Soil Geography and Land Use

Soil Geography and Land Use

HENRY D. FOTH
Michigan State University

JOHN W. SCHAFER
Iowa State University

John Wiley & Sons
New York   Chichester   Brisbane   Toronto

Library of Congress Cataloging in Publication Data:

Foth, H D
  Soil geography and land use.

  Includes index.
  1. Soil geography. 2. Land use. I. Schafer,
John W., joint author. II. Title

S591.F68 1980   631.4'7   79-27731
ISBN 0-471-01710-8

Printed in the United States of America

10 9 8 7 6 5 4 3 2 1

Preface

Since World War II, there has been an outpouring of publications dealing with the characterization, genesis, classification, and geography of soils throughout the world. The completion of the FAO *Soil Map of the World* and the publication of *Soils of Canada* and *Soil Taxonomy* in the United States are only a few examples of the excellent work being done in this area. Although these publications and many others like them have greatly increased our knowledge and understanding of soils, they have addressed themselves mostly to a professional audience and present much detail about a limited geographic area. A need was created for a book that synthesized and organized this knowledge to make it readily available to students, teachers, and others interested in the land. *Soil Geography and Land Use* satisfies that need.

This is a soils textbook with a worldwide perspective. To write a soil geography book with a world view requires many compromises. We have given major emphasis to soils of North America and secondary emphasis to soils of other continents. Our knowledge of world soils is naturally limited; therefore we hope that persons with greater knowledge than ours will inform us of any errors so that they can be corrected in the second edition.

We are indebted to many people for information, photographs, diagrams, and time spent with us in the field. We are also grateful to the authors of the many publications of which we made extensive use. We recognize in particular the work of Dr. Guy D. Smith for his contribution to the development of *Soil Taxonomy,* which is used extensively in the book.

Henry D. Foth
John W. Schafer

Contents

Soil Geography and Land Use

1

Soil Classification

Classification schemes of natural objects seek to organize knowledge so that the properties and relationships of the objects can be most easily remembered and understood for some specific purpose. Ultimately, soil taxonomy and other soil classification schemes seek to make knowledge useful in satisfying the human needs that depend on soils. This chapter briefly considers the nature and categories of *Soil Taxonomy,* which was published by the U.S. Department of Agriculture in 1975, and the Canadian System of Soil Classification, published in 1978. Understanding of this chapter is important for understanding the remainder of this book.

Diagnostic Horizons in Soil Taxonomy

Diagnostic soil horizons are used to classify soils into orders—the highest category. Diagnostic horizons formed at the surface are called diagnostic surface horizons, or *epipedons;* six horizons are recognized: mollic, umbric, anthropic, ochric, histic, and plaggen. Subsurface diagnostic horizons form below the surface of the soil and, in some cases, just immediately below a layer of leaf litter. Diagnostic subsurface horizons may be exposed by truncation of the soil. The diagnostic subsurface horizons include cambic, argillic, natric, spodic, oxic, and agric.

The Diagnostic Epipedons

Mollic Epipedon   A dark-colored surface horizon is characteristic of soils of the steppes in the Americas, Asia, and Europe. In virtually all of these soils there is a relatively thick, dark-colored, humus-rich surface horizon or horizons in which bivalent cations are dominant on the exchange complex. The structure is moderate to strongly developed, so that the soil is *soft* even when dry. This horizon is called the *mollic* horizon from the word mollify, which means "to soften." The mollic epipedon is believed to have formed mainly by the underground decomposition of organic residues in the presence of considerable calcium. The organic residues are partly roots and partly materials from the surface that have been carried underground by animals.

The mollic horizon is defined on the basis of its morphology and has the following properties.

1.   Soil structure strong enough so that the soil is soft when dry.
2.   Dark color with Munsell color value darker than 3.5 moist and 5.5 dry and with chroma less than 3.5 moist.
3.   Base saturation is 50 percent or more.
4.   Organic carbon content is 2.5 percent (slightly more than 4 percent organic matter) or more in the upper 18 centimeters. The horizon is mainly mineral, however, and not organic in nature.
5.   Thick, 25 centimeters or more unless underlain by lithic or other such horizon in which 10 centimeters is minimum thickness. In other cases thickness must be 18 centimeters or more and comprise more than one-third of the soil above pedogenic lime, and so on, that occurs at depths less than 75 centimeters.
6.   Contains less than 250 parts per million of P_2O_5 soluble in 1 percent citric acid.

The mollic horizon also has quite favorable water content and temperature for plant growth. The mollic epipedon of a Mollisol is shown in Fig. 1-1.

Umbric Epipedon   Many soils have thick, dark-colored epipedons high in organic matter that cannot be distinguished by the eye from mollic horizons, but they have base saturation less than 50 percent. These horizons are called umbric, coined from the Latin *umbric,* meaning shade (hence dark). The umbric horizon is not required to meet the soft when dry criteria of mollic horizons. Umbric horizon

Figure 1-1 Soil developed under grass with dark-colored mollic epipedon (chernozemic A in the Canadian soil classification).

development seems favored by high rainfall, as in the mountains of western Oregon and Washington.

Anthropic Epipedon   The anthropic horizon is made by humans and resembles the mollic horizon in color, structure, and organic matter content. Anthropic horizons develop where humans have, over a long period of time, disposed of bones and shells high in phosphorus and calcium, or used the soil for irrigated crops. The phosphorus content is over 250 parts per million expressed as P_2O_5. The addition of shells and bones usually results in over 50 percent base saturation; if less than 50 percent, the base saturation is considerably higher than adjacent soils. Anthropic horizons occur mostly in kitchen middens in Europe, United States, and South America.

Plaggen Epipedon   The plaggen epipedon is also made by humans; it is 50 centimeters or more and is produced by long-continued manuring. In the early Middle Ages and continuing until this century farmers on sand soils in northwestern Europe collected highly organic materials from heaths and forested areas and used them as bedding in barns. The material collected was soddy and contained some mineral soil particles, mainly quartz sand. Cattle and sheep grazed on outlying pastures and dropped their manure in the barns. The manure from the animals plus the sod used for bedding was spread on the cultivated land. The organic matter slowly decomposed and the quartz sand in the manure accumulated on the fields at a rate of about 1 millimeter per year. Thus, in about 1000 years, plaggen epipedons about 1 meter thick developed. The plaggen horizons commonly overlie buried Spodosols or Podzols. The word plaggen comes from the German *plaggen,* meaning sod.

The color of the plaggen epipedon and its organic carbon content depend on the source of materials used for bedding. If the sod was cut from the heath, the epipedon tended to be black or very dark gray, rich in organic matter, and with a large carbon/nitrogen ratio. Sod from forested areas produced brown plaggen epipedons with less organic matter and a low carbon/nitrogen ratio. A diagram of the formation of plaggen epipedons is given in Fig. 1-2.

Ochric Epipedon   An ochric epipedon is too light colored, low in organic matter, or thin to be a mollic, umbric, anthropic, or plaggen epipedon or it is both hard and massive when dry. Ochric epipedons are common in youthful soils and soils developed under forest vegetation. Ochric comes from the Greek *ochros,* meaning pale.

Histic Epipedon   The epipedons discussed thus far are mainly composed of mineral material with only a relatively small amount of organic matter. The histic epipedon occurs in soils that are saturated with water for 30 consecutive days or more at some time in most years unless the soils are artifically drained. The horizon is organic in nature and typically over 20 centimeters thick. The organic carbon may vary from 12 percent or more in soils where the mineral fraction contains no clay

Figure 1-2   Diagram of the development of plaggen epipedons.

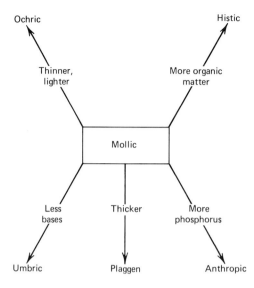

Figure 1-3 Diagram showing the relationship between several epipedons to the mollic horizon.

to 18 percent or more organic carbon if the mineral fraction is 60 percent or more clay. Histic is derived from the Greek *histos,* meaning tissue. Muck and peat soils characteristically have histic epipedons. A diagram showing the relationship of histic, ochric, anthropic, plaggen, and umbric horizons to mollic epipedons is given in Fig. 1-3.

Diagnostic Subsurface Horizons

Cambic Horizon The word cambic is derived from the Latin *cambiare,* meaning to change. The cambic horizon is a subsurface horizon that has been little altered in soils where the parent material has the texture of very fine sand or loamy very fine sand or is finer in texture. It is common in young soils where the major physical changes in the subsoil have caused a destruction of the orginal rock structure, including stratification of sediments and aggregation of soil particles into peds. Chemically, some alteration has occured, but it is usually minimal. Cambic horizons may have lost bases, carbonates, and some iron and aluminum oxide. Gains in organic matter may have occurred but not as a result of illuviation. Cambic horizons have features that represent genetic soil development but without mineral accumulation by illuviation and extreme weathering. Clay skins are typically absent, and the clay content does not increase with increasing depth or increases very gradually.

Argillic Horizon The word argillic comes from the Latin *argilla,* meaning clay. The argillic horizon is an illuvial horizon, where layer-lattice silicate clays have

Figure 1-4   Natric horizon between depths of 20 and 30 centimeters (solonetzic B in the Canadian soil classification).

accumulated to a significant extent. An eluvial horizon usually occurs above the argillic horizon, and the clay content must increase abruptly as the argillic horizon is approached. In most soils the increase in clay content must be at least 20 percent greater in the argillic horizon than in the eluvial horizon above. For example, 25 percent clay in the eluvial horizon would require at least 30 percent clay in the argillic horizon. The illuviation of clay is typically associated with the presence of clay skins. The nature and the development of argillic horizons are related to time. Since argillic horizons take thousands of years to develop, their presence indicates a fairly stable land surface and that clay translocation is dominant over processes that destroy clay or mix soil horizons. It is likely that more soils in the world have argillic horizons than any other kind of subsoil horizon.

Natric Horizon   The natric horizon is a special kind of argillic horizon. Natric horizons have all the properties of argillic horizons but, in addition, are 15 percent or more sodium saturated. Their formation is favored where leaching results in the accumulation of sodium on the cation-exchange complex. The word natric is derived from the Latin *natrium,* meaning sodium. Natric horizons commonly have a characteristic prismatic or columnar structure and low water permeability (see Fig. 1-4).

Spodic Horizon   The spodic horizon is an illuvial horizon enriched with amorphous iron and aluminum oxides and amorphous organic matter. Spodic horizons form mostly in acid, quartzitic sands in areas of high annual precipitation from the tropics to the tundra. Spodic (rhymes with odd) is derived from the Greek *spodos,* meaning wood ash. Wood ash has been associated with Podzols in Russia, where farmers thought the prominent white-colored A2 horizon resulted from the ashes of burning the forest. Now the word wood ash is associated with a horizon that typically underlies an ashy-white layer and it commonly has a reddish-brown color. Many soils with spodic horizons have a thin A2 (albic) horizon and a very thin or missing A1 horizon, so that plowing usually results in a thorough mixing of the spodic horizon with the layers above. The relationship of illuvial spodic, argillic, and natric horizons to the non-illuvial cambic horizon is shown in Fig. 1-5.

Oxic Horizon   The oxic horizon is a mineral subsurface horizon in an advanced stage of weathering; it is at least 30 centimeters thick. Oxic horizons consist of a mixture of hydrated oxides of iron or aluminum or both, with variable amounts of 1:1 lattice clay (kaolinite). In addition there may be some highly insoluble accessory minerals such as quartz present. The word oxic comes from the French word *oxide.*

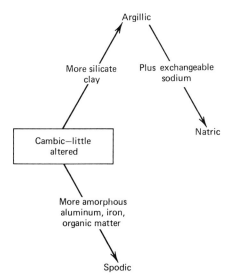

Figure 1-5 Diagram showing the relationship between several illuvial horizons to the non-illuvial cambic horizon.

The fine earth fraction of oxic horizons has few or no primary minerals that can weather and release bases, iron or aluminum. The clay has a cation-exchange capacity of 16 milliequivalents or less per 100 grams. The low CEC and amount of weatherable minerals distinguish the oxic horizon from the cambic horizon. The extreme degree of weathering means that soils with oxic horizons occur on very old and stable geomorphic surfaces. These surfaces are likely to be mid-Pleistocene or older. Soils with oxic horizons have no reserve of bases beyond those held on the exchange complex and in plant tissue. None of the soils in the continental United States or Canada have oxic horizons, which are found in some humid, tropical region soils.

Agric Horizon   The word agric comes from the Latin *ager,* meaning field. The agric horizon is an illuvial horizon formed under cultivation that contains significant amounts of illuvial clay, silt, and humus. When a field is brought under cultivation, the soil fauna and flora change drastically. The plow layer is mixed periodically and, in effect, a new cycle of soil formation is started. After long-continued cultivation changes occur immediately below the plow layer. The large pores in the plow layer and the absence of vegetation immediately after plowing permit a turbulent flow of muddy water to the base of the plow layer. This muddy water can later enter wormholes or fine cracks between peds; the suspended materials are deposited as lamellae and eventually fill wormholes and other cavities.

Agric horizons develop only after long periods of cultivation and are not recognized in United States. Some have formed in loessial soils in northwestern Europe and are concentrated along Roman roads, where agriculture has been carried on the longest. Some of the agric horizons have been excavated and used for making bricks in landscapes where clay is not readily available.

Diagnostic Horizons in the Canadian Soil Classification

Several diagnostic horizons or layers used in *The Canadian System of Soil Classification* are comparable to those in *Soil Taxonomy.* These are chernozemic A horizon for mollic, O horizon for histic, podzolic B horizon for spodic, and solonetzic B horizon for natric. The diagnostic horizons are discussed in relation to whether they are mineral or organic.

Mineral Diagnostic Horizons

Chernozemic A Horizon   This a dark-colored A horizon that is at least 10 centimeters thick with the same color requirements as the mollic; in addition, must be one color value darker than the C horizon (see Fig. 1-1). As an Ap horizon it must be at least 15 centimeters thick and meet the color requirements. The organic C is

1 to 17 percent and the carbon-nitrogen ratio is less than 17. Characteristically, the structure is well developed, so the soil is not hard and massive when dry. Base saturation is more than 80 percent,with calcium as the dominant exchangeable cation. Soils with chernozemic A horizons are restricted to soils with mean annual soil temperatures of 0°C or higher and soil moisture regimes drier than humid. On the basis of moisture and base saturation, many mollic horizons in Illinois would not qualify as chernozemic A; however, mollic epipedons on the Great Plains would mostly qualify.

Duric Horizon   This is a strongly cemented horizon that does not meet the requirement of a podzolic B. The upper surface is usually the most cemented. Air-dry clods do not slake when immersed in water.

Fragipan   This is a loam subsurface horizon of high bulk density and very low organic matter content. Hard when dry and brittle when moist.

Ortstein   This is strongly cemented Bh, Bhf, and Bf horizons at least 3 centimeters thick that occur in more than one-third of the exposed face of the pedon. (See Table 1 of appendix for meaning of lowercase suffixes.) Colors are generally reddish brown to very dark reddish brown.

Placic Horizon   This horizon is a thin layer (usually 5 millimeters or less thick) or a series of thin layers that are irregular or involuted, hard, impervious, often vitreous, and dark reddish brown to black. May be cemented with various complexes of iron, aluminum, manganese, and organic matter.

Podzolic B Horizon   This is a subsurface horizon at least 10 centimeters thick with a texture coarser than clay. The moist crushed color is either black or the hue is 7.5 YR or redder or 10YR near the upper boundary and becomes redder with depth. The chroma is higher than 3, or the value is 3 or less. Amorphous material is indicated by brown to black coatings on some mineral grains or by brown or black microaggregates.

Chemically, the podzolic B is of one of two types. Bh is a podzolic B horizon very low in iron and contains more than 1 percent organic carbon. Pyrophosphate-extractable iron content is less than 0.3 percent, and the ratio of organic carbon to pyrophosphate-extractable iron is 20 or more.

Bf and Bhf are podzolic B horizons that contain appreciable iron as well as aluminum. Organic carbon content is more than 0.5 percent. Pyrophosphate-extractable iron is at least 0.3 percent, and/or the ratio of organic carbon to pyrophosphate-extractable iron is less than 20.

Solonetzic B Horizon   This is a subsurface horizon with prismatic or columnar primary structure that breaks into blocky secondary structure (see Fig. 1-4). Peds are extremely hard when dry. The ratio of exchangeable calcium to sodium is 10 or less. It includes Bn and Bnt horizons.

Lithic Layer   This is a consolidated bedrock layer (R) within the control section below a depth of 10 centimeters. The upper surface of a lithic layer is a lithic contact.

Mull   This is a zoogenous, forest humus form consisting of an initiate mixture of well-humified organic matter and mineral soil with crumb or granular structure that makes a gradual transition to the horizon underneath. Soil mixing by fauna, mostly earthworms, prevents the organic debris from accumulating as a distinct layer, as in mor or moder. Organic matter content is usually 5 to 20 percent, and the carbon-nitrogen ratio is 10 to 15. It is one kind of Ah horizon.

Organic Diagnostic Horizons

Organic horizons are typically at the soil surface in organic soils but also occur as buried horizons in other soils. Organic carbon is more than 17 percent (about 30 percent organic matter) by weight.

O Horizon   This organic horizon is developed mainly from mosses, rushes, and woody materials. The Of is composed largely of fibric material or is readily identifiable as to botanical origin. The Om horizon or layer is mesic in degree of decomposition and has been partially altered physically and biochemically. The Oh consists of humic material that is at an advanced stage of decomposition. Oco is coprogenous earth, which is limnic material that occurs in some Organic soils.

L, F, and H Horizons   Whereas O horizons have a major component of organic matter from mosses, only a minor component of mosses exists in L, F, and H layers. The L layer is organic in nature and characterized by an accumulation of organic matter derived mainly from leaves, twigs, and woody materials that are relatively unaltered. The F layer is characterized by an accumulation of materials similar to those in L layers but partially decomposed. The H layer is sufficiently decomposed that the original botancial structures or origins are indiscernible.

Soil Temperature Regimes in the United States and Canada

Conventional classification of temperature and moisture have relevance for above-soil environment and above-ground plant growth. Soil temperature and soil moisture regimes or soil climates are more relevant for growth of roots, soil fauna, and microorganisms and for frost action. The soil temperature and soil moisture regimes characterize the changes in soil temperature and moisture over time that are relevant for biological activity and soil genesis. The regimes are also used in soil classification.

The Control Section

Extreme and rapid changes in temperature and moisture content may occur in the immediate surface soil and have little significance for biological activity or soil genesis. The control section is the subsurface portion used to characterize the regimes. For the soil temperature regime, the depth is arbitrarily set at 5 to 100 centimeters in the United States. In Canada the temperature at 50 centimeters depth is considered standard for soil classification.

Soil Temperature Regimes in the United States

Soil temperature classes or regimes in the United States are defined according to the mean annual soil temperature (MAST) in the control section (5 to 100 centimeters). Average root zone temperatures of freezing and below freezing characterize the pergelic soil temperature regime where permafrost is common. Low soil temperature and short growing season result in low biomass production. The temperature regimes, with their mean annual soil temperature ranges and general characteristics, are given in Table 1-1.

The soil temperature regimes are roughly parallel to the air temperature thermal regimes. Approximate equivalents are pergelic-polar, cryic-subarctic, mesic-midlatitude, thermic-tropical, and hyperthermic-equatorial.

Soil Temperature Regimes in Canada

All of the soils of Canada are mesic or cooler, and it is important to make further subdivisions of temperature classes toward the cold end of the range compared to the United States. The soil temperature regimes and classes used in Canada are given in Table 1-2.

An arctic climate or extremely cold soil temperature regime characterizes 30.5 percent of Canadian soils, in which biological activity may occur only briefly once in 5 years; another 31.5 percent of the soils have a subarctic climate or a very cold temperature regime, which also greatly restricts biological activity. Thus, only 38 percent of the soils have enough thermal energy for sustaining productive vegetation, and only 34 percent of this land does not have severe moisture limitations (subhumid or humid). Grain crops (wheat), forage, and forest trees are the major economic plants.

Table 1-1 Definitions and Features of Soil Temperature Regimes in the United States

Temperature Regime	Mean Annual Temperature in Root Zone, 5 to 100 centimeters		Characteristics and Some Locations
	C	F	
Pergelic	<0	<32	Permafrost and ice wedges common. Tundra of northern Alaska (and Canada) and high elevations of middle and northern Rocky Mountains
Cryic and frigid [a]	0–8	32–47	Cool to cold soils of northern Great Plains of United States (and southern Canada) where spring wheat is the dominant crop. Forested regions of New England
Mesic	8–15	47–59	Midwestern and Great Plains regions where corn and winter wheat are common crops
Thermic	15–22	59–72	Coastal plain of southeastern United States where temperatures are warm enough for cotton. Central valley of California
Hyperthermic	Over 22	Over 72	Citrus areas of Florida peninsula, Rio Grande Valley of Texas, southern California, and low elevations in Puerto Rico and Hawaii. Tropical climates and crops

[a]Frigid soils have warmer summers than cryic soils; both have same MAST. Frigid soils have more than 5°C temperature difference between mean winter and mean summer temperatures at a depth of 50 centimeters (or lithic or paralithic contact, if shallower).

Soil Moisture Regimes in the United States

The amount of water available to plants is affected by soil as well as climate. For example, wet soils exist in deserts where soils have impermeable layers and receive run-on water from surrounding higher land or springs. Gravelly soils in humid regions may be droughty because little water is retained. The soil property that expresses the order of soil moisture changes over time is the *soil moisture regime*. In addition to the supply of water for plants, the soil moisture regime expresses the availability of water for weathering and leaching and indicates whether or not the root zone lacks oxygen because of water saturation.

Table 1-2 Definition and Characteristics of Soil Temperature Classes in Canada

Extremely cold
>MAST[a] $< -7°C$
>Continuous permafrost usually occurs below the active layer within 1 meter of the surface
>No significant growing season, <15 days $>5°C$
>Remains frozen within the lower part of the control section
>Cold to very cool summer, MSST[b] $<5°C$
>No warm thermal period $>15°C$

Very cold
>MAST $-7-2°C$
>Discontinuous permafrost may occur below the active layer
>Soils with aquic regimes usually remain frozen within part of the control section
>Short growing season, <120 days $>5°C$
>Degree-days $>5°$ C are <550
>Moderately cool summer, MSST 5–8°C
>No warm thermal period $>15°C$

Cold
>MAST 2–8°C
>No permafrost
>Undisturbed soils are usually frozen in some part of the control section for a part of the dormant season
>Soils with aquic regimes may remain frozen for part of the growing season
>Moderately short to moderately long growing season 140–220 days $>5°C$
>Degree-days $>5°C$ are 550–1250
>Mild summer, MSST 8–15°C
>An insignificant or very short, warm thermal period, 0–50 days $>15°C$
>Degree-days $>15°C$ are <30

Cool
>MAST 5–8°C
>Undisturbed soils may or may not be frozen in part of the control section for a short part of the dormant season
>Moderately short to moderately long growing season, 170–220 days $>5°C$
>Degree-days $>5°C$ are 1250–1700
>Mild to moderately warm summer, MSST 15–18°C
>Significant very short to short warm thermal period, >60 days $>15°C$
>Degree-days $>15°C$ are 30–220

Mild
>MAST 8–15°C
>Undisturbed soils are rarely frozen during the dormant season
>Moderately long to nearly continuous growing season, 200–365 days $>5°C$
>Degree-days$>5°C$ are 1700–2800
>Moderately warm to warm summer, MSST 15–22°C
>Short to moderately warm thermal period 90–180 days $>15°C$
>Degree-days $>15°C$ are 170–670

From The Canadian System of Soil Classification, *1978, and reproduced by permission of the Minister of Supply and Services Canada, Ottawa.*
[a]*MAST mean annual soil temperature*
[b]*MSST mean summer soil temperature*

The Soil Moisture Control Section

Soil moisture regimes are based on moisture conditions in the *soil moisture control section*. The upper boundary of the moisture control section is the depth to which 2.5 centimeters of water will moisten dry soil (tension over 15 bars not air dry) in 24 hours. The lower boundary is the depth of penetration of 7.5 centimeters of water in dry soil in 48 hours. These depths are exclusive of large cracks open at the soil surface. For many loamy soils the moisture control section is located between depths of 20 to 60 centimeters.

Aquic Soil Moisture Regime

Soils with aquic (L. *aqua,* water) moisture regime are wet and are virtually free of dissolved oxygen because the soil is saturated. Very commonly the level of groundwater will fluctuate with the season. In some cases, as in tidal marshes, the water table is at or close to the soil surface all the time. Drainage is needed to grow plants that require an aerated root zone. Most of the world's rice, however, is grown on aquic soils. Subsoil colors in mineral soils are frequently gray, indicating reducing conditions, or mottled, indicating alternating reducing and oxidizing conditions. Aquic soils are unsuited for home-building sites (see Fig. 1-6).

Figure 1-6 Aquic soils with water table at or near the soil surface in wet seasons are poor building sites. Basements may flood, and septic tank effluent may contaminate shallow water supplies.

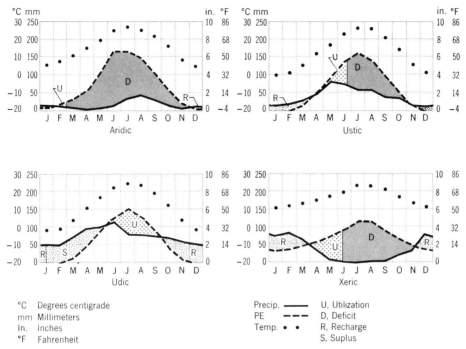

°C	Degrees centigrade		Precip. ———	U, Utilization
mm	Millimeters		PE – – –	D, Deficit
In.	Inches		Temp. • •	R, Recharge
°F	Fahrenheit			S, Suplus

Figure 1-7   Climatic data and soil water balance for moisture regimes. (Available water capacities (AWC), in millimeters, are aridic, 50; udic, 394; ustic, 342; and xeric, 196. (From *Soil Taxonomy,* 1975.)

Aridic Soil Moisture Regime

The driest soils have aridic (L. *aridus,* dry) moisture regimes (also called *torric,* meaning hot and dry). The moisture control section is dry in all parts more than one-half of the growing season and is not moist (SMT less than 15 bars) in some parts for as long as 90 consecutive days during a growing season in most years. Most soils with aridic moisture regimes are in arid or desert regions with widely spaced shrubs and cacti. A crop cannot be matured without irrigation.

Some climatic data and the soil-water budget typical of an aridic soil moisture regime are given in Fig. 1-7. Rainfall is low most months, and the potential evapotranspiration is very high in summer relative to the precipitation. Consequently little soil moisture recharge or water storage occurs, and the little water that is stored is quickly used in late winter or early spring. The result is a lack of water for plant growth most of the time and a large water deficit. Many of the native plants have an unusual capacity to endure a high degree of desiccation without serious injury. Grazing is the dominant land use, and crop production requires irrigation.

Weathering occurs when soils are moist, but there is little or no leaching. Soluble salts commonly accumulate in a zone that marks the average depth of moisture penetration. Some soils with aridic moisture regimes exist in the semiarid regions if they are shallow over rock or have very low infiltration rates.

Udic Soil Moisture Regime

Udic (L. *udus,* humid) means humid. Soils with udic moisture regimes have a moisture control section that is not dry in any part as much as 90 cumulative days in most years. Udic moisture regimes are common in soils of humid climates that have well-distributed rainfall. The amount of summer rainfall plus stored soil water is approximately equal to or exceeds the amount of evapotranspiration (see Fig. 1-7). Forests and tall grass prairies are typical vegetation on udic soils. If precipitation exceeds the amount of evapotranspiration each month, the moisture regime is called *perudic.*

In most years there is surplus water, and leaching occurs. Significant amounts of plant nutrients are removed from the soil, and the soils develop acidity. Sufficient leaching to produce soil acidity and low natural fertility are typical of forest soils with udic moisture regimes. Short droughts occur occasionally. Irrigation is not widespread and is used for specialty crops and special benefits, as in frost control. Udic soils are common from the east coast to the western boundaries of Minnesota, Iowa, Missouri, Arkansas, and Louisiana.

Ustic Soil Moisture Regime

The ustic (L. *ustus,* burnt, implying dryness) soil moisture regime is intermediate between aridic and udic regimes. The concept is one of limited available soil moisture, but soil moisture is available for significant plant growth when other conditions are favorable for growth. Compared to the aridic regime, significantly more water storage occurs between fall and spring, and summer rainfall is greater (see Fig. 1-7). The water deficit is much less than the aridic regime and much greater than the udic regime. Ustic soil mositure regimes are common on the Great Plains, where water is available for wheat production in winter, spring, and early summer in most years. Droughts are not uncommon. Sorghum is grown because it can interupt growth when water is lacking and grow again if more rainfall occurs. Corn requires irrigation. Wheat and sorghum are the major dry land crops, and grazing is important. Surplus water is rare, and soils are unleached. Native vegetation was mainly mid-, short, and bunch grasses.

Xeric Soil Moisture Regime

Soils in areas with Mediterranean climates typically have xeric (Gr. *xeros,* dry) soil moisture regimes. Winters are cool and moist, and summers are hot and dry. The precipitation occurs in the cool months when evapotranspiration is low and is effec-

Figure 1-8 Landscape with soils having xeric moisture regime in California. Crops are irrigated during long dry summer and hills are brown from dried grass. Landscape is green with plants in winter due to winter rains.

tive for weathering and leaching (see Fig. 1-7). Surplus water may occur. Pastures and crops are well supplied with moisture in the winter and the landscape is green. A large water deficit occurs in summer, when hilly grasslands are brown (Fig. 1-8). Xeric soils of the Central Valley of California are used for a wide variety of crops, including vines, fruits, nuts, vegetables, seeds, and agricultural produce. Xeric soils of the Palouse in Washington and Oregon are used for winter wheat and peas.

Summary Statement

Soil moisture regimes express changes in soil moisture over time as a function of soil and climate. Aquic soils are characterized by water saturation and oxygen deficiency. They occur in any climatic zone. Aquic soils that have been drained represent one of the world's most productive agricultural soils. Aquic soils, however, are unsuited for septic tank filter fields unless properly drained.

The central concept of the aridic, ustic, and udic regimes is one of differences in leaching frequency with summer maximum rainfall (see Table 1-3). These regimes generally encompass the rainfall classes from arid to semiarid to subhumid to humid. Land use without irrigation ranges from grazing to grazing and wheat plus sorghum and cotton to corn, soybeans, and cotton.

The xeric soil moisture regime is characterized by winter rainfall maximum when leaching and weathering are effective. A long period of soil dryness occurs in summer. Soils with xeric regimes occur in areas with Mediterranean climate.

Table 1-3 Characteristics of Soil Moisture Regimes

Saturated Soils	Unsaturated Moisture Control Sections		
	Leaching Frequency	Time of Maximum Rainfall	
		Summer	Winter
Aquic	Never	Aridic	
	Some years	Ustic	Xeric
	Every year	Udic	

Soil Moisture Regimes in Canada

Two soil moisture regimes for water-saturated soils and six regimes for moist, unsaturated soils are used in Canadian soil classification. The water deficit is used to characterize the moisture regime of moist, unsaturated soils and indicates the extent to which plant growth is limited by soil moisture shortage. This shortage is also expressed with the climatic moisture index (CMI), which expresses the growing season precipitation as a percentage of the potential water used by annual crops when water is readily available from the soil. It is calculated as follows.

$$\text{CMI} = \frac{P}{P + SM + IR} \times 100$$

where
  P = growing season precipitation
SM = water available to crops that is stored in the soil at the beginning of the growing season
 IR = irrigation requirements or water deficit for the growing season
The regimes and a brief description are given in Table 1-4.

Ninety-two percent of Canadian soils are in perhumid (25.8) and humid (66.2) moisture classes. Of these soils, 26 percent are cryoboreal and generally too cold for most agricultural crops, but they are productive for forests. The agriculture is mainly on the 18 percent with boreal soil temperature regime and 4 percent with mesic regime (see Table 1-5). Since Canada is a large country, exceeded only in size by the USSR, there is a relatively large amount of land suited for grain and forage crops.

Categories of Soil Taxonomy

There are six categories in the American soil classification, the *order* is the most general one. The names and number of members in each category in the United States are given in Table 1-6.

Table 1-4   Soil Moisture Regimes of Canadian Soils

Regime	Characteristics	Water Deficit,[a] centimeters	CMI
	Saturated soils		
Aqueous	Free water on surface continuously	0	—
Aquic	Saturated only part of the time	0	—
	Moist unsaturated soils		
Prehumid	No Significant water deficit during growing season	Less than 2.5	Over 84
Humid	Very slight deficit during growing season	2.5–6.5	74–84
Subhumid	Significant deficit during growing season	6.5–13	59–73
Semiarid	Moderately severe deficit during growing season	13–19	46–58
Subarid	Severe deficit during growing season	19–38 (boreal) 19–51 (mesic)	25–45
Arid	Very severe deficit during growing season	over 38 (boreal) over 51 (mesic)	Over 25

Adapted from The Canadian System of Soil Classification, *1978.*
[a]Amount of additional water plants could use during growing season, considering stored soil water or irrigation requirement.

Table 1-5   Extent of Soil Climatic Classes in Canada

Class and Subclass		Arctic	Subarctic	Cryoboreal	Boreal	Mesic	Subclass Total	Percent of Total
				100 square kilometers				
d.	Perhumid	—	7,216	10,782	4,141	250	22,388	25.8
e.	Humid	26,180	19,577	10,077	606	911	57,352	66.2
f.	Subhumid	—	546	3,524	313	62	4,447	5.2
g.	Semiarid	—	—	409	1,181	5	1,618	1.8
h.	Subarid	—	—	—	914	—	914	1.0
	Class total	26,180	27,340	24,794	7,180	1,227	86,720	
	Percent	30.2	31.5	28.6	8.3	1.4	100	

Reproduced from Field Soil Water Regime, Soil Science Society of America Special Publication 5, *1973, Table 8, page 209, by permission of the Soil Science Society of America.*

Table 1-6 Categories in Soil
Taxonomy and Number of Members in
Each Category in United States

Category	Number
Order	10
Suborder	44
Great group	184
Subgroup	975
Family	5,202
Series	12,156

*From Classification of Soil Series in the United
States, SCS, USDA, 1977.*

Orders

The order is the highest category and there are 10 orders, each ending in sol (L.
solum, meaning soil). The orders, along with their derivations, meanings, and
approximate equivalents in great soil groups, are given in Table 1-7.

Soil orders are differentiated by the presence or absence of diagnostic horizons
or features that are marks in the soil of differences in the degree and kind of
dominant soil-forming processes that have gone on. Entisols are very recent soils
(see Table 1-7). Vertisols are soils high in clay that become "inverted" because of
alternate swelling and shrinking. Inceptisols are either young soils with just the
beginning of genetic horizon development or older soils where horizonation has
disappeared. Aridisols are soils of arid regions. Mollisols are the grassland soils
with thick, "soft," dark-colored surface horizons (mollic epipedons). Spodosols
have spodic horizons (and are comparable to Podzols). Alfisol is derived from
pedalfer—a word first used to refer to humid region soils leached of lime and with
a tendency for aluminum and iron to accumulate in the subsoil. Ultisols are
extremely leached soils, very low in bases. Oxisols are the red tropical soils rich in
oxides of iron and aluminum and also 1:1 clays; that is, they have oxic horizons.
The Histosols are bog soils composed mainly of plant tissue. Data on the area and
relative abundance of the soil orders in the world are given in Table 1-8.

Suborders

Soils in the orders are classified into suborders mainly on the basis of soil moisture
and temperature regimes and certain physical and chemical properties associated
with differences in parent materials and soil-forming processes. The suborder name
has two syllables. The last syllable is the formative element. The first syllable con-
notes additional diagnostic properties of the soils in the suborders. Thus, an Entisol
that has an aquic moisture regime is called an Aquent (L. *aqua,* water, and *ent,*

Table 1-7 Soil Orders, Formative Syllables with Derivations and Meanings, and Approximate Equivalents with Great Soil Groups Used in 1949 Soil Classification System

Order	Formative Syllable	Derivation	Meaning	Approximate Equivalents
1. Entisol	ent	Coined syllable	Recent soil	Azonal soils and some Low Humic Gley soils
2. Vertisol	ert	L. *verto*, turn	Inverted soil	Grumusols
3. Inceptisol	ept	L. *inceptum*, beginning	Inception, or young soil	Ando, Sol Brun Acide, some Brown Forest, Low Humic Gley, and Humic Gley soils
4. Aridisol	id	L. *aridus*, dry	Arid soil	Desert, Reddish Desert, Sierozem, Solonchak, some Brown and Reddish Brown Soils, and associated Solonetz
5. Mollisol	oll	L. *mollis*, soft	Soft soil	Chestnut, Chernozem, Brunizem (Prairie), Rendzinas, some Brown, Brown Forest, and associated Solonetz and Humic Gley soils
6. Spodosol	od	Gk. *spodos*, wood ash	Ashy (Podzol) soil	Podzols, Brown Podzolic soils, and Ground-Water Podzols
7. Alfisol	alf	Coined syllable	Pedalfer (Al-Fe) soil	Gray-Brown Podzolic, Gray Wooded, Noncalcic Brown, Degraded Chernozem, and associated Planosols and Half-Bog soils
8. Ultisol	ult	L. *ultimus*, last	Ultimate (of leaching)	Red-Yellow Podzolic, Reddish-Brown Lateritic (of United States), and associated Planosols and Half-Bog soils
9. Oxisol	ox	F. *oxide*, oxide	Oxide soils	Laterite soils, Latosols
10. Histosol	ist	G. *histos*, tissue	Tissue (organic soils)	Bog soils

Table 1-8   Areas of Soil Orders and Relative Abundance in the World

Soil Order	Area, Billions of Hectares	Percentage of World Total	Rank
Alfisol	1.73	13.2	2
Aridisol	2.47	18.8	1
Entisol	1.09	8.3	6
Histosol	0.12	0.9	10
Inceptisol	1.17	8.9	3
Mollisol	1.13	8.6	4
Oxisol	1.12	8.5	5
Spodosol	0.56	4.3	8
Ultisol	0.73	5.6	7
Vertisol	0.24	1.8	9
Soils of mountain areas	2.80	21.3	
Grand total	13.16	100	

Complied by the Soil Geography Unit of USDA, mimeograph, 1971.

from Entisol). Fluvents are Entisols formed in very young sediments (L. *fluvius,* river, and *ent,* from Entisol). The formative elements, with their derivations and connotations for the suborders, are given in Table 1-9. Figure 1-9 is a map showing the world distribution of the most extensive orders and suborders. The legend of the figure also gives abbreviated definitions of the orders and suborders.

Great Groups

The great group name includes the suborder plus a prefix. The prefixes used for the great groups are given in Table 1-10. An Aquent with a cryic temperature regime is a Cryaquent. A Natrargid is an Aridisol with an argillic horizon that has high exchangeable sodium or an Aridisol with a natric horizon. By comparing formative elements in Table 1-10 with those in Table 1-9, one can see that some of the same elements are used to form both great group and suborder names.

Subgroups

Great groups are separated into three kinds of subgroups: typic, intergrade, and extragrade. The name of a subgroup is formed by placing one or more adjectives before the name of the relevant great group.

A *typic* subgroup represents the central concept of its great group. A soil in a typic subgroup, however, is not necessarily more extensive than the other kinds of soil in the same great group.

An intergrade subgroup has the definitive properties of the great group whose name it carries as a substantive. It also has some of the properties of another taxon or more than one other taxon—an order, a suborder, or a great group. The adjective or adjectives in the intergrade subgroup name are formed from the names of the other taxon or taxa. A Fluvaquentic Haploxeroll is a Haploxeroll that intergrades to a Fluvaquent.

Table 1-9 Formative Elements for Suborders, Including Derivations and Connotations

Formative Element	Derivation	Connotation
Alb	L. *albus*, white	Presence of albic horizon
And	Modified from ando	Andolike
Aqu	L. *aqua*, water	Aquic moisture regime
Ar	L. *arare*, to plow	Mixed horizons
Arg	Modified from argillic horizon; L. *argilla*, white clay.	Presence of argillic horizon
Bor	Gr. *boreas*, northern	Cool
Ferr	L. *ferrum*, iron	Presence of iron
Fibr	L. *fibra*, fiber	Least decomposed stage
Fluv	L. *fluvius*, river	Floodplain
Fol	L. *folia*, leaf	Mass of leaves
Hem	Gr. *hemi*, half	Intermediate stage of decomposition
Hum	L. *humus*, earth	Presence of organic matter
Ochr	Gr. base of *ochros*, pale	Presence of ochric epipedon
Orth	Gr. *orthos*, true	The common ones
Plagg	Modified from Ger. *plaggen*, sod	Presence of plaggen epipedon
Psamm	Gr. *psammos*, sand	Sand texture
Rend	Modified from Rendzina	High carbonate content
Sapr	Gr. *sapros*, rotten	Most decomposed stage
Torr	L. *torridus*, hot and dry	Torric moisture regime
Ud	L. *udus*, humid	Udic moisture regime
Umbr	L. *umbra*, shade	Presence of umbric epipedon
Ust	L. *ustus*, burnt	Ustic moisture regime
Xer	Gr. *xeros*, dry	Xeric moisture regime

From Soil Taxonomy, *1975.*

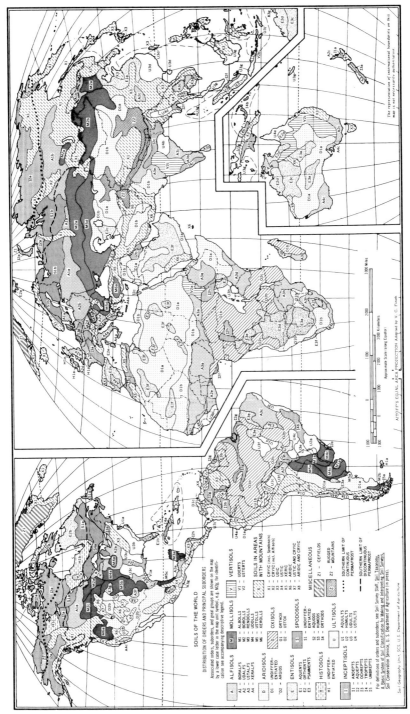

Figure 1-9 Soils of the world. Distribution of orders and principal suborders and great groups.

DISTRIBUTION OF ORDERS AND PRINCIPAL SUBORDERS AND GREAT GROUPS
1:50,000,000

A **ALFISOLS**—Soils with subsurface horizons of clay accumulation and medium to high base supply; either usually moist or moist for 90 consecutive days during a period when temperature is suitable for plant growth.

    A1 **Boralfs**—cold.
        A1a with Histosols, cryic temperature regimes common
        A1b with Spodosols, cryic temperature regimes

    A2 **Udalfs**—temperate to hot, usually moist.
        A2a with Aqualfs
        A2b with Aquolls
        A2c with Hapludults
        A2d with Ochrepts
        A2e with Troporthents
        A2f with Udorthents

    A3 **Ustalfs**—temperate to hot, dry more than 90 cumulative days during periods when temperature is suitable for plant growth.
        A3a with Tropepts
        A3b with Troporthents
        A3c with Tropustults
        A3d with Usterts
        A3e with Ustochrepts
        A3f with Ustolls
        A3g with Ustorthents
        A3h with Ustox
        A3j Plinthustalfs with Ustorthents

    A4 **Xeralfs**—temperate or warm, moist in winter and dry more than 45 consecutive days in summer.
        A4a with Xerochrepts
        A4b with Xerorthents
        A4c with Xerults

D **ARIDISOLS**—Soils with pedogenic horizons, usually dry in all horizons and never moist as long as 90 consecutive days during a period when temperature is suitable for plant growth.

    D1 **Aridisols**—undifferentiated.
        D1a with Orthents
        D1b with Psamments
        D1c with Ustalfs

    D2 **Argids**—with horizons of clay accumulation.
        D2a with Fluvents
        D2b with Torriorthents

E **ENTISOLS**—Soils without pedogenic horizons, either usually wet, usually moist, or usually dry.

    E1 **Aquents**—seasonally or perenially wet.
        E1a Haplaquents with Udifluvents
        E1b Psammaquents with Haplaquents
        E1c Tropaquents with Hydraquents

    E2 **Orthents**—loamy or clayey textures, many shallow to rock.
        E2a Cryorthents
        E2b Cryorthents with Orthods
        E2c Torriorthents with Aridisols
        E2d Torriorthents with Ustalfs
        E2e Xerorthents with Xeralfs

    E3 **Psamments**—sand or loamy sand textures.
        E3a with Aridisols
        E3b with Orthox
        E3c with Torriorthents
        E3d with Ustalfs
        E3e with Ustox
        E3f shifting sands
        E3g Ustipsamments with Ustolls

H **HISTOSOLS**—Organic soils.

    H1 **Histosols**—undifferentiated.
        H1a with Aquods
        H1b with Boralfs
        H1c with Cryaquepts

Figure 1-9 (*Continued*)

I **INCEPTISOLS**—Soils with pedogenic horizons of alteration or concentration but without accumulations of translocated materials other than carbonates or silica; usually moist or moist for 90 consecutive days during a period when temperature is suitable for plant growth.

I1 **Andepts**—amorphous clay or vitric volcanic ash or pumice.
I1a Dystrandepts with Ochrepts

I2 **Aquepts**—seasonally wet.
I2a Cryaquepts with Orthents
I2b Haplaquepts with Salorthids
I2c Haplaquepts with Humaquepts
I2d Haplaquepts with Ochraqualfs
I2e Humaquepts with Psamments
I2f Tropaquepts with Hydraquents
I2g Tropaquepts with Plinthaquults
I2h Tropaquepts with Tropaquents
I2j Tropaquepts with Tropudults

I3 **Ochrepts**—thin, light-colored surface horizons and little organic matter.
I3a Dystrochrepts with Fragiochrepts
I3b Dystrochrepts with Orthox
I3c Xerochrepts with Xerolls

I4 **Tropepts**—continuously warm or hot.
I4a with Ustalfs
I4b with Tropudults
I4c with Ustox

I5 **Umbrepts**—dark-colored surface horizons with medium to low base supply.
I5a with Aqualfs

M **MOLLISOLS**—Soils with nearly black, organic-rich surface horizons and high base supply; either usually moist or usually dry.

M1 **Albolls**—light gray subsurface horizon over slowly permeable horizon; seasonally wet.
M1a with Aquepts

M2 **Borolls**—cold.
M2a with Aquolls
M2b with Orthids
M2c with Torriorthents

M3 **Rendolls**—subsurface horizons have much calcium carbonate but no accumulation of clay.
M3a with Usterts

M4 **Udolls**—temperate or warm, usually moist.
M4a with Aquolls
M4b with Eutrochrepts
M4c with Humaquepts

M5 **Ustolls**—temperate to hot, dry more than 90 cumulative days in year.
M5a with Argialbolls
M5b with Ustalfs
M5c with Usterts
M5d with Ustochrepts

M6 **Xerolls**—cool to warm, moist in winter and dry more then 45 consecutive days in summer.
M6a with Xerorthents

O **OXISOLS**—Soils with pedogenic horizons that are mixtures principally of kaolin, hydrated oxides, and quartz and are low in weatherable minerals.

O1 **Orthox**—hot, nearly always moist.
O1a with Plinthaquults
O1b with Tropudults

O2 **Ustox**—warm or hot, dry for long periods but moist more than 90 consecutive days in the year.

O2a with Plinthaquults
O2b with Tropustults
O2c with Ustalfs

Figure 1-9 (*Continued*)

S **SPODOSOLS**—Soils with accumulation of amorphous materials in subsurface horizons; usually moist or wet.

  S1 **Spodosols**—undifferentiated.
  S1a cryic temperature regimes; with Boralfs
  S1b cryic temperature regimes; with Histosols

  S2 **Aquods**—seasonally wet.
  S2a Haplaquods with Quartzipsamments

  S3 **Humods**—with accumulations of organic matter in subsurface horizons.
  S3a with Hapludalfs

  S4 **Orthods**—with accumulations of organic matter, iron, and aluminum in subsurface horizons.
  S4a Haplorthods with Boralfs

U **ULTISOLS**—Soils with subsurface horizons of clay accumulation and low base supply; usually moist or moist for 90 consecutive days during a period when temperature is suitable for plant growth.

  U1 **Aquults**—seasonally wet.
  U1a Ochraquults with Udults
  U1b Plinthaquults with Orthox
  U1c Plinthaquults with Plinthaquox
  U1d Plinthaquults with Tropaquepts

  U2 **Humults**—temperate or warm and moist all of year; high content of organic matter.
  U2a with Umbrepts

  U3 **Udults**—temperate to hot; never dry more than 90 cumulative days in the year.
  U3a with Andepts
  U3b with Dystrochrepts
  U3c with Udalfs
  U3d Hapludults with Dystrochrepts

  U3e Rhodudults with Udalfs
  U3f Tropudults with Aquults
  U3g Tropudults with Hydraquents
  U3h Tropudults with Orthox
  U3j Tropudults with Tropepts
  U3k Tropudults with Tropudalfs

  U4 **Ustults**—warm or hot; dry more than 90 cumulative days in the year.
  U4a with Ustochrepts
  U4b Plinthustults with Ustorthents
  U4c Rhodustults with Ustalfs
  U4d Tropustults with Tropaquepts
  U4e Tropustults with Ustalfs

V **VERTISOLS**—Soils with high content of swelling clays; deep, wide cracks develop during dry periods.

  V1 **Uderts**—usually moist in some part in most years; cracks open less then 90 cumulative days in the year.
  V1a with Usterts

  V2 **Usterts**—cracks open more than 90 cumulative days in the year.
  V2a with Tropaquepts
  V2b with Tropofluvents
  V2c with Ustalfs

X **Soils in areas with mountains**—Soils with various moisture and temperature regimes; many steep slopes; relief and total elevation vary greatly from place to place. Soils vary greatly within short distances and with changes in altitude; vertical zonation common.
  X1 Cryic great groups of Entisols, Inceptisols, and Spodosols.
  X2 Boralfs and cryic great groups of Entisols and Inceptisols.
  X3 Udic great groups of Alfisols, Entisols, and Ultisols; Inceptisols.

Figure 1-9 (*Continued*)

X4   Ustic great groups of Alfisols, Inceptisols, Mollisols, and Ultisols.

X5   Xeric great groups of Alfisols, Entisols, Inceptisols, Mollisols, and Ultisols.

X6   Torric great groups of Entisols; Aridisols.

X7   Ustic and cryic great groups of Alfisols, Entisols, Inceptisols, and Mollisols; ustic great groups of Ultisols; cryic great groups of Spodosols.

X8   Aridisols, torric and cryic great groups of Entisols, and cryic

great groups of Spodosols and Inceptisols.

Z MISCELLANEOUS

Z1   Icefields.

Z2   Rugged mountains—mostly devoid of soil (includes glaciers, permanent snowfields and, in some places, areas of soil).

. . . Southern limit of continuous permafrost.

-- Southern limit of discontinuous permafrost.

An extragrade subgroup has aberrant properties that do not represent inter-grades to any known kind of soil. Hard rock, for example, is not considered to be soil. Consequently, a soil with underlying hard rock at a depth of less than 50 centimeter is placed in a lithic subgroup. A permanently frozen layer (permafrost) below the soil is the basis for placing a soil in a pergelic subgroup. These subgroups, in a sense, are made up of intergrades to "not soil." A soil at the base of a slope may accumulate sediments slowly, in time producing only a very thick, dark-colored A1 horizon. Such a soil is placed in a cumulic subgroup because there is no known thick soil that consists of only an A1 horizon.

Table 1-10   Formative Elements in Names of Great Groups

Formative Element	Derivation	Connotation
Acr	Modified from Gr. *akros*, at the end	Extreme weathering
Agr	L. *ager*, field	An agric horizon
Alb	L. *albus*, white	An albic horizon
And	Modified from ando	Andolike
Arg	Modified from argillic horizon; L. *argilla*, white clay.	An argillic horizon
Bor	Gr. *boreas*, northern	Cool
Calc	L. *calcis*, lime	A calcic horizon
Camb	L. *cambiare*, to exchange	A cambic horizon
Chrom	Gr. *chroma*, color	High chroma
Cry	Gr. *kryos*, icy cold	Cold
Dur	L. *durus*, hard	A duripan
Dystr, dys	Modified from Gr. *dys*, ill; dystrophic, infertile	Low base saturation
Eutr, eu	Modified from Gr. *eu*, good; eutrophic, fertile	High base saturation

Table 1-10   (*Continued*)

Formative Element	Derivation	Connotation
Ferr	L. *ferrum*, iron	Presence of iron
Fluv	L. *fluvus*, river	Floodplain
Frag	Modified from L. *fragilis*, brittle	Presence of fragipan
Fragloss	Compound of fra(g) and gloss	See the formative elements frag and gloss
Gibbs	Modified from gibbsite	Presence of gibbsite in sheets or nodules
Gyps	L. *gypsum*, gypsum	Presence of a gypsic horizon
Gloss	Gr. *glossa*, tongue	Tongued
Hal	Gr. *hals*, salt	Salty
Hapl	Gr. *haplous*, simple	Minimum horizon
Hum	L. *humus*, earth	Presence of humus
Hydr	Gr. *hydor*, water	Presence of water
Luv	Gr. *louo*, to wash	Illuvial
Med	L. *media*, middle	Of temperate climates
Nadur	Compound of na(tr) and dur	See the formative elements natr and dur
Natr	Modified from *natrium*, sodium	Presence of natric horizon
Ochr	Gr. base of *ochros*, pale	Presence of ochric epipedon
Pale	Gr. *paleos*, old	Excessive development
Pell	Gr. *pellos*, dusky	Low chroma
Plac	Gr. base of *plax*, flat stone	Presence of thin pan
Plagg	Modified from Ger. *plaggen*, sod	Presence of plaggen epipedon
Plinth	Gr. *plinthos*, brick	Presence of plinthite
Psamm	Gr. *psammos*, sand	Sand texture
Quartz	Ger. *quarz*, quartz	High quartz content
Rhod	Gr. base of *rhodon*, rose	Dark red color
Sal	L. base of *sal*, salt	Presence of salic horizon
Sider	Gr. *sideros*, iron	Presence of free iron oxides
Sombr	F. *sombre*, dark	A dark horizon
Sphagn	Gr. *sphagnos*, bog	Presence of sphagnum
Sulf	L. *sulfur*, sulfur	Presence of sulfides or their oxidation products
Torr	L. *torridus*, hot and dry	Torric moisture regime
Trop	Modified from Gr. *tropikos*, of the solstice	Humid and continually warm
Ud	L. *udus*, humid	Udic moisture regime
Umbr	L. base *umbra*, shade	Presence of umbric epipedon
Ust	L. base of *ustus*, burnt	Ustic moisture regime
Verm	L. base of *vermes*, worm	Wormy, or mixed by animals
Vitr	L. *vitrum*, glass	Presence of glass
Xer	Gr. *xeros*, dry	A xeric moisture regime

From Soil Taxonomy, *1975.*

Families (of Mineral Soils)

A complete family name consists of a subgroup name preceded by a few (usually three) modifiers that narrow the range of properties enough to permit general statements about use and management of the soils. Modifiers in a family name of mineral soils represent the following differences, listed according to their sequence in the family name.

Particle-size class.
Mineralogy class.
Calcareous and reaction classes.
Soil temperature class.
Soil depth class.
Soil slope class.
Soil consistence class.
Coatings.
Class of permanent cracks.

Series

The series is the name of an individual soil; as a rule, they are abstract names. The name is usually taken from a place near the one where the series was first recognized. The name of a series carries no meaning to people who have no other information about the soil. The series, however, often carries a clue to the geographic location of the soil, as in the case of Houston soils in Texas and San Joaquin soils in the San Joaquin valley in California.

Classification of Abac Series—An Example

The Abac series is classified as follows: loamy, mixed (calcareous), frigid, shallow Typic Ustorthents. This name provides the following information about the Abac soils.

They have no significant pedogenic horizons and only moderate to small amounts of organic matter (from *ent*).
They are on recent slopes, subject to erosion (from *orth*).
They are well drained and are dry for a significant part of year, but they are dry for less than one-half of the growing season (from *ust*).
There is soft rock, probably calcareous, within 50 centimeters of the surface (from *shallow* and *calcareous*).
The soils are warm in summer and frozen in winter (a snow cover is unlikely

for soils in an ustic great group), and the mean annual soil temperature is between 0 and 8°C (from *frigid*).

The mineralogy is not dominated by any one mineral; it is a mixture of several, including some $CaCO_3$, but less than 40 percent (from *mixed, calcareous*).

The clay percentage is less than 35, and the texture is finer than loamy sand (from *loamy*).

Abac soils, having such a combination of properties, are not likely to be cultivated. They can produce a moderate amount of forage for grazing in late spring and early summer.

The Soil Type—Not a Part of Soil Taxonomy

The soil type has been the lowest category in all previous classification systems that were used in the United States. Types have been distinguished within series on the basis of texture of the soil surface, a single characteristic. At first, the distinction involved the texture of the soil when texture meant a combination of particle size, structure, and consistence. In recent years, when texture has been defined in terms of particle size distribution alone, the texture of the plow layer or of its equivalent in virgin soils was used to distinguish types within a series. Because the significance of the texture or particle size distribution of the plow layer is mainly pragmatic and the texture classes can be made more useful if adjusted to fit circumstances at a given time, the soil type has not been retained as a category of this system. This mention is made to explain its disappearance as a category. The texture of the plow layer will commonly be shown in the soil name in published soil surveys as heretofore but will be considered as a part of the *phase* name instead of the name of a taxon.

Simplified Key for Orders in Soil Taxonomy

A key is used to place soils in the various categories of the classification system. The organic soils are keyed out first. A very abbreviated key to the orders is as follows.

Organic soils	HISTOSOLS
Soils with spodic horizons	SPODOSOLS
Soils with oxic horizons	OXISOLS
Inverted, clayey soils	VERTISOLS
Soils with aridic soil moisture regime	ARIDISOLS

Mesic or warmer soils with argillic horizons and base saturation less than 35 percent	ULTISOLS
Soils with mollic epipedons	MOLLISOLS
Soils with argillic horizons and base saturation 35 percent or more	ALFISOLS
Soils with slightly altered horizons	INCEPTISOLS
All other soils	ENTISOLS

The Entisol order is keyed out last and includes all soils not placed in one of the other orders; consequently, it is an order with very diverse soils.

Categories in the Canadian Soil Classification

The Canadian System of Soil Classification was published in 1978. The system has five categories: order, great group, subgroup, family, and series.

Orders

Taxa at the order level are based on properties of the pedon that reflect the nature of the soil environment and the effects of the dominant soil-forming processes. Nine soil orders are recognized.

Chernozemic soils, soils of the Chernozemic order, are generally characterized by well- to imperfectly drained soils with chernozemic A horizons. They are the dominant soils of the grassland plains or steppes. *Solonetzic* soils have solonetzic B horizons and develop on saline parent materials in some areas of the semiarid and subhumid Interior Plains in association with Chernozemic soils and, to a lesser extent, with Luvisolic and Gleysolic soils. *Luvisolic* soils are characterized by light-colored eluvial horizons underlain by illuvial horizons in which silicate clay has accumulated (Bt). Most Luvisols developed under forest, and some exist in the forest-grassland transition zones. Soils of the *Podzolic* order are Podzols and have a podzolic B horizon. Podzols typically develop under forest or heath in cool to very cold climates from medium to coarse-textured and acid or low-lime parent materials. *Brunisolic* soils are weakly developed soils that are comparable to Inceptisols in United States. *Regosolic* soils are younger than Brunisols and have no genetic horizons. *Gleysolic* soils have features indicative of periodic or prolonged water saturation and reducing conditions. Gleysols have wide distribution. Soils of the *Organic* order are composed mostly of organic materials and include soils commonly known as peat, muck, or bog. *Cryosolic* soils are both mineral and organic soils with permafrost within at least 2 meters of the soil surface.

Great Groups

Great groups are soil taxa formed by the subdivision of the orders. Each group carries with it the differentiating criteria of the order and, in addition, properties that reflect differences in the strengths of dominant processes or a major contribution of a process in addition to the dominant one. For example, in Luvic Gleysols, the dominant process is gleying (Gleysols), but clay translocation is also a major process (Luvisols). The orders and great groups are given in Table 1-11 along with approximate equivalents in American and FAO systems.

Subgroups, Families, and Series

Soils of the subgroups carry the differentiating criteria of the order and great group to which they belong. Subgroups are differentiated on the basis of the kind and arrangement of horizons that indicate conformity to the central concept of the great group or intergrading to another order or some additional feature as an ortstein layer. The family differentiates soils of subgroups on the basis of parent material characteristics such as texture and mineralogical composition and on the basis of soil climatic factors and soil reaction (pH). Series within a family are differentiated on the basis of detailed features of the pedon. Pedons belonging to a series have similar kinds and arrangements of horizons whose color, texture, structure, consistence, thickness, reaction, and composition fall within a narrow range.

Simplified Key for Orders in the Canadian Soil Classification

One of the most recent major changes in *The Canadian System of Soil Classification* was the introduction of the Cryosolic order in 1978. Cryosols are soils with permafrost and are keyed out first. Organic soils keyed out second, whereas they are keyed first in United States. A very abbreviated key to the soil orders is as follows.

Soils with permafrost	CRYOSOLIC
Organic soils	ORGANIC
Soils with podzolic B	PODZOLIC
Soils water saturated all or part of the time	GLEYSOLIC
Soils with solonetzic B	SOLONETZIC
Soils with chernozemic A	CHERNOZEMIC
Other soils with Bt horizons	LUVISOLIC
Other soils with minimal horizon development, cambic or weak Bt or podzolic B	BRUNISOLIC
All other soils	REGOSOLIC

Table 1-11   Correlation of Canadian Orders and Great Groups with Nearest
Equivalents in Soil Taxonomy and the FAO Classification

Canadian	Soil Taxonomy	FAO
Chernozemic	Boroll	Kastanozem, Chernozem, Rendzina, Phaeozem
Brown	Aridic Boroll	Kastanozem (aridic)
Dark Brown	Typic Boroll	Kastanozem (typic)
Black	Udic Boroll, Rendoll	Chernozem, Rendzina
Dark Gray	Boralfic Boroll, Alboll	Greyzem
Solonetzic	Natric great groups of Mollisol and Alfisol	Solonetz
Solonetz	Natric great groups of Mollisol and Alfisol	Solonetz
Solodized Solonetz	Natric great groups of Mollisol and Alfisol	Solentz
Solod	Glossic Natriboroll, Natriboroll	Solodic Planosol
Luvisolic	Boralf, Udalf	Luvisol
Gray Brown	Hapludalf or Glossudalf	Albic Luvisol
Gray	Boralf	Albic Luvisol, Podzoluvisol
Podzolic	Spodosol	Podzol
Humic	Cryaquod, Humod	Humic Podzol
Ferro-Humic	Humic Cryorthod, Humic Haplorthod	Orthic Podzol
Humo-Ferric	Cryorthod, Haplorthod	Orthic Podzol
Burnisolic	Inceptisol	Cambisol
Melanic	Cryochrept, Eutrochrept, Hapludoll	Cambisol, Eutric Cambisol
Eutric	Cryorchrept, Eutrochrept	Eutric Cambisol
Sombric	Umbric Dystrochrept	Dystric Cambisol
Dystric	Dystrochrept, Cryorchrept	Dystric Cambisol
Regosolic	Entisol	Regosol, Fluvisol
Regosol	Entisol	Regosol
Humic Regosol	Entisol	Regosol, Fluvisol
Gleysolic	Aquic suborders	Gleysol, Planosol
Humic	Aquoll, Humaquept	Gleysol
Gleysol	Aquent, Fluvent, Aquept	Gleysol
Luvic Gleysol	Argialboll, Argiaquoll, Aqualf	Planosol
Organic	Histosol	Histosol
Fibrisol	Fibrist	Histosol
Mesisol	Hemist	Histosol

Table 1-11   (*Continued*)

Canadian	Soil Taxonomy	FAO
Humisol	Saprist	Histosol
Folisol	Folist	Histosol
Cryosolic	Pergelic subgroups	Gelic
Turbic	Pergelic Ruptic subgroups	Cambisol, Regosol, Fluvisol
Static	Pergelic subgroups	—
Organic	Pergelic Histosol	Gelic Histosol

Adapted from The Canadian System of Soil Classification, *1978.*

References

Baier, W., and A. R. Mack, "Development of Soil Temperature and Soil Water Criteria for Characterizing Soil Climates in Canada," in *Field Soil Water Regime,* Soil Sci. Soc. Am. Spec. Pub. 5, pp. 195–212, 1973.

Canada Soil Survey Committee, *The Canadian System of Soil Classification,* Can. Dept. Agric. Publ. 1646, Supply and Services, Canada, Ottawa, 1978.

Edelman, E. H., *Soils of the Netherlands,* North Holland Pub., Amsterdam, 1950.

Simonson, R. W., "Soil Classification in the United States," *Science, 137:*1027–1034, 1962.

Soil Conservation Service, *Classification of Soil Series of the United States* (mimeo), USDA, Washington, D.C., 1977.

Soil Survey Staff, *Soil Taxonomy,* USDA Agriculture Handbook 436, Washington, D.C., 1975.

Thorp, J., and G. D. Smith, "Higher Categories of Soil Classification: Order, Suborder and Great Soil Groups," *Soil Sci., 67:*117–126, 1949.

2

Entisols of the United States

Central Concept and Suborders

The central concept of Entisols is that of soils with little or no evidence of the development of pedogenic horizons. Many Entisols have ochric epipedons, and some of the sands have albic horizons. Histic epipedons are found in some Entisols in coastal marshes.

There are several reasons why horizon development is so limited. The time has been too short in many Entisols and many have formed on steep actively eroding slopes. On floodplains and some other surfaces, new deposits of material frequently occur. Some Entisols, however, have formed in old materials that consist mostly of quartzic sand that does not alter to form horizons. A few Entisols have been created by deep plowing or land leveling, which destroys and mixes existing pedogenic horizons.

Entisols rank sixth out of 10 soils in abundance in the world, and great variations exist in this order. Entisols may have any moisture or temperature regime, parent

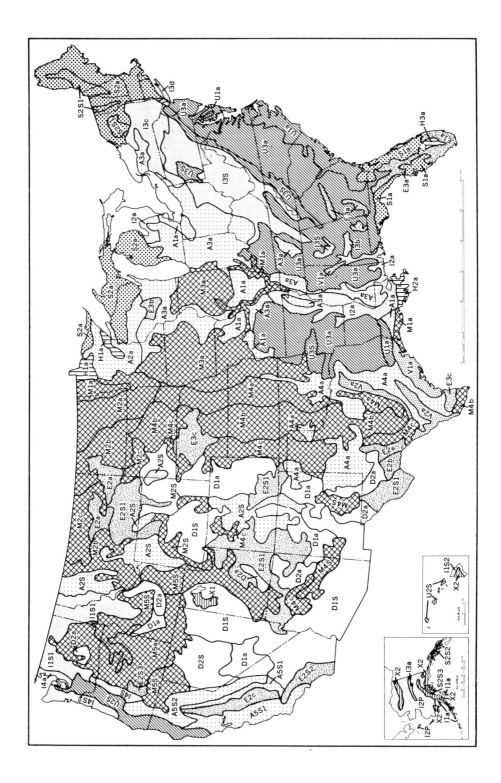

Figure 2-1 General soil map of the United States. (From *Soil Taxonomy*, 1975.)

ALFISOLS

Aqualfs

A1a—Aqualfs with Udalfs, Haplaquepts, Udolls; gently sloping.

Boralfs

A2a—Boralfs with Udipsamments and Histosols; gently and moderately sloping.

A2S—Cryoboralfs with Borolls, Cryochrepts, Cryorthods, and rock outcrops; steep.

Udalfs

A3a—Udalfs with Aqualfs, Aquolls, Rendolls, Udolls, and Udults; gently or moderately sloping.

Ustalfs

A4a—Ustalfs with Ustochrepts, Ustolls, Usterts, Ustipsamments, and Ustorthents; gently or moderately sloping.

Xeralfs

A5S1—Xeralfs with Xerolls, Xerorthents, and Xererts; moderately sloping to steep.

A5S2—Ultic and lithic subgroups of Haploxeralfs with Andepts, Xerults, Xerolls, and Xerochrepts; steep.

ARIDISOLS

Argids

D1a—Argids with Orthids, Orthents, Psamments, and Ustolls; gently and moderately sloping.

D1S—Argids with Orthids, gently sloping; and Torriorthents, gently sloping to steep.

Orthids

D2a—Orthids with Argids, Orthents, and Xerolls; gently or moderately sloping.

D2S—Orthids, gently sloping to steep, with Argids, gently sloping; lithic subgroups of Torriorthents and Xerorthents, both steep.

ENTISOLS

Aquents

E1a—Aquents with Quartzipsamments, Aquepts, Aquolls, and Aquods; gently sloping.

Orthents

E2a—Torriorthents, steep, with borollic subgroups of Aridisols; Usterts and aridic and vertic subgroups of Borolls; gently or moderately sloping.

E2b—Torriorthents with Torrerts; gently or moderately sloping.

E2c—Xerorthents with Xeralfs, Orthids, and Argids; gently sloping.

E2S1—Torriorthents; steep, and Argids, Torrifluvents, Ustolls, and Borolls; gently sloping.

E2S2—Xerorthents with Xeralfs and Xerolls; steep.

E2S3—Cryorthents with Cryopsamments and Cryandepts; gently sloping to steep.

Psamments

E3a—Quartzipsamments with Aquults and Udults; gently or moderately sloping.

E3b—Udipsamments with Aquolls and Udalfs; gently or moderately sloping.

E3c—Ustipsamments with Ustalfs and Aquolls; gently or moderately sloping.

HISTOSOLS

Histosols

H1a—Hemists with Psammaquents and Udipsamments; gently sloping.

H2a—Hemists and Saprists with Fluvaquents and Haplaquepts; gently sloping.

H3a—Fibrists, Hemists, and Saprists with Psammaquents; gently sloping.

INCEPTISOLS

Andepts

I1a—Cryandepts with Cryaquepts,

Figure 2-1 *(Continued)*

Histosols, and rockland; gently or moderately sloping.

I1S1—Cryandepts with Cryochrepts, Cryumbrepts, and Cryorthods; steep.

I1S2—Andepts with Tropepts, Ustolls, and Tropofolists; moderately sloping to steep.

Aquepts

I2a—Haplaquepts with Aqualfs, Aquolls, Udalfs, and Fluvaquents; gently sloping.

I2P—Cryaquepts with cryic great groups of Orthents, Histosols, and Ochrepts; gently sloping to steep.

Ochrepts

I3a—Cryochrepts with cryic great groups of Aquepts, Histosols, and Orthods; gently or moderately sloping.

I3b—Eutrochrepts with Uderts; gently sloping.

I3c—Fragiochrepts with Fragiaquepts, gently or moderately sloping; and Dystrochrepts, steep.

I3d—Dystrochrepts with Udipsamments and Haplorthods; gently sloping.

I3S—Dystrochrepts, steep, with Udalfs and Udults; gently or moderately sloping.

Umbrepts

I4a—Haplumbrepts with Aquepts and Orthods; gently or moderately sloping.

I4S—Haplumbrepts and Orthods; steep, with Xerolls and Andepts; gently sloping.

MOLLISOLS

Aquolls

M1a—Aquolls with Udalfs, Fluvents, Udipsamments, Ustipsamments, Aquepts, Eutrochrepts, and Borolls; gently sloping.

Borolls

M2a—Udic subgroups of Borolls with Aquolls and Ustorthents; gently sloping.

M2b—Typic subgroups of Borolls with Ustipsamments, Ustorthents, and Boralfs; gently sloping.

M2c—Aridic subgroups of Borolls with Borollic subgroups of Argids and Orthids, and Torriorthents; gently sloping.

M2S—Borolls with Boralfs, Argids, Torriorthents, and Ustolls; moderately sloping or steep.

Udolls

M3a—Udolls, with Aquolls, Udalfs, Aqualfs, Fluvents, Psamments, Ustorthents; Aquepts, and Albolls; gently or moderately sloping.

Ustolls

M4a—Udic subgroups of Ustolls with Orthents, Ustochrepts, Usterts, Aquents, Fluvents, and Udolls; gently or moderately sloping.

M4b—Typic subgroups of Ustolls with Ustalfs, Ustipsamments, Ustorthents, Ustochrepts, Aquolls, and Usterts; gently or moderately sloping.

M4c—Aridic subgroups of Ustolls with Ustalfs, Orthids, Ustipsamments, Ustorthents, Ustochrepts, Torriorthents, Borolls, Ustolls, and Usterts, gently or moderately sloping.

M4S—Ustolls with Argids and Torriorthents; moderately sloping or steep.

Xerolls

M5a—Xerolls with Argids, Orthids, Fluvents, Cryoboralfs, Cryoborolls, and Xerorthents; gently or moderately sloping.

M5S—Xerolls with Cryoboralfs, Xeralfs, Xerorthents, and Xererts; moderately sloping or steep.

SPODOSOLS

Aquods

S1a—Aquods with Psammaquents, Aquolls, Humods, and Aquults; gently sloping.

Orthods

S2a—Orthods with Boralfs, Aquents, Orthents, Psamments, Histosols, Aquepts, Fragiochrepts, and

Figure 2-1 (*Continued*)

Dystrochrepts; gently or moderately sloping.
S2S1—Orthods with Histosols, Aquents, and Aquepts; moderately sloping or steep.
S2S2—Cryorthods with Histosols; moderately sloping or steep.
S2S3—Cryorthods with Histosols, Andepts and Aquepts; gently sloping to steep.

ULTISOLS

Aquults
U1a—Aquults with Aquents, Histosols, Quartzipsamments, and Udults; gently sloping.

Humults
U2S—Humults with Andepts, Tropepts, Xerolls, Ustolls, Orthox, Torrox, and rock land; gently sloping to steep.

Udults
U3a—Udults with Udalfs, Fluvents, Aquents, Quartzipsamments, Aquepts, Dystrochrepts, and Aquults; gently or moderately sloping.

U3S—Udults with Dystrochrepts; moderately sloping or steep.

VERTISOLS

Uderts
V1a—Uderts with Aqualfs, Eutrochrepts, Aquolls, and Ustolls; gently sloping.

Usterts
V2a—Usterts with Aqualfs, Orthids, Udifluvents, Aquolls, Ustolls, and Torrerts; gently sloping.

Areas With Little Soil

X1—Salt flats.
X2—Rock land (plus permanent snowfields and glaciers).

Slope Classes

Gently sloping—Slopes mainly less than 10 percent, including nearly level.
Moderately sloping—Slopes mainly between 10 and 25 percent.
Steep—Slopes mainly steeper than 25 percent.

---

material, or vegetation. The only features common to all Entisols is the near absence of pedogenic horizons and the mineral nature of the soil. A very brief characterization of the suborders is as follows.

Aquents—wet, aquic moisture regime.

Arents—disturbed pedogenic horizons.

Fluvents—irregular distribution of organic matter with depth.

Psamments—loamy sand or coarser throughout.

Orthents—other Entisols.

The general distribution of Entisols in the United States is given in Fig. 2-1. About 66 percent of the Entisols are Orthents (see appendix Table 2) and located in the western United States, where limited rainfall, steep slopes, and the recent origin of parent material are major factors contributing to their existence (E2 areas in Fig. 2-1). Psamments are the dominant Entisols in the Sand Hills of north-central Nebraska and southern South Dakota, in Minnesota, Wisconsin, and Michigan, and the coastal plain of the southeastern United States. Aquents are common in the southern one-third of the Florida peninsula and in marshy areas along the Atlantic coast.

Aquents

Aquents are Entisols with aquic or peraquic moisture regimes. They usually have bluish and gray subsoil colors. They are common where soils are saturated at least part of the year in tidal marshes along the coast, deltas, margins of lakes, and floodplains (see Fig. 2-2).

About 0.2 percent of the soils in the United States are Aquents. Aquents occur as minor inclusions in many landscapes and are extensive only in landscapes in the southeastern United States. The major area of Aquents in the United States is the lower coastal plain in southern Florida (see area Ela of Fig. 2-1). The parent materials are mainly marine sands and marl. Drainage is needed for crop production, and groundwater is readily available for irrigation during dry seasons. The hyperthermic temperature regime with limited frost hazard makes soils of southern Florida valuable for production of winter vegetables, citrus, and improved pasture. The major associated soils are Aquods and Aquolls.

Arents

Arents are created by activities of humans. Arents are Entisols that do not have horizons because they have been deeply plowed, spaded, or otherwise mixed. Some of these soils result from cuts and fills, as in land leveling to prepare land for

Figure 2-2   Aquent landscape on the lower coastal plain of the southeastern United States. Some Aquents in coastal marshes are bathed with saltwater, contain sulfides, and are Sulfaquents.

Figure 2-3   Kanima soils, Udalfic Arents, developed under hardwood trees and grass in parent material from strip mining of coal in Tulsa and Wagoner counties, Oklahoma. The soils are well suited for wildlife habitat. (Photo USDA-SCS.)

irrigation, and some are created by intensive bombardment during a war. Arents are not common in the United States, but their area is increasing as a result of larger tractors that make earth movement easier.

Two kinds of Arents are currently recognized in the United States. Hapludollic Arents have a udic moisture regime and have fragments of a mollic epipedon within the upper meter of soil. Udalfic Arents have a udic moisture regime and fragments of an argillic horizon that has 35 percent or more base saturation within the upper meter of soil. Only two series have been recognized by the USDA up to 1977, and both are in Oklahoma. They are Barge soils (Hapludollic Arents), developed from materials from river dredging, and Kanima (Udalfic Arents), developed on shaly material from strip mining of coal (see Fig. 2-3).

Some undesirable effects of deep plowing can occur. Deep plowing makes seedbed preparation difficult if the subsoil material is clayey. In an experiment designed to measure the effect of creating a 50-cemtimeter-thick layer of "topsoil" high in fertility, it took three plowings over a period of 7 years to obtain plant populations and corn yields on deeply plowed plots equal to those of check plots. Yields on the Arent eventually surpassed those of the Udalf in the eighth year. Where soils have calcareous subsoils, deep plowing may create plant nutrient deficiencies of iron or zinc. This may occur when soils with ustic and aridic moisture regimes are leveled for gravity flow of irrigation water.

Fluvents

Fluvents are formed in recent water-deposited sediments on floodplains, fans and deltas of rivers, and small streams (Fig. 2-4). Fluvents do not occur in backwater swamps where drainage is poor because Fluvents cannot have an aquic moisture regime. Sediments are young; they are usually only a few decades or hundreds of years old in humid regions. Fluvents are frequently flooded and receive new sediments, which results in a building of the soil from the "bottom up." The sedimentation commonly causes stratification. Most alluvial sediments come from eroding soils or stream banks that contain appreciable organic matter. The organic matter moves with and is deposited largely in association with clay, so that the finer-textured layers have more organic matter than coarser-textured layers. Thus Fluvents are characterized by an irregular decrease in organic matter with increasing depth

Figure 2-4   Udifluvents along a river in Massachusetts used for agriculture and subject to flooding except where protected by levees. Note that the road and buildings on the right are on higher land and are not subject to flooding. (USDA photograph.)

that is associated with stratification. The irregular decrease in organic matter with increasing depth is the basis for the definition of Fluvents. Some Fluvents have similar texture and decreasing organic carbon content with increasing depth. These Fluvents must have over 0.2 percent organic carbon at depth of 1.25 meters.

Fluvents can have any vegetation and moisture regime and any temperature regime except pergelic. The subgroups include Cryofluvents, Xerofluvents, Ustifluvents, Torrifluvents, Tropofluvents, and Udifluvents.

No major area of Fluvents is shown on the generalized soil map of the United States (see Fig. 2-1). Most soils formed from alluvium along the Mississippi River are Inceptisols. Fluvents are restricted to recent sediments because the organic matter decomposes and the irregular decrease in organic matter content disappears over time in the absence of frequent deposition of new materials. Fluvents cannot have an aquic moisture regime because of rapid reduction and movement of iron to form a cambic horizon in most wet soils in a large river system such as the Mississippi; the soils are classified as Aquepts.

As the Mississippi River has wandered about over a long period of time Fluvents, in the absence of frequent deposition, have become Inceptisols. When Fluvents are left on terraces because of the cutting down of a river, they may evolve into soils typical of the region, such as Alfisols or Mollisols.

Alluvial soils have had a history of fertility and productivity and are capable of supporting large populations. Most Chinese spend their lives on flat alluvial plains of varying size in close association with Fluvents. About one-fourth of the world's people are supported from soils developed from alluvium, and about one-half of these soils are most likely Fluvents or closely associated soils. Eroding materials carried by rivers are commonly calcareous and laden with nutrients. The periodic rejuvenation of soils by flooding has been symbolized by the flooding of the Nile. The major hazard in their use is flooding.

Psamments

Psamments are Entisols that, below the plow layer or a depth of 25 centimeters, whichever is deeper, have a sand or loamy sand texture in all subhorizons to a depth of 1 meter unless a lithic contact occurs at shallower depths. Psamments have low water-holding capacity and rapid permeability. They are subject to wind erosion when devoid of vegetation and have low support for wheeled vehicles.

Psamments form in sand dunes, cover sands, levees, beaches, and sandy parent materials that were sorted in an earlier geologic cycle. Psamments occur under any climate or vegetation and on surfaces of virtually any age, from recent historic to Pliocene or older. The Psamments of old stable surfaces consist of quartz sand, which cannot form diagnostic horizons of accumulation of clays or sexquioxides. Groundwater is more than 50 centimeters deep and is usually much deeper, so that soils do not have an aquic moisture regime. The three most extensive great groups in the United States are Ustipsamments, Udipsamments, and Quartzipsamments.

Ustipsamments of the Nebraska Sand Hills

Ustipsamments have formed mostly in windblown sands in areas with ustic moisture regime. About 5 million hectares are located in north-central Nebraska and 500,000 hectares in southern South Dakota (area E3c of Fig. 2-1). The Sand Hills occupy almost one-fourth of the state of Nebraska, and smaller areas are located in northeastern Colorado.

Environmental Setting   The Sand Hills landscape is a monotonous succession of dunes, swales with some narrow elongated dry valleys, scattered shallow lakes, and infrequent streams, as illustrated in Fig. 2-5. No semblance of order is apparent in the arrangement of the hills over much of the area, although in some parts, particularly in the center and along the eastern border, a nearly east-west orientation of ridges and valleys is noticeable. The Sand Hills are frequently referred to as sand dunes, but this is not always fitting. Although they are mainly the product of wind action on the disintegrated material of the Ogallala formation, some of the

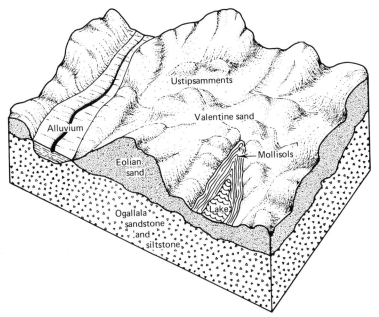

Figure 2-5   Diagram showing relationship of soils to parent materials and topography in the Nebraska Sand Hills. (After Sherfey, 1965.)

Figure 2-6 Ustipsamments in the Nebraska Sand Hills occur in a dunelike landscape and are covered with mid- and tall grasses that protect the soil from wind erosion.

larger dunes still contain cores of the old formation. Wind has moved, sifted, and redeposited the loose material in the form of hills and ridges over 70 meters high in the western sector that decrease to swells on the eastern spurs. The soils of the hills are loose sand, but those in the depressions are chiefly loamy sand and, infrequently, sandy loam.

Mid- and tall grasses cover the area. On the drier sites at higher elevation the mid grasses are dominant; in the depressions where the water supply is greater, tall grasses are dominant. The grasses protect the sand from blowing (see Fig. 2-6).

Normally, precipitation ranges from 46 to 56 centimeters a year; the frost-free season is 140 to 150 days. Little or no surface runoff occurs in the Sand Hills. Precipitation readily infiltrates the sand, and the excess water is transmitted to the groundwater. Lakes are formed on the groundwater level in many of the depressions. The Sand Hills area is important for groundwater recharge of the area to the south. It is estimated that 1.5 million hectare-meters infiltrate annually to the zone of saturation.

Soils of the Nebraska Sand Hills The Sand Hills region is unique in that a single soil series, Valentine, occupies about 95 percent of the area. Valentine soils have a thin, moderately dark-colored ochric epipedon and sand or loamy sand tex-

Table 2-1   Selected Properties of Valentine Soils

Horizon	Depth, Centimeters	Fine Sand	Clay	Organic Carbon	CEC, milliequivalents 100 grams	Ca	Mg	Na	K	pH
A1	0–10	59	4	0.78	4.6	3.3	0.7	0.1	0.5	7.3
AC	10–16	61	3	0.44	3.8	2.6	0.8	0.1	0.3	7.4
C1	16–35	61	3	0.28	3.0	2.4	0.3	0.1	0.2	7.0
C2	35–150	62	3	0.13	2.8	2.0	0.5	0.1	0.2	6.9

From Soil Conservation Service, *1966, soil number 554 Nebr.-16.*

ture; fine sand is the dominant fraction (see Table 2-1). The sands are noncalcareous and mostly quartz with about 20 percent feldspar; potassium feldspar is dominant. Clay content is about 5 percent or less. Only 12 centimeters of available water can be held in the upper 150 centimeters of soil. The low clay and organic matter contents result in a low cation-exchange capacity that is entirely base saturated. Soil pH of all horizons is near 7.

A considerable amount of the precipitation passes through the soil and substratum to the water table and shallow groundwater lakes. Soils with the water table within the root zone of the grasses occur throughout the area surrounding the numerous small lakes. These soils are subirrigated, produce much more biomass than Valentine soils, and are Aquolls. They occupy only 1 or 2 percent of the landscape.

Land Use   The Sand Hills is one of the world's greatest beef-producing areas. Over 90 percent of the land is in pasture for beef cattle. The numerous small lakes provide watering sites; where there are no lakes, windmills pump water from the shallow groundwater (see Fig. 2-7). The major management problem is that of controlled grazing to maintain a vigorous grass cover to protect the land from wind erosion and get maximum production from the land.

All of the food that plants use is manufactured in the leaves. Cutting or grazing off the tops of plants reduces food for growth of both tops and roots. A vigorous root system for range plants can only be maintained by limited or controlled grazing. Maximum production of forage and beef occurs when about one-half of the forage is grazed each year. Overgrazing also results in an undesirable change in forage composition. Grazing results in the decline of some of the original climax species (decreasers) and an increase in some species of the original climax vegetation (increasers). When grazing results in less than 70 percent of the original climax vegetation species, however, there is a rapid increase in invaders, which are primarily weeds and plants of low forage quality (see Fig. 2-8).

Figure 2-7 Nearly all of the land in the Nebraska Sand Hills is used to produce forage for beef cattle. Water is provided by numerous small lakes or windmills that pump water from shallow groundwater.

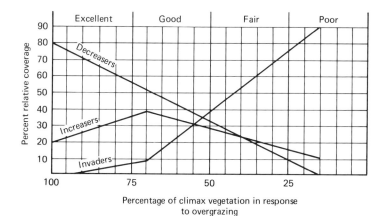

Figure 2-8 Four range condition classes based on percentage of climax vegetation and behavior of climax range plants under increasingly intensive grazing. (From Sherfey, 1965.)

A small amount of land with Aquolls surrounding the small lakes is naturally subirrigated and is used for hay production. This small but important land use provides forage for range cattle during winter blizzards, when the cattle cannot be on the range. Some hay is also cut from the range land and is stacked on the range to provide forage during winter months. Only one-half percent of the land is suited for continuous cultivation. Some areas of finer-textured soil along streams and rivers are used for crops such as corn, rye, and alfalfa. Irrigation is increasing on this small cropland acreage. The high permeability of the soil makes furrow irrigation impossible. Experiments are being conducted to study the use of pivot sprinklers for corn production by killing grass in strips with herbicides. The grass between the corn rows protects the soil from wind erosion, and the irrigation offers the possibility of increasing the forage as well as the corn production.

Cattle ranching began in the Sand Hills about 1870. Homesteading was difficult because 160 acres was not large enough to support a family on the sandy soils, which were unsuited for crop production. In 1904 Representative Kincaid of Nebraska persuaded Congress to pass the Kincaid Act, which increased homestead allotments to 640 acres. Even these larger grants proved inadequate, except when they included some alluvial land. Crops planted in the sand withered, and wind erosion became severe. These settlers, known as Kinkaiders, or "soddies," built homes of sod and "wore their lives out" trying to make farming pay. After the residents became convinced that the land was only suited for ranching, the cattle industry grew rapidly. The grass cover is probably better today than when the Indians frequently used fire to herd buffalo. Large landholdings needed for ranching keep the urban population low. Perhaps the conditions in no other area of comparable size in the Great Plains are as similar to those that existed when settlement began.

Ustipsamments of Southern Texas

The Ustipsamments of Texas are located in southern Texas along the Gulf coast south of Corpus Christi (area E3c, Fig. 2-1). The aeolian sand parent material originated from sediments of barrier islands, lagoons, and bays. The sand fraction is mineralogically similar to the sand in the Nebraska Sand Hills. Many of the sandy soils of this area have a sandy clay loam B horizon generally between 1 and 2 meters deep and are Arenic Paleustalfs. The IIB horizon or argillic horizon is a paleosol that makes the soil less droughty and causes water to pond in some of the slight depressions. Grazing is the dominant land use.

Udipsamments of Central Wisconsin

Udipsamments have a udic moisture regime and occupy about 30 percent of a five-county area in central Wisconsin (see area E3b of Fig. 2-1). The soils occur mainly on moderate to gentle slopes; they developed from acid, sandy, glacial outwash.

The sand fraction is of mixed mineralogy and is similar to that of the Ustipsamments of Nebraska. Plainfield sand is representative, and the sand fraction contains about 25 percent weatherable minerals. Clay content is about 5 percent or less, soils are quite acid, and base saturation is less than 10 percent in the C horizon. Natural fertility is low. Native vegetation consists of jack pine and scrub oak. Rainfall is moderate and barely adequate for crops and pastures in normal years. Droughts are common; they plagued early settlers of the area. Water is abundant in the glacial sediments, and large pivot irrigation systems are used to irrigate truck crops and corn. Cranberries are produced on some of the bogs in the area.

Outcrops of extremely quartzose sandstone occur in the area. Boone soils have developed from the weathered sandstone and contain only 1 or 2 percent weatherable minerals. Boone soils are considered the least fertile soils in Wisconsin and are Quartzipsamments—the great group to be discussed next.

Quartzipsamments of the Southeastern United States

Quartzipsamments are old soils and have a sand fraction that is 95 percent or more quartz, zircon, tourmaline, rutile, or other insoluble minerals that do not weather to liberate iron or aluminum or bases. As a result, clays are not synthesized, spodic horizons do not form, and the soils are very infertile. The sand grains are mainly uncoated. The major areas are located in the southeastern United States (areas E3a in Fig. 2-1) and a landscape is shown in Fig. 2-9.

The cover sands are thick and have been intensely weathered. They originated from weathering of rocks in the Piedmont and the Appalachian mountains and have gone through one or more weathering cycles before coming to rest at their present location. The water table is generally deeper than 2 meters, and the soils naturally support drought-tolerant oaks, pines, shrubs and grasses. Two extensive series include Lakeland with thermic soil temperature regime and Lake with hyperthermic soil temperature regime. The soils are acid, infertile, and droughty.

Even though the soils seem to have serious limitations for general farming, they are highly prized for citrus production. The Quartzipsamments of the southern Central Florida Ridge are choice citrus sites because the soils are well drained and have warm temperatures (see Fig. 2-10). Overhead sprinklers are common and air fans are used to reduce frost injury. Fertilizers and lime are used to increase soil fertility. Florida produces over two-thirds of the nations oranges and grapefruit and nearly all of the limes, tangelos, tangerines, and temples. A major problem of the citrus industry is the conversion of desirable sites to urbanization and other uses. This is forcing citrus production away from Quartzipsamments to more poorly drained soils, where shallow water tables require careful water management. About 25 percent of the deep sandy soils of the Central Florida Ridge have a significant increase in clay content within 2 meters depth. These soils are Paleudults, which are also well suited for citrus production.

Most of the land containing Quartzipsamments in the southeastern United States is not irrigated, is in forest and native grass, and is used for grazing by cattle. The soils are ideal building sites and have a high potential for disposal of septic tank effluent.

Use of Asphalt Barriers to Increase Water Retention

Thick sands retain little water for plant growth and allow most of the natural rainfall to percolate below the root zone in humid regions. The result is inefficient use of natural precipitation for plant growth. Soil scientists at Michigan State University experimented with the installation of asphalt barriers in sands to increase water retention capacity.

The barrier holds up rain or irrigation water and, in effect, creates a free water surface; soil moisture tension approaches zero at the immediate top of the barrier. Water films from water underneath the barrier are broken by the barrier and cannot exert a downward pull to cause water to move downward from soil above the barrier. Thus, at the barrier surface, the soil is saturated after the soil has been thoroughly wetted, and excess water drains laterally off from the barrier. Above the barrier the water content of the soil is a function of height above the barrier.

Figure 2-9   Gentle slopes mainly covered with pine forest and some pasturelands are typical of Quartzipsamments of the southeastern United States.

Figure 2-10 Well-drained Quartzipsamments with hyperthermic temperature regime on the central Florida ridge are prized for citrus production. (Photograph courtesy Larry Jackson, University of Florida.)

The effect has been to double the available water retention capacity of the soil above the barrier. This means that in the spring of the growing season the barriered soils contain two times more water in the soil above the barrier. It also means that with each significant rain, more water will be retained throughout the growing season. The increased yields of vegetables has been sufficient to pay for the cost of the barrier in about 3 years. The life of the barrier is expected to be at least 15 years.

Orthents

Fluvents are becoming thicker due to deposition of eroded material, while many Orthents remain thin or become thinner because of erosion. Orthents occur mostly on recent erosional surfaces as a result of geologic erosion or accelerated erosion by agriculture or other means. No diagnositc horizons are present. Orthents are very diverse, because all Entisols not in one of the other suborders are in Orthents. Orthents may occur in any climate and under any vegetation. The subgroups are Cryorthents, Torriorthents, Xerorthents, Troporthents, Udorthents, and Ustorthents.

Orthents With R Horizons

Perhaps the best way to conceptualize Orthents is to contrast Orthents deep to hard rock (with C horizon) with Orthents shallow to rock, as shown in Fig. 2-11. Many Orthents with R horizons occur on steep mountainous areas in Montana, Wyoming, Colorado, Arizona, Utah, New Mexico, and western Texas (see areas E2S1 of Fig. 2-1).

Cryorthents with rock close to the surface are found at the higher elevations of the Rocky Mountains. The soils occur where glaciers left little debris or where slopes are very steep. Even though the soils are rocky and shallow, deep plant rooting may occur (Fig. 2-12). Forestry, grazing, mining, and recreation are important land uses. Orthents shallow to rock in southern California are Xerorthents; they are developed in residuum weathered from the underlying bedrocks and occur on steep slopes. Large acreages of these soils are covered with brush.

Figure 2-11   Cryorthent shallow to granite rock in the Rocky Mountains.

Orthents With C Horizons On Actively Eroding Slopes

Orthents deep to hard rock are important in an Entisol-Mollisol landscape in southwestern Iowa that occupies 5 percent of the state (see Fig. 2-13). The Orthents have developed from thick loess in a strip along the Missouri River floodplain. The loess on the ridge tops is about 12 to 18 meters thick and rests on top of the highly dissected Kansan till plain (see Fig. 2-14). Some of the loess hills are similar to sand dunes. The soils have a udic moisture regime and are Udorthents.

The Hamburg soils on the steepest slopes are used for permanent pasture, and forage yields are low. Some of the Ida soils, also Udorthents, are used for cropping in association with the Monona soils, which are weakly developed Mollisols. Corn is an important crop, and erosion is a major problem. Care must be exercised to prevent large gullies that prohibit the use of large machinery.

Figure 2-12 Ponderosa pine roots have penetrated deeply in this Ustorthent formed from sandstone.

Figure 2-13   Location of the Monona-Ida-Hamburg soil association area in Iowa, where Udorthents (Ida, Hamburg) are important soils. (From Oschwald et. al., 1965.)

Soil properties vary little with depth except for a slight accumulation of organic matter in the upper 25 centimeters and a slight loss of carbonate and slight concentration of clay in the upper 12 centimeters. The surface soil is still calcareous, because erosion about keeps pace with removal of carbonates by leaching. This means that the loss of the surface soil by erosion during cropping exposes subsoil material that is quite similar to the original surface soil. The effect of erosion on crop yields is mainly due to the loss of organic matter and a reduced supply of nitrogen from mineralization.

Large areas of Orthents deep to hard rock exist in southwestern North Dakota and east-central Montana. These Orthents have developed in residuum from the weathering of sandstones and shales. Mollisols are the major associated soils where wheat is produced by summer fallowing. Some Orthents in central Oregon have developed from pumice, with particles mainly greater than 2 millimeters in diameter. These Orthents have a cryic temperature regime and are used mostly for forests.

Orthents With C Horizons on Nonactively Eroding Slopes

Many Entisols with C horizons do not occur on actively eroding slopes but in valleys and basins, where there is an absence of frequent sedimentation. The result is the formation of Orthents with C horizons in alluvial parent material and not the formation of Fluvents. Orthents deep to rock are minor inclusions of landscapes in the southwest, but the major areas occur in California. The largest area is the Central Valley, composed of the Sacramento Valley in the north of the San Joaquin Valley in the south, shown as area E2c Fig. 2-1. These soils have formed in sediments from the Sierra Nevada to the east and in sediments of the coastal ranges to the west. Different mineralogy exists in the soils in the eastern and western parts of the valley. Soil temperature regime is thermic. Sediments of the Imperial Valley were put in place by the Colorado River. Soils of the Imperial Valley have hyperthermic temperature regime and are adapted to production of many tropical crops, including dates, cotton, and citrus.

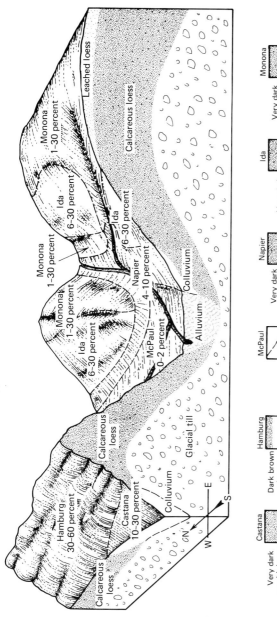

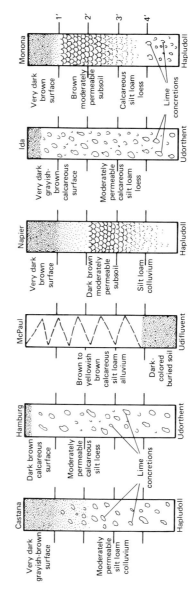

Figure 2-14 Diagram showing the relationship of Udorthents (Ida, Hamburg) to parent material and topography and relationship of Udorthents to associated soils and their properties in southwestern Iowa. (From Oschwald et al., 1965.)

Irrigated Agriculture on Orthents

Orthents of the basins of the southwest and west are used intensively for agriculture if irrigation water is available. Land surfaces are nearly level, and water can be distributed by low-cost gravity systems (see Fig. 2-15). The earliest major irrigation project in the United States occurred with the diversion of water from the Colorado River 100 kilometers through Mexico to the Imperial Valley in 1901. A flood on the Colorado River washed out the waterworks and, for about 2 years (1905 to 1907), the Colorado River flowed into the Imperial Valley and created the Salton Sea. New water control works were constructed; the most recent, the All American Canal, was completed in 1942 and is located entirely in the United States. Water for irrigation in the San Joaquin and Sacramento valleys is highly dependent on runoff water collected in reservoirs in the mountains of eastern and northern California. Water collected in reservoirs behind the Shasta and Oroville dams of northern California is transported hundreds of kilometers to serve the needs of agriculture and urban centers as far south as Los Angeles.

California ranks first in the nation in value of agricultural products and in the production of over 40 crops. Eight of the top 10 counties based on agricultural income are in California. The key to this agricultural productivity is irrigation of nearly level Orthents—the central valley being the largest irrigated area in United States. One of the major problems in the use of Orthents for agriculture, however, is salt accumulation in soils and development of saline soils.

Figure 2-15 Typical basin landscape of the southwestern United States, where nearly level Orthents with C horizons are the dominant soils.

Figure 2-16 Saline soil in the Imperial Valley caused by upward movement of water from a shallow water table and salt accumulation as water evaporated. Salt accumulation is greatest on tops of ridges.

Effect of Salt on Plants The salt in saline soils causes ion toxicities and increases osmotic pressure of the soil solution. The increased osmotic pressure reduces the uptake of water by germinating seeds and by roots. A common symptom is wilting. An increase in salinity produces the same effect on water uptake as that produced by increased soil mosisture tension. In fact, the effects seem to be additive. For example, if a plant wilts when the SMT is 20 bars and the osmotic pressure of the soil solution is 1 bar, the plant will wilt if the soil moisture tension is 1 bar and the osmotic pressure of the water is 20 bars.

Salinity Control Irrigation of land in arid region basins results in the buildup of water tables. When the water table rises to about 1 meter of the soil surface, water moves upward by capillarity at a rate sufficient to deposit salt on top of the soil from the evaporation of water (Fig. 2-16). It is essential in these cases that drainage systems are installed, with tile laid at least 1 or 2 meters below the surface to remove drainage water. Thus we see that salinity control and maintenance of permanent irrigated agriculture in arid regions are dependent on drainage for salt removal. Archaeological studies in Iraq showed that the Sumerians grew about one-half wheat and one-half barley 3500 B.C. One thousand years later, only one-sixth of the grain was the less-salt-tolerant wheat and, by 1700 B.C., the production of wheat was abandoned. This decline in wheat production and increase in more tolerant barley coincided with soil salinization. Records show that there was wide-spread land abandonment and that salt accumulation in soils was a factor in the demise of the Sumerian civilization.

For permanent agriculture there must be a favorable salt balance in the soil. The salt added to soils in irrigation water must be balanced by the removal of salt by leaching. The fraction of the irrigation water that must be leached through the root zone to control soil salinity has been defined as the *leaching requirement.* The leaching requirement is directly related to the salt content of the irrigation water. A leaching requirement of 0.25 or 25 percent, means that 25 percent of the water applied should leach through the root zone and carry with it the salt contained in the original irrigation water. Obviously, the drainage water contains a higher concentration of salt than the irrigation water.

Environmental Consequences of Irrigation   About 60 percent of the water diverted for irrigation in the United States is evaporated or consumed. The remainder appears as irrigation return flow (includes drainage water) and is returned to the rivers. Therefore the salt concentration of river waters is increased. A 2- to 7-fold increase in salt concentration is common for many rivers. The Colorado, one of the largest rivers, experiences a 21-fold salt increase between Grand Lake in northwestern Colorado and the Imperial Dam. Large increases in salt concentrations have become harmful to freshwater fish, and salts decrease water quality for downstream users. There is no inexpensive method to remove the salt at the present time.

References

Allison, L. R., "Salinity in Relation to Irrigation," in *Advances in Agronomy,* Vol. 16, pp. 139–180, Academic, New York, 1964.

Anonyomous, "Facts About California Agriculture," *AXT-46,* Cooperative Extension Service, University of California, 1974.

Austin, M. E., *Land Resource Regions and Major Land Resource Areas,* USDA Handbook 296, Washington, D.C., 1965.

Bower, C. A., "Salinity of Drainage Waters," in *Drainage for Agriculture,* Jan Van Schilfgaarde, Ed., American Society of Agronomy, Madison, Wis., 1974, pp. 471–487.

Buol, S. W., Ed., "Soils of the Southern States and Puerto Rico," *Southern Cooperative Series Bulletin 174,* 1973.

Buol, S. W., F. D. Hole, and R. J. McCracken, *Soil Genesis and Classification,* Iowa State Press, Ames, 1973.

Cantor, L. M., "The California Water Plan," *Jour. Geog.,* 68:366–371, 1969.

Criddle, W. D., and H. R. Haise, "Irrigation in Arid Regions," in *Soil,* USDA Yearbook, pp. 359–367, Washington, D.C., 1957.

Erickson, A. E., C. M. Hansen, and A. J. M. Smucker, "The Influence of Subsurface Asphalt Barriers on the Water Properties and the Productivity of Sand Soils," *9th Int. Cong. Soil Sci. Trans.,* 1:331–337, 1968.

Foth, H. D., C. M. Hansen, A. E. Erickson, and L. S. Robertson, "Effect of Deep Plowing on the Productivity of the Conover Loam," *Mich. Agr. Exp. Sta. Quart. Bull.,* 49:4–11, 1966.

Franki, G. E., R. N. Garcia, B. F. Hajek, D. Arriage, and J. C. Roberts, *Soil Survey of Nueces County, Texas,* USDA, Washington, D.C., 1965.

Gersmehl, P. J., "Soil Taxonomy and Mapping," *Annals Assoc. Am. Geog.,* 67:419–428. 1977.

Gray, F., and M. H. Roozitalab, "Benchmark and Key Soils of Oklahoma," *MP 97 Okla. Agr. Exp. Sta.,* 1976.

Higbee, E., *American Agriculture,* Wiley, New York, 1958.

Hole, F. D., *Soils of Wisconsin,* University of Wisconsin Press, Madison, 1976.

Johnson, R. W., J. F. Brasfield and F. S. Merrill, *Soil Survey Supplement Seminole County Florida,* USDA, Soil Conservation Service, Washington, D.C., 1975.

Keech, C. F., and R. Bentall, "Dunes on the Plains, The Sand Hills Region of Nebraska," *Resource Report 4,* Conservation and Survey Division, University of Nebraska, Lincoln, 1971.

Kellogg, C. E., and A. C. Orvedal, "Potentially Arable Soils of the World, and Critical Measures for their Uses," in *Advances Agronomy,* in Vol. 21, pp. 109–170, Academic, New York, 1969.

Jacobsen, T., and R. M. Adams, "Salt and Silt in Ancient Mesopotamian Agriculture," *Science,* 128:1251–1258, 1958.

Jensen, P. N., "Conservation Ranching in the Nebraska Sandhills," *Jour. Soil Water Con.,* 24:26–27, 1969.

Law, J. P., and J. L. Witherow, "Irrigation Residues," *Jour. Soil Water Con.,* 26:54–56, 1971.

Matelski, R. P., "Great Soil Groups of Nebraska," *Soil Sci.,* 88:228–239, 1959.

Oschwald, W. R., F. F. Riecken, R. I. Dideriksen, W. H. Scholtes, and F. W. Schaller, *Principal Soils of Iowa,* Special Report No. 42, Department of Agronomy, Iowa State University, 1965.

Pierce, J. Bell, Ed., *Agricultural Growth in an Urban Age,* Institute of Food and Agricultural Sciences, University of Florida, 1975.

Polone, D. J., *Soil Survey of Wagoner County, Oklahoma,* USDA, Washington, D.C., 1976.

Rivers, E. D., C. Godfrey, and G. W. Kunze, "Physical, Chemical and Mineralogical Properties of the Lakeland Soil Series in Texas," *Soil Sci., 96:*395–403, 1963.

Russell, J. S., and H. F. Rhoades, "Water Table as a Factor in Soil Formation," *Soil Sci., 82:*319–328, 1956.

Sherfey, L. E., C. Fox and J. Nishimura *Soil Survey of Thomas County, Nebraska,* Washington, D.C., USDA, 1965.

Smith, F. B., R. G. Leighty, R. E. Caldwell, V. W. Carlisle, L. G. Thompson, Jr., and T. C. Mathews, "Principal Soil Areas of Florida," *Fla. Ag. Exp. Sta. Bull. 717,* 1967, reprinted 1973.

Soil Conservation Service, *Distribution of Principal Kinds of Soils: Orders, Suborders and Great Groups,* Sheets, 85 and 86, U.S. Geol. Survey, Washington, D.C., 1969.

Soil Conservation Service, *Soil Survey Laboratory Data and Descriptions for Some Soils of Nebraska,* USDA, Washington, D.C., 1966.

Soil Survey Staff, *Soil Taxonomy,* USDA Agriculture Handbook 436, Washington, D.C., 1975.

Storie, R. E., "Soil Regions of California Illustrated by Twenty-four Dominant Soil Types," *Soil Sci. Soc. Am. Proc., 11:*425–430, 1946.

The Argicultural Commissioner's Reports, "Gross Values of California's Agricultural Production," *California Crop and Livestock Reporting Service,* 1974, 1975.

Thorne, W., and H. B. Peterson, "Salinity in United States Waters," in *Agriculture and the Quality of Our Environment,* American Association of Advancement Science, Washington, D.C., 1967.

Tuan, Yi-Fu, *China,* Aldine, Chicago, 1969.

United States Salinity Laboratory Staff, *Diagnosis and Improvement of Saline and Alkali Soils, USDA Handbook* 60, Washington, D.C., 1969.

3

Inceptisols of the United States

Central Concept and Suborders

Inceptisols are *humid* region soils that have one or more pedogenic horizons as contrasted to Aridisols, which are *dry* region soils with one or more pedogenic horizons. Inceptisols show great variation in age, but all show little horizon differentiation. They are related to Entisols in that they show more pedogenic development. Inceptisols can form in almost any climate from the polar to the tropical, except where dryness prevails. Vegetation is likewise very diverse. Most Inceptisols are on relatively youthful geomorphoric surfaces of the late Pleistocene or Holocene (recent) age. The Inceptisol order includes soils of great diversity, as does the Entisol order.

The central concept of Inceptisols is that of humid region soils with altered horizons but that retain some weatherable minerals. Inceptisols are without diagnostic illuvial horizons but may have ochric, umbric, cambic, fragipan, or duripan

horizons. Most Inceptisols have an ochric or umbric epipedon over a cambic horizon. Drainage ranges from very poor to well drained. Few Inceptisols are thick
sands, since most of the deep sands are Psamments. Inceptisols are important soils
on the tundra, in river valleys, on mountain slopes, and where recent volcanic ash
has been deposited. General distribution of Inceptisols in the United States is
shown in Fig. 2-1 (and world distribution in Fig. 1-11).

Inceptisols are the second most extensive soils in the United States. The suborders include Andepts (recent volcanic ash), Aquepts (wet), Ochrepts (well drained),
Plaggepts (plaggen horizon), Tropepts (tropical), and Umbrepts (umbric horizon).
All suborders are extensive in the United States except Plaggepts and Tropepts
(see appendix Table 2).

Aquepts

Aquepts are wet Inceptisols, and most have an aquic moisture regime or are
drained. Surface horizons tend to be dark colored, and subsoils show evidence of
restricted drainage. Mottled and rusty colors are within 50 centimeters of the soil
surface. Most of the Aquepts have formed in sediments of Wisconsin age or
younger on floodplains, very flat plains, or depressions. They are extensive in two
very contrasting environments in North America: on the broad, nearly level alluvial and lacustrine plains (Haplaquepts) and on the tundra (Cryaquepts). Aquepts
of warm regions are the most important soils for rice production in the world.

Aquepts of the Mississippi River Valley

Haplaquepts are extensive soils on the broad, nearly level floodplains of the Mississippi River south of its confluence with the Ohio River and on the floodplains of
the Pearl, Alabama, and Tombigee rivers in Mississippi and Alabama, as shown in
Fig. 2-1 (areas I2a). Most of these soils have developed from clayey alluvium of
slack-water areas a considerable distance from the main channel (see Fig. 3-1).
Fluvents are common close to the rivers on stratified levee sediments. Soil temperature regime is thermic. On some of the longest exposed levee areas there has
been significant clay translocation and leaching to form Aqualfs.

Crittenden County, Arkansas, is entirely within the bottomlands of the Mississippi River; it has a complex soil pattern, because the channel has wandered about
in the past depositing clayey sediments over coarser levee materials and coarser
levee materials over formerly slack-water sediments. Fluvents comprise about 7
percent of the county and Aquepts comprise nearly 60 percent. Sharkey is the
most extensive series and occupies 37 percent of the land. The native vegetation
of the slack-water areas where Sharkey soils develop is forest, including bald
cypress. Sharkey clay has 1 percent sand or less in all horizons and no evidence of

Figure 3-1 Aquepts in the foreground floodplain, with the Mississippi River about 60 kilometers away. Cleared land is intensely cropped, and trees occupy the wettest areas.

clay migration; the clays are mainly montmorillonitic. There has been sufficient leaching to remove the original carbonates and develop soil acidity (see Table 3-1). There is a gradual decline in organic matter with increasing depth. The Sharkey soil is in the very fine family of Vertic Haplaquepts. There is not sufficient churning in the soils for them to be classified as Vertisols.

Cotton was the major crop of the early economy on the soils of the Mississippi River valley. Today cotton ranks first and soybeans second in acreage. Wheat, corn, alfalfa, and pasture are important throughout the area. In southern Louisiana sugarcane is important. It is grown on ridges or beds for improved drainage and aeration of the root zone.

Table 3-1   Selected Properties of Sharkey Silty Clay

Horizon	Depth, centimeters	Percent Clay	Silt	pH	Percent Base Saturation	Percent Organic Matter
Ap	0–12	57	42	6.1	74	3.4
A12	12–20	58	41	6.3	74	2.5
B21g	20–44	61	39	6.4	72	2.0
B22g	44–64	65	35	6.6	77	1.6
B23g	64–90	76	24	6.6	78	1.4
B24g	90–120	65	35	6.9	82	1.2

From Gray, 1974.

Figure 3-2 Rice harvest on Haplaquepts. Note the contour ridges used to distribute water. Fields are drained before harvest to allow use of harvesters.

About 21 percent of the rice produced in the United States is grown on Sharkey and similar soils. The low water permeability and nearly level surface make the soils well suited for rice production, as shown in Fig. 3-2. Arkansas is the major rice-producing state, and one of the two major rice-growing areas in the United States is closely associated with Haplaquepts in the states of Arkansas, Mississippi, and Louisiana (see Fig. 3-3). Rice production by states is given in Table 3-2.

Aquepts of the Huron Lake Plain

An area in eastern lower Michigan is dominated by Aquepts (see area I2a Fig. 2-1). The soils have formed on a broad, nearly level lacustrine plain. Most of the soils

Figure 3-3 The major rice-producing areas in the United States. The northern area includes large acreages of Haplaquepts, while Alfisols and Vertisols are the dominant soils for rice along the Gulf Coast. (Adapted from Flack and Slusher, 1977.)

have dark-colored A horizons and are Mollic Haplaquepts. The principal associated soils are Aquolls. Temperature regime is mainly mesic. Some Spodosols (Podzols) have developed in the sands of old beach ridges.

The Huron Lake plain is an important cash crop area; about two-thirds of the land is in crops (see Fig. 3-4). Michigan ranks first among the states in dry bean (navy) production; the production is concentrated on the lake plain. Michigan produces about 70 percent of the fine eastern soft white winter wheat east of the Rocky Mountains. The climate of this area is cool, moist in the summer, and conducive to production of this soft, low-protein wheat, which is used for making pastry flour. Other important crops include soybeans, sugar beets, potatoes, canning crops, and corn. Corn production in the northern part of the area, however, is limited by low temperature, because it is on the mesic-cryic temperature regime boundary. The fine-textured soils have severe limitations for building sites and septic drain fields.

Table 3-2   Rice Production by States, 1976

State	1000 Hectares Harvested
Arkansas	340
Louisiana	243
Texas	217
California	163
Mississippi	58
Other	7
Total	1028

From Flack and Slusher, 1977.

Figure 3-4 Aquepts on the lake plain of eastern Michigan are naturally fertile and are highly productive for agricultural crops when properly drained. Light areas indicate where tile drainage has been installed.

Aquepts of Cold Places—Pergelic Cryaquepts

Aquepts are more common in cold places because evapotranspiration is less if all other factors are the same. The presence of permafrost also increases soil wetness. The permafrost is impermeable to water, since any cracks that form are sealed by the refreezing of percolating water. Permafrost causes soil wetness even on slopes and inhibits eluviation. Marshes and swamps are common where permafrost exists. Taking into consideration that relative humidity is high and some moisture is added to the soil by condensation, only 10 centimeters of annual precipitation produce Aquepts on the tundra of the Arctic coastal plain near Point Barrow in Alaska. Climatewise the polar area is considered a desert.

Aquepts with annual temperature above 0°C and lower than 8°C are Cryaquepts. Cryaquepts with permafrost are Pergelic Cryaquepts and are the most extensive soils in Alaska, as shown in Fig. 3-5. The soils have formed under a wide variety of parent materials and mainly under tundra or boreal forest (taiga) vegetation. Smaller areas of Pergelic Cryaquepts occur above the timberline in the central and northern Rocky Mountains (see Fig. 3-6). The timberline-tundra boundary is the approximate boundary of permafrost.

Pedoturbation Caused by Frost   Pedoturbation is the process of soil mixing. Two sets of processes are at work in Pergelic Cryaquepts. One produces soil horizons. The other destroys these same soil horizons. One process is solifluction, which causes soil to move slowly downslope enmasse with a churning effect.

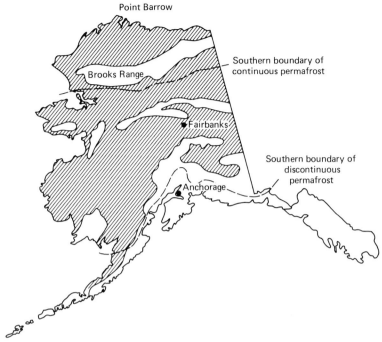

Figure 3-5 Lined area shows location where Aquepts are the dominant soils in Alaska. Most of the soils are Pergelic Cryaquepts, since permafrost is extensive.

Figure 3-6 Tundra landscape above timberline in the central Rocky Mountains, where Pergelic Cryaquepts are important soils.

About a 9 percent increase in volume occurs when water freezes and in the fall the soil freezes from the surface downward. The increase in volume caused by the freezing of water between a frozen surface layer and permafrost creates a pressure that is released where the surface soil is weakest. The soil is pushed up, and mounds, or frost boils, are formed. In other cases large stones are forced to the surface to create stone rings or nets.

Properties of Pergelic Cryaquepts   Soil development in Pergelic Cryaquepts is dominated by the influence of wetness and low temperature. Typically, a spongy mat of organic matter rests on top of the mineral soil, as shown by the 011 and 012 horizons in Table 3-3. The organic matter mat insulates the soil from solar radiation and accentuates the difference between summer air and soil temperatures. The result is that a small biomass production causes large accumulations of organic matter in soils because decomposition is very slow. Many of the soils intergrade to Histosols. As in all Aquepts, the gray and mottled colors are evidence of wetness. The thin mineral soil or active layer rests directly on the permafrost that represents the average annual depth of thawing in summer.

Low soil temperature favors physical rather than chemical weathering. Soil texture is related to the grain size of the parent rocks. The content of silt and sand is high and that of clay is low, as shown in Table 3-4. Evidence of clay migration is absent and there is no illuvial B horizon. Carbon nitrogen ratios are high, indicative of the low intensity of organic matter decomposition. Ratios over 20 are common. The clay and organic matter contribute to a significant cation-exchange capacity

Table 3-3   Horizons, Depth, and Characteristics of a Pergelic Cryaquept Located Midway between Anchorage and Fairbanks, Alaska

Depth, centimeters	Horizon	Characteristics
23		
	011	Raw sedge and hypnum moss peat
8		
	012	Black, partly decomposed organic matter
0		
	Cg[a]	Dark gray, loam, mottled, few roots
18		
	Cgf[b]	Dark gray, loam, frozen with ice lens up to 2 centimeters thick
33		

Adapted from Allan, et al., 1969.
[a]*Gleyed.*
[b]*Gleyed and with permafrost (frozen).*

Table 3-4 Selected Properties of Mineral Horizons of a Pergelic Cryaquept
Located Midway Between Anchorage and Fairbanks, Alaska

Depth, centi- meters	Horizon	Percent			Carbon/ Nitrogen Ratio	CEC milli- equiv- alent per 100 grams	Percent Base Saturation	pH
		Silt	Sand	Clay				
0–10	C1g	42	41	17	30	14	79	5.5
10–18	C2g	40	45	15	30	11	83	5.6
18–33	C3gf	44	44	12	25	10	71	5.9

Reproduced from Soil Science Society of America Proceedings, *Volume 33, 1969, by permission of Soil Science Society of America.*

that is typically over 50 percent base saturated. Base saturation and pH increase with increasing soil depth. The acid nature of soil and base unsaturation is evidence of some leaching that is mainly lateral; the nutrients eventually leave the soil ecosystem as surface runoff. Many of the soils have buried organic horizons (Ob) caused, in some cases, by burial by solifluction and in other cases by cryoturbation.

Land Use on Cryaquepts   Land on Cryaquepts in Alaska is used mainly by a sparse population that lives by hunting and fishing. Some Eskimos collect wild berries and have small garden plots for vegetables. In Lapland similar lands are used for reindeer herding. Eskimos in Alaska harvest caribou, but the management of large herds has never developed.

About 600,000 caribou spend the winter in the spruce forest in the protected valleys on the southern slopes of the Brooks Range. In the spring they migrate north over the Brooks Range to the coastal plain for summer grazing on the tundra. Fifteen families that lived in the village of Anaktuvuk Pass near the migration trail in the Brooks Range were studied in 1962. It was found that about 90 caribou per year were required per family for their livelihood. Russian and U.S. nuclear testing prior to 1962 had released radioactive cesium that occurred as fallout throughout the world. Radioactive cesium that landed on lichens, the major food of caribou, was transferred to the caribou and then to the Eskimos. The Eskimos had about 100 times more radioactivity in their bodies than persons in the lower 48 states. Eskimos in other villages that depended mainly on fish had much lower radioactivity in their bodies. The Eskimos have been assimilated into Western society, and snowmobiles and motor boats are common. There are few, if any, Eskimos that live in traditional igloos, and so on.

Agriculture is confined mainly to several river valley areas and to Cryochrepts and Cryofluvents, and Cryaquepts. The largest agricultural area is 5100 hectares in the Matanuska Valley in the Palmer area north of Anchorage. The second largest is the 3400 hectares near Fairbanks, in the Tanana River region. It is estimated that about 400,000 hectares in Alaska are suitable for cultivation, which is less than 0.3 percent of the land area. Considerably more land is used for pasture and range, and this acreage could be expanded. There are about 85,000 hectares of range- and pasturelands concentrated on Kodiak Island and in Homer County along the Cook Inlet south of Anchorage.

Pergelic Cryaquepts are called Cryosols in Canada. These soils are extensive in areas rich in minerals and oil. Exploration, development, and tourism are increasing traffic. Off-road vehicle traffic may cause denudation of vegetation and stimulate erosion and permafrost melting (thermokarst).

Ochrepts

Ochrepts are well-drained soils that are light colored and have brownish subsoils. They are extensive on moderate and steep actively eroding slopes of mountain regions. The hardness of the underlying rocks and erosion on moderate to steep slopes account for soil youthfulness. The soils have formed on a wide variety of sedimentary, metamorphic, and igneous rocks as well as on glacial till in the northern section of the Appalachian region. Temperature regimes are mainly mesic and cryic, with some thermic in the southernmost states. An area of Ochrepts occurs in Alaska in the region of the Yukon and Tanana rivers. Most Ochrepts had or now support forest vegetation, and some have a tundra vegetation. As a group, Ochrepts present many difficulties for farming, including steep slopes, low temperature, low soil fertility and root-restricting fragipans.

The Ochrepts in the eastern United States are represented mainly by three great groups: Fragiochrepts that have fragipans, Dystrochrepts that are low in base saturation, and Eutrochrepts that are high in base saturation. The Dystrochrepts are the most extensive and will be discussed first.

Dystrochrepts of the Central Appalachian Region

Dystrochrepts are the most extensive Inceptisols of the eastern United States and extend from Pennsylvania to Alabama (see areas I3s in Fig. 2-1). Most of them have developed in the residuum from the weathering of sandstone and shale. On the steeper slopes rock or R horizons occur at shallow depths. Climate is humid or perhumid, and the soils are intensely leached. Most of the parent materials were originally acid, and base saturation is low. Dystrochrepts must have less than 60 percent base saturation at depths of 25 to 75 centimeters below the surface. Some properties of a Dystrochrept are given in Table 3-5.

The B horizon lacks evidence of clay accumulation, and it is a cambic horizon. The soil is quite acid throughout; perhaps recycling of bases or addition of lime for farming accounts for the pH of 6.2 in the Ap horizon. Base saturation between the

Table 3-5   Selected Properties of a Dystrochrept

Depth, centimeters	Horizon	Percent Clay	Texture	pH	Percent Base Saturation
0–25	Ap	14	Loam	6.2	56
25–45	B21	13	Loam	5.4	48
45–65	B22	19	Loam	4.8	42
65–90	C	22	Loam	4.7	43
90+	R(weathered shale)				

From Long, 1975.

depths of 25 and 75 centimeters is well below the maximum limit of 60 percent. The soil developed from the weathering of shale.

Most of the Dystrochrepts are covered with mixed conifer-hardwood forests. Settlers cleared some of the land, and much of this land is no longer cultivated (see Fig. 3-7). Only a small percentage of the Dystrochrepts are now used for agriculture, which is restricted to more favorable soils on gentler slopes in the stream and

Figure 3-7   Typical Appalachian valley dominated by hardwood forest and some abandoned fields now used for pasture.

Figure 3-8 Steep slopes, where bedrock is close to the surface and high rainfall creates many rockslides and landslides in the Appalachian region in which Dystrochrepts are the dominant soils.

river valleys. In the north the associated soils are largely Alfisols and Umbrepts, and in the south they are Ultisols. When traveling through the region one encounters limestone valleys where topography is gentler, soils are mainly Alfisols and agriculture is important. The steep slopes along highways result in frequent rock- and landslides (see Fig. 3-8). Most of the nation's coal is mined in this region.

Fragiochrepts of Southern New York and Northern Pennsylvania

Fragiochrepts are extensive in southern New York and northern Pennsylvania (see area I3c Fig. 2-1). The soils have a fragipan at about the 50-centimeter depth that underlies a cambic B horizon, as shown in Fig. 3-9. Usually there is perched groundwater above the pan at some time during the year. Few roots penetrate the pan which contributes to shallow rooting. Trees exposed to strong winds are susceptible to wind throw, and such areas may have a microrelief due to soil disturbance by falling trees.

Most Fragiochrepts are loamy-textured soils developed from glacial sediments. Since the area was glaciated, the slopes are more gentle than in the Dystrochrept area to the south. Soil temperature regime is mesic, and soil water regimes are udic and perudic. A large acreage is in mixed-hardwood forest. Hay, pasture, and grain for dairy cattle are the major crops. Potatoes, fruits, and truck crops are locally important. The soils have severe limitations for homesites and septic drain fields because the low permeability of fragipan results in a seasonal shallow water table.

Eutrochrepts of Alabama

Eutrochrepts are well-drained Ochrepts that are base rich. Base saturation between depths of 25 and 75 centimeters must be 60 percent or more. Parent materials generally were calcareous or basic sedimentary rocks. Some Eutrochrepts have

Figure 3-9 Fragipan (labeled x) in a Fragiochrept in southern New York. The fragipan has vertical, light-colored streaks that are the surfaces of large prisms; it is slowly permeable to water and impermeable to roots. (Scale in inches.)

cambic horizons that are calcareous. The most extensive area in the United States is in south-central Alabama (see area I3b Fig. 2-1). These soils developed in the residum of weathered chalk and marl. The associated soils on smooth upland surfaces are Vertisols.

Soil temperature regime is thermic, and about one-half of the land is used for agriculture. Much of the land is in pasture. Soybeans and corn are the principal crops. Soil erosion is a critical problem in cultivated areas. About 40 percent of the area is in forest.

Andepts

Andepts were formerly called Ando soils, which generally meant soils developed from volcanic ash (see Fig. 3-10). The unique feature of Andepts is the amorphous nature of the mineral fraction. The soils inherit amorphous material from the pyroclastic parent material that readily weathers in a humid climate to allophane—an amorphous aluminum silicate clay mineral.

Andepts occur mostly on or near mountains that have active volcanoes from the equator to the high latitudes. They are extensive in the mountains from the southern tip of South America to Alaska and the Aleutian Islands. Vegetation varies

Figure 3-10 Andepts have formed from ash on the slopes of Diamond Head, a volcanic cone near Honolulu.

with climate and includes bamboo, tundra, tropical rain forest, coniferous forest, grasslands, and bracken ferns. Repeated ashfall results in buried soils in many cases. Typical properties of Andepts include:

1. Amorphous nature of mineral fraction due to inherited materials and formation of allophane from weathering.
2. High organic matter content resulting from protection by the amorphous mineral fraction.
3. High cation-exchange capacity.
4. High phosphorus fixing capacity and low phosphorus availability.
5. Low bulk density, commonly 0.4 to 0.8 grams per cubic centimeter.

Andepts may represent a transient soil (depending on time frame) in a given landscape because: (1) erosion may remove the ash and other amorphous materials and expose crystalline parent material for soil development, and (2) the Andepts may be transformed into other soils as Ultisols or Oxisols as the mineral fraction becomes more crystalline (allophane to montmorillonite to kaolinite, etc).

Andepts of Hawaii

The Hawaiian archipelago consists of a series of many volcanic islands that rise over 10,000 meters above the floor of the Pacific Ocean. The islands are strung out in a northwest–southeast direction. The ages of the islands decrease in a southeasterly direction. Because eruptions of ash and cinders are more common during the late stages of volcanism, ash and Andepts are more extensive on the younger islands. On the oldest islands (e.g., Kauai), ash has been removed by erosion, and modern soils have formed largely in the basaltic lava. Many soils are Ultisols and Oxisols. The large island of Hawaii is the youngest island and the only one with active volcanoes; Andepts are extensive on Hawaii. A few of the Andepts have formed in alluvium derived from ash and cinders.

Andepts and Climate Relationships. Hawaii is located in the northeast trade winds belt, and rainfall is closely related to elevation. Rainfall increases from about 75 centimeters over the open ocean to 750 centimeters at the 1000-meter level on windward slopes. Rainfall decreases above the 1000-meter elevation because the winds flow around the mountains instead of over the top. At any given elevation, the rainfall is less on the leeward than windward slopes. These rainfall differences account for most of the differences in Andepts on Hawaii.

Some Aridisols have developed from volcanic ash on Hawaii in areas where soil moisture regime is aridic. As rainfall increases beyond that which gives rise to Aridisols, the soils show greater evidence of weathering and leaching, and areas

Figure 3-11   Eutrandept showing molliclike epipedon and consolidated ash or tuff at a depth of 75 centimeters. Little difference between A and B horizons is typical of Andepts.(Scale in feet.)

with ustic soil moisture regimes have Eutrandepts (base rich). Eutrandepts have a base saturation of 50 percent or more at a depth between 25 and 75 centimeters. Some have calcium layers, or calcic horizons, in the subsoil. More of the ashy materials have been weathered as compared to Aridisols formed from volcanic ash. Natural vegetation is savannah or open forest, and mollic epipedons are common (see Fig. 3-11). If the soils had developed from crystalline parent materials, the soils would be mostly Ustolls. Eutrandepts on gentle slopes are mainly cultivated, and most of the other are used for grazing cattle. The Parker Ranch, one of the world's largest cattle ranches, is mainly on Eutrandepts.

Dystrandepts (low bases) have developed on Hawaii under udic soil moisture regimes. Base saturation is less than 50 percent in all horizons between the depths of 25 and 75 centimeters. Although the soil moisture regime is udic, there are occasional periods of soil drying and wetting. The drying and wetting cycles result in an amorphous fraction that does not change irreversibly after drying into aggregates. Fern and forests are the native vegetation. Dystrandepts are more acid than Eutrandepts and have greater phosphorus-fixing capacity. Weathering has been intense enough to weather most of the original ashy material. The mineral fraction is mainly amorphous allophane and oxides of iron and aluminum. Sugarcane is an important crop on Dystrandepts on Hawaii.

Hydrandepts (hyd stands for water) develop where the soil moisture regime is perudic. Rainfall is well distributed throughout the year, and the soil is near field capacity most of the time. Leaching is an almost continuous process. Base saturation is generally lower than in Dystrandepts, but the distinguishing feature is the irreversibility of soil aggregates on drying. A summary of the relationship among Andepts, soil moisture regimes, soil properties is given in Table 3-6.

In the absence of wetting and drying cycles during development of Hydrandepts, the amorphous fraction of mineral and organic matter congeals into aggregates that do not take up water again once they have been dried. This happens to exposed soil that dries along roadcuts. Soils are also thixotropic. The soil is remarkably uniform from top to bottom and feels smeary. The soil is the mineral counterpart of an organic soil. Natural vegetation is tree ferns and forests. The supply of nutrients is very low, but occasional ashfalls seem to maintain the base supply in some. The distribution of Andepts on Hawaii is given in Fig. 3-12.

Land Use on Andepts on Hawaii. Sugarcane is the principal crop of the island of Hawaii; the production is concentrated on Hydrandepts and some Dystrandepts at low elevations, where sunshine and high temperature are the most suitable for high yields. The coastal area along northeast Hawaii is the major sugar-growing area of the state. Hydrandepts at elevations too high for sugarcane (too cloudy and cool) are mainly forested. Pasture is the dominant use of Dystrandepts at higher elevations and of Eutrandepts, regardless of elevation. Smaller acreages of truck crops, coffee (along the west coast), and macadamia nuts are also important crops grown on Andepts on the island of Hawaii.

Table 3-6 Relationships Among the Great Groups of Andepts

Soil Property	Great Group		
	Eutrandept	Dystrandept	Hydrandept
Soil moisture regime	Ustic	Udic	Perudic
Percent base saturation	Fifty or more	Less than 50	Less than 50
Clays	Dry reversibly	Dry reversibly	Dry irreversibly

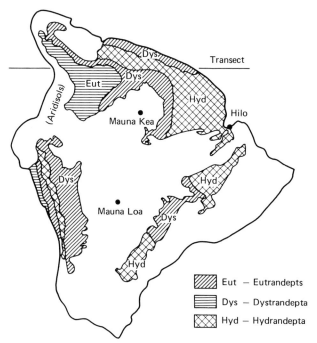

Figure 3-12   Distribution of Andepts on Hawaii. Note that soils along the transect east to west are Dystrandepts (udic) at low elevation and Hydrandepts (perudic) on the higher slopes of Mauna Kea. Eutrandepts (ustic) occur to the leeward of the mountains, with Aridisols (aridic) near sea level on the leeward side.

Cryandepts of Alaska

Cryandepts cover nearly all well-drained, nonmountainous areas of the Aleutian Islands, the Alaska peninsula, and the Kodiak Island group. Low temperature and frequent rain result in an almost constantly moist soil environment. Tall grasses, forbs, and shrubs are the dominant vegetation. The two largest areas of pasture- and rangeland in Alaska occur on Andepts on Kodiak Island and near Homer, south of Anchorage.

Ashfall has been frequent and, in 1912, the Andepts on Kodiak Island were buried by about 25 centimeters of ash. Selected properties of the Kodiak silt loam, a representative Cryandept, are given in Table 3-7. The recent ashfall (C horizon) shows little weathering compared to the underlying horizons. The high organic matter content and cation-exchange capacity and the low base saturation and soil pH are obvious from the table.

Table 3-7   Selected Properties of Kodiak Silt Loam—A
Cryandept

Horizon	Depth, centimeters	Percent Organic Carbon	CEC	Percent Base Saturation	pH
C(ash)	0–25	—	2	30	5.8
		Buried soil			
A11b	25–37	23+	102	6	4.9
A12b	37–50	12+	110	3	5.1
ACb	50–65	—	52	4	5.4
IIC	65+	3	94	3	5.3

Adapted from Simonson and Rieger, 1967. Used by permission of Soil Science Society of America.

Most Andepts in Alaska are covered with grass vegetation that is believed to be the vegetation under which they developed. The area of grassland is decreasing because spruce forests are expanding. Some Cryandepts invaded by spruce forest in Alaska have been converted to Spodosols or Podzols.

Cryandepts of the Pacific Northwest

The Cryandepts of the Pacific Northwest are located in Washington, Idaho, and Montana (see area I1S1 of Fig. 2-1). The soils occur on steep mountain slopes and are mostly in forest. The principal associated soils are Cryumbrepts in the Cascade Mountains and Cryorchrepts in the northern Rocky Mountains. Small areas of Cryorthods are present.

Umbrepts

Umbrepts are Inceptisols with dark-colored epipedons high in organic matter content. They differ from Ochrepts mainly in having greater precipitation that results in greater accumulation of organic matter. Most Umbrepts have an umbric epipedon. Some have a mollic epipedon with an underlying cambic horizon that has less than 50 percent base saturation. Others have an anthropic horizon. Umbrepts are the dark reddish or brownish, freely drained, organic matter-rich Inceptisols of the humid middle to high latitudes. Most occur in hilly or mountainous regions of high precipitation and under coniferous forest. Umbrepts in the United States are most extensive near the Pacific Ocean in Oregon and Washington (see areas I4S and I4a

Figure 3-13   Douglas fir is the major crop growing on Umbrepts of the coastal ranges in Oregon and Washington. Clear cutting is popular because regeneration of Douglas fir requires light intensity greater than that under the shade of mature trees.

of Fig. 2-1). Umbrepts are also important on well-drained sites in association with Cryaquepts in Alaska and high elevations of the middle and northern Rocky Mountains. Umbrepts are a minor component of the soil associations of the high rainfall areas of the southern Appalachian Mountains.

Umbrepts of Oregon and Washington

The Umbrepts of Oregon and Washington have a mesic soil temperature regime and udic or xeric soil moisture regimes. Fog is frequent close to the Pacific Ocean. Many of the soils are on steep slopes that are densely covered with coniferous forest and lumbering is the principal industry in the area (see Fig. 3-13). Umbrepts in the Puget Sound Valley (including Seattle) developed from glacial deposits and occur on gentle slopes of dissected terraces. Although they are mostly forested, a wide variety of crops are grown such as deciduous fruits, berries, vegetables, seed crops, and grain.

References

Allan, R. J., J. Brown, and S. Rieger, "Poorly Drained Soils With Permafrost in Interior Alaska," *Soil Sci. Soc. Am. Proc., 33:*599–605, 1969.

Anonymous, "Fallout in the Food Chain," *Time,* September 13, 1963, p. 63.

Austin, M. E., *Land Resource Regions and Major Land Resource Areas of the United States,* USDA Handbook 296, Washington, D.C., 1965.

Bridges, E. M., *World Soils,* Cambridge University Press, London, 1970.

Brown, D. A., V. E. Nash, A. G. Caldwell, L. J. Bartelli, R. C. Carter, and O. R. Carter, "A Monograph of the Soils of the Southern Mississippi River Valley Alluvium," *Southern Cooperative Series Bull. 178,* Ark. Agr. Exp. Sta., Fayetteville.

Brown, J., "Tundra Soils Formed Over Ice Wedges, Northern Alaska," *Soil Sci. Soc. Am. Proc., 31:*686–691, 1967.

Buol, S., *Soils of the Southern States and Puerto Rico,* Southern Cooperative Series Bulletin No. 174, Agr. Exp. Sta. of the Southern States and Puerto Rico and the USDA, 1973.

Canada Department of Agriculture, *Soils of Canada,* Ottawa, 1977.

Douglas, L. A., and J. C. F. Tedrow, "Tundra Soils of Arctic Alaska," *7th Intern. Cong. Soil Science, 4:*291–304, 1960.

Flack, K. W., and D. F. Slusher, "Soils Used for Rice Culture in the United States," *Mimeo, USDA Soil Conservation Service,* Washington, D.C., 1977.

Gersmehl, P. J., "Soil Taxonomy and Mapping," *Annals Assoc. Am. Geog., 67:*419–248, 1977.

Gray, J. L., and D. V. Ferguson, *Soil Survey of Crittenden County, Arkansas,* USDA, Washington, D.C., 1974.

Kellogg, C. E., and I. J. Nygard, *Exploratory Study of the Principal Soil Groups of Alaska,* USDA Monograph No. 7, 1951.

Long, R. S., *Soil Survey of Franklin County, Pennsylvania,* USDA and Pennsylvania State University and Pennsylvania Department of Environmental Resources, Washington, D.C., 1975.

Lytle, S. S., *The Morphological Characteristics and Relief Relationships of Representative Soils in Louisiana,* La. Agr. Exp. Sta. Bull 631, Baton Rouge, 1968.

McCall, W. W., *Soil Classification in Hawaii,* Hawaii Coop. Ext. Service Circular 476, Honolulu, 1973.

Michigan Department of Agriculture, *Michigan Food Facts,* Department of Agriculture, Michigan, 1975.

National Cooperative Soil Survey of United States, National Soil Survey Committee of Canada and FAO, *Soil Map of the World,* II: North America, Rome, 1975.

Nimlos, T. J., and R. C. McConnell, "Alpine Soils in Montana," *Soil Sci., 99:*310–321, 1965.

Palmer, H. E., W. C. Hanson, B. I. Griffin, and W. C. Roesch, "Cesium-137 In Alaskan Eskimos," *Science, 142:*64–65, 1963.

Retzer, J. L., "Alpine Soils of the Rocky Mountains", *Jour. of Sci., 7:*22–32, 1956.

Retzer, J. L., "Present Soil-Forming Factors and Processes in the Arctic and Alpine Regions," *Soil Sci., 99:*38–44, 1965.

Retzer, J. L., *Soil Survey of Fraser Alpine Area, Colorado,* USDA and Colo. Agr. Exp. Sta., Washington, D.C., 1962.

Sato, H. H., et al., *Soil Survey of the Island of Hawaii, State of Hawaii,* USDA and Hawaii Agr. Exp. Sta., Washington, D.C., 1973.

Simonson, R. W., and S. Rieger, "Soils of the Andept Suborder in Alaska," *Soil Sci. Soc. Am. Proc., 31:*692–699, 1967.

Smiley, T. L, and J. H. Zumberge, "Polar Deserts," *Science, 174:*79, 1971.

Soil Conservation Service, USDA, "Distribution of Principal Kinds of Soils: Orders, Suborders, and Great Groups," *National Atlas Sheet 85 and 86,* Department of the Interior, Washington, D.C., 1969.

Soil Survey Division of Bureau of Soils, "Soils of the United States, in *Soils and Men,* USDA Yearbook, Washington, D.C., 1938. pp. 1019–1161.

Soil Survey Staff, *Soil Taxonomy,* USDA Agriculture Handbook 436, Washington, D.C., 1975.

Tedrow, J. C. F., "Concerning Genesis of the Buried Organic Matter in Tundra Soil," *Soil Sci. Soc. Am. Proc., 29:*89–90, 1965.

Tedrow, J. F. C., *Soils of the Polar Landscapes,* Rutgers, New Brunswick, 1977.

Tedrow, J. C. F., J. V. Drew, D. E. Hill, and L. S. Douglas, "Major Genetic Soils of the Arctic Slope of Alaska," *Jour. Soil Sci., 9:*33–45, 1958.

Uehara, G., H. Ikawa, and H. H. Sato, *Guide to Hawaii Soils,* Hawaii Agr. Exp. Sta. Mesc. Pub. 83, 1971.

U.S. Department of Agriculture, Department of the Interior, and the state of Alaska, *Alaska Conservation Needs Survey,* Washington, D.C., 1968 (mimeo).

Wulforst, J. P., J. A. Phillips, W. E. Hanna, P. Puglia, and L. Crandall, *Soil Survey of Wyoming County, New York,* Soil Conservation Service and Cornell University Agr. Exp. Sta., Washington, D.C., 1974.

Vertisols of the United States

Central Concept

The central concept of Vertisols is clayey soils with deep, wide cracks at some time during the year. The word Vertisol is derived from the Latin *verto,* meaning to turn. Repeated expansion and shrinking with wetting and drying causes Vertisols to be slowly inverted. Two conditions required for their development are parent material high in content of expanding clay or rocks that weather to form parent materials high in content of expanding clay, and alternate wetting and drying of the soil. Vertisols comprise about 1 percent of the land in the United States and are generally found within 45 degrees of the equator. The Canadian Soil Classification System does not contain an order equivalent to Vertisols.

Genesis of Vertisols

The genesis of Vertisols is illustrated in Fig. 4-1. Deep, wide cracks, sometimes over 1 meter deep, develop during the dry season. Dry surface soil materials fall

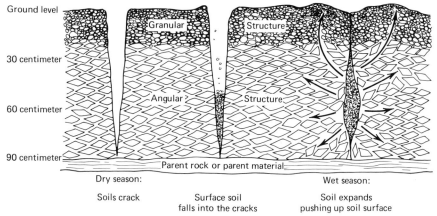

Figure 4-1   Diagram of evolution of Vertisols. Soils crack in dry season, and surface material falls into cracks that later expand when wetted by rain that falls quickly to the bottom of the cracks. The soil then wets from the bottom up. Uneven wetting of the soil creates pressures that result in a wedge-shaped structure and upward movement of soil between peds. (Adapted from Buol, 1966.)

into the cracks, and rains wash material into the cracks. Nearly all the rain after a dry season may quickly move to the bottom of the cracks, and the soil then wets from the bottom upward, although it may also wet slowly from the surface downward. Expansion during wetting exerts pressure in all directions and causes soil movement hortizontally and upward. Unequal drying of the soil from the bottom up and accompanying soil movement result in the formation of wedge-shaped aggregates or polyhedrons, as shown in Fig. 4-1. The surfaces of the polyhedrons are polished because of soil movement and are called *slickensides* (see Fig. 4-2).

Surface soil material that falls into deep cracks causes a downward movement of the surface soil. Pressures created during expansion cause a slow movement of soil upward between cracks. The result is a microrelief of alternating depressions and ridges (Fig. 4-3). The microrelief is called *gilgai*. Soil in the depressions may be colored black to a depth of as much as 2 meters, while the soil on nearby adjacent ridges is brownish or yellow-brown. Plowing fields causes a color pattern related to the microrelief. Because of the slow churning or inversion of the soils, they are called "self-swallowing" soils. Vertisols without gilgai have developed in Oregon in only 550 years.

Alternating wet and dry seasons are necessary for development of Vertisols, and the native vegetation on most of them was grasses. Some Vertisols supported trees. Desert shrubs were the native vegetation on Vertisols that developed in desert basins that received occasional run-on water. Parent materials include marine clays, shales, limestone, chalk, and basalt. Vertisols have previously been called Grumusols and Rendzinas in the United States. In other countries they are referred to by many names, including Tirs, Black Cotton, Black Earth, and Regur.

Figure 4-2　Large, shiny slickenside surface that is partially exposed in Houston Black clay near Temple, Texas. (USDA-SCS photograph.)

Figure 4-3　Gilagi microrelief on Vertisols. (USDA-SCS photograph.)

Dominant Properties of Vertisols

Vertisols are mineral soils in regions with mesic or warmer soil temperature regimes that have:

1. Thirty percent or more clay in all horizons to a depth of 50 centimeters or more.

2. Cracks that are at least 1 centimeter wide at the 50-centimeter depth at some time during the year.

3. Wedge-shaped structure in subsoil whose long axes are tilted 10 to 60 degrees from the hortizontal.

4. Gilgai, or at some depth between 25 centimeter and 1 meter, slickensides or wedge-shaped structural peds.

The self-swallowing nature of Vertisols results in profiles that are quite homogenous vertically; however, the soils are cyclic and properties change horizontally. Selected properties of the Houston Black Clay are given in Table 4-1. There is a high content of clay that is uniformly distributed vertically. The range of clay content in Vertisols is about 30 to 80 percent. The clay is dominantly montmorillonite. The organic matter content decreases slowly with increasing depth. Decreasing content of organic matter with increasing soil depth maybe associated with little change in color. Black color of moist soil may occur as deep as 1 meter in microdepressions. On microknolls, as compared to microdepressions, the A horizons are much thinner and lighter in color, and color changes more rapidly with depth.

There is high cation-exchange capacity reflecting both the high clay content and high exchange capacity of the montmorillonitic clay. The gradual decrease in cation-exchange capacity with depth is associated with decreasing content of organic matter. The impermeability of the underlying parent material and somewhat limited rainfall prevented the removal of the calcium carbonate. There is evidence, however, of some downward movement of carbonate. The calcium carbonate content of all horizons is high, reflecting the high content of lime in the

Table 4-1   Selected Properties of Houston Black Clay

Horizon	Depth, centimeter	Percent Clay	Percent Organic Matter	CEC, milliequivalents per 100 grams	Percent $CaCO_3$
A11	0–45	58	4.1	64	17
A12	45–100	58	2.1	58	20
AC	100–150	58	1.0	53	26
C	150–195	58	0.4	47	32

Data from Kunze and Templin, 1956.

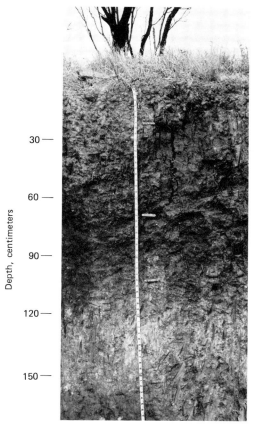

Depth, centimeters

30 —
60 —
90 —
120 —
150 —

Figure 4-4   Vertisol (Pellustert) showing strong structure in the surface soil, deep soil crack, and slickensides near bottom of crack. (USDA-SCS photograph.)

parent material. In the Houston all horizons are alkaline, especially on the micro-knolls, although some Vertisols have a pH as low as 5 in the surface horizon. A Vertisol profile is shown in Fig. 4-4.

Bulk density of soil between cracks is high and is 2 grams per cubic centimeter in some cases. The structure of the surface soil varies widely. Some Vertisols have a surface mulch of fine and medium granules that is 5 to 19 centimeters thick. If destroyed by cultivation, the mulch may reform in a single wetting and drying cycle. Other Vertisols have surface horizons with massive structure and are very hard when dry. The type of surface soil structure is very important for tillage operations, especially for seedbed preparation.

Unique Water Relationships

The moisture relationships in Vertisols are unique because of their extensive cracking and low permeability when wet. Water from rains that occur when cracks are open can quickly move to the bottom of the cracks with little or no runoff. After cracks close, however, the low permeability results in all or almost all runoff of rains, and erosion is a serious problem on sloping land. During the dry season, the cracks also permit the escape of soil moisture from deep in the soil along the crack surfaces. Experimental studies have shown that evaporation from soil cracks may equal or exceed loss of water by evaporation from the soil surface. The formation of cracks also prunes plant roots.

Vertisol Suborders

Vertisol suborders are Torrerts, Uderts, Usterts, and Xererts, depending on the number of days cracks are open in most years, as shown in Table 4-2.

Torrerts

Torrerts are Vertisols of dry climates and tend to form in closed depressions that receive run-on water. Cracks may remain open all year. There is a short rainy period and, if the cracks close, they remain closed for less than 60 consecutive days. The infrequent wetting results in limited or no development of gilgai.

Uderts

Uderts are Vertisols of humid climates and are usually moist. Cracks open and close one or more times in most years. Cracks are open less than 90 cumulative days in most years. Most of these soils had a grass vegetation at the time of settlement, and some had an open or closed hardwood forest. Forested Uderts have a gray- or dark gray-colored surface horizon.

Table 4-2   Number of Days Cracks are Open in Most Years of Vertisol Suborders

Suborder	Days Cracks Open
Torrerts	About 365
Uderts	Less than 90 cumulative or less than 60 consecutive
Usterts	Over 90 cumulative
Xererts	More than 60 consecutive

Usterts

Usterts in the United States develop in areas with low summer rainfall and high summer temperature. Cracks are open for at least 90 cumulative days and are closed for more than 60 or more consecutive days in most years. Usterts that occur outside the United States in monsoon climates have two rainy and dry seasons per year.

Xererts

Xererts occur in areas with a Mediterranean climate. Winters are cool and rainy, and summers are dry and warm. Cracks open and close quite regularly each year and remain open for more than 2 of the 3 months following the summer solstice in more than 7 out of 10 years. Temperature regimes are thermic or mesic. Xererts are moderately extensive in California. Small grains and grazing are the dominant dry land use. A wide variety of crops is grown if irrigated. Salt accumulation from irrigation, however, is difficult to remove by leaching because of low soil permeability. Where "good" water is available for flushing, salting is not a problem on Vertisols.

Vertisols of Texas

Texas has the largest area of Vertisols in North America; they occupy 10.4 percent of Texas land. The suborders are mainly Usterts and Uderts, with some Torrerts. Locations of major areas are shown in Fig. 4-5.

Usterts of the Blackland Prairies

Perhaps the best-known area of Vertisols in the United States is the Blackland Prairie in north-central Texas, where Dallas is located. This area also extends into southern Oklahoma. The soils developed primarily from soft, calcareous sedimentary rocks such as marl and chalk. East of the Vertisols on sandy marine sediments the soils are mainly Alfisols. The Vertisols occur on gently sloping to nearly level uplands of a dissected plain. The most extensive soil is the Houston Black Clay.

Land Use.   Nearly all of the land is in farms, about two-thirds of it is cropland, and one-third is pasture. The landscape is one of prosperous agriculture. Soils are Usterts, with a thermic soil temperature regime. Climate, soil, and presence of cotton root rot organisms are the major factors affecting the kind of crops grown. Tillage operations too soon after rains or irrigation result in puddling and soil compaction. Bedding and rebedding (throwing soil up into ridges) several times during

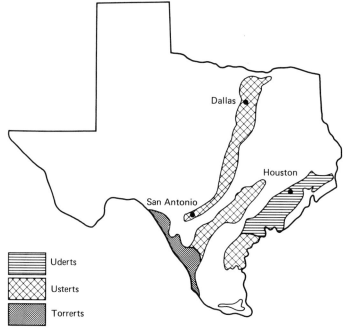

Figure 4-5   Major areas of Vertisols in Texas. (Adapted from Godfrey, 1973 and National Atlas Sheets 85 and 86, 1969, U.S. Department of Interior.)

the fall and winter allow the soil to go through several cycles of wetting and drying, which results in a fine, granular soil. Planting is on the tops of the beds that dry quickly at the surface and thus warm up faster than the lower, wetter, unbedded soil. Early planting enables crops to make the best use of moisture before the dry summer season and helps cotton resist cotton root rot disease. Cotton, grain sorghum, and small grain, with or without sweet clover, make up the basic crop rotation. Cotton requires a warm season and is quite drought resistant. The importance of cotton has been declining because of acreage controls, erosion, and root rot disease. The soils generally are quite fertile. They have high exchangeable potassium and respond to nitrogen and phosphorus. Iron chlorosis occurs on some shrubs, fruits, and grasses.

Usterts of Southern Texas

Another area of Usterts is located in southern Texas (see Fig. 4-5). These Vertisols developed mainly in calcareous clays on nearly level to gently undulating plains. The soil temperature regime is hyperthermic. These soils are quite dry and are used mainly for grazing and for production of some cotton and grain sorghum. Grain sorghum is adapted to the ustic soil moisture regime because it can interrupt its growth when water is limited and continue its growth when water becomes available again. In the southernmost and driest part of the area Torrerts, not Usterts, are common.

Uderts of The Gulf Coast Prairie

Vertisols of the Texas gulf coast have developed with annual precipitation ranging from 70 to 140 centimeters. The soils have developed on a nearly level coastal plain just inward from a narrow strip of marsh, where soils are largely Aquolls and Aquepts. Parent materials are mainly calcareous marine clays.

The greatest rainfall is in the eastern part of the gulf coast, and soils are the most leached and acid. Beaumont is a dominant series and is acid to considerable depth, with a pH 5 or less in the surface horizons. In the central part of the area where rainfall is less than where Beaumont soils occur, Lake Charles is the dominant series and is acid in the upper horizons and calcareous in the lower horizons. Beaumont and Lake Charles are Uderts and have thermic soil temperature regimes. Victoria clay is the dominant series in the westernmost part of the area and is an Ustert; the soil is calcareous throughout as are Usterts of the Blacklands. Victoria soils have hyperthermic temperature regimes. The differences in the Vertisols along the gulf coast are caused mainly by precipitation variation.

Drainage is a major problem in the use of Uderts for agriculture on the gulf coast. Land is mounded up to provide elevated walkways for cattle on pastures. Minor flooding sometimes occurs from hurricanes. Urban, industrial, and recreational developments are rapidly increasing. Groundwater removal in the Houston area is causing subsidence of the land surface and increasing flooding hazard from hurricanes.

The nearly level surface of the gulf coast Uderts and impermeability make the soils well suited to rice production (see Fig. 4-6). Texas ranks third in the United States in rice production; about 11 percent of the rice produced in the United States is grown on Uderts in Texas. Because they churn during periods when they dry out, the soils have to be releveled frequently to obtain even distribution of irrigation water for rice production. Rice acreage expansion has slowed in recent years because of limited groundwater supplies for irrigation. Other important crops include cotton, corn, grain sorghum, and tame pasture.

Vertisols of Lesser Extent

Vertisols require high clay content parent material and alternating wet and dry seasons. As a result, there are numerous locations where small isolated areas of Vertisols develop. Vertisol series are recognized in at least 16 states.

Figure 4-6   Rice harvest on Lake Charles clay in Jefferson County, Texas. (USDA-SCS photograph.)

Uderts of Mississippi and Alabama

Uderts have developed in west central Alabama and east-central Mississippi in residum weathered from marl and chalk (see area Vla of Fig. 2-1). Average annual precipitation is about 130 centimeters; it is 65 to 75 centimeters during the frost-free seasons, which lasts about 200 to 230 days. The major associated soils are the Eutrochrepts on the more sloping and eroded lands. Principal crops are cotton, corn, and soybeans.

Usterts of Arizona

About 400,000 hectares of Vertisols occur in north-central and eastern Arizona. The soils developed mainly from weathered ultrabasic volcanic rocks; a brown color is dominant, and the soil moisture regime is ustic. The soils are Chromusterts (chrom for high chroma or light color). Most of the Vertisols are uncultivated and are used for cattle and sheep ranges. The vegetation cover is dominated by grasses.

Vertisols of The West Coast, Hawaii, and Puerto Rico

Torrerts occur in the coastal ranges of California that developed from shaly parent materials. The soils are used mainly for grazing and dryland grains. A considerable area of Xererts occur in southwestern Oregon, where some Xererts support pear orchards.

Usterts are found on some of the valley floors in Hawaii, where montmorillonitic clays produced by the weathering of basalt are deposited by eroding waters. In southwestern Puerto Rico Usterts are found in the Lahas Valley, which developed from clays originated from the weathering of the surrounding limestone hills.

Engineering Problems on Vertisols

Although Vertisols have a high bearing capacity because of their density, heavy cracking and soil movement endanger the stability of building foundations. Several characteristics of Vertisols, such as slickensides and gilgai, indicate that the soil undergoes strong pressures and is subject to churning. Changes in volume with wetting and drying cause significant changes in surface elevation. Vertical soil movement of concrete bench marks penetrating to a depth of 1.5 meters has been recorded to be 25 millimeters and 38 millimeters for bench marks buried to depth of 1 meter. Bench marks buried deeper than 2.5 meters moved less than 5 millimeters vertically. As a result of contraction and expansion of soil, mainly in the

Figure 4-7 These houses built on Vertisols have suffered foundation failure and downhill movement.

upper 1.5 meters. concrete and brick structures frequently crack and are twisted out of shape (see Fig. 4-7). Building foundations on Vertisols need to be deep enough to be placed in soil that remains relatively moist and not subject to extremes of wetting and drying.

Foundations for roads are also a problem on Vertisols. Since road foundations may not be laid at great depth because of high cost, the construction must provide good drainage to prevent excessive swelling and churning. The same drainage precautions apply to construction of airport runways. Machinery is difficult to move when soils are wet.

On slopes there is danger of landslides because of an excess of water that lubricates the soil in the wet season, allowing entire hillsides to give way. Structures built on slopes risk total destruction. The downward movement may be very gradual on gentle slopes, with trees and posts leaning downhill.

Because of high plasticity, Vertisols can be used to only a limited extent for building dams. The soil erodes very easily, and care should be taken to minimize cracking and the formation of runnels in the banks. Vertisols can be used safely only if they are mixed with less plastic or nonplastic material, the amount determined by the resulting infiltration capacity of the mixture. The impermeability of Vertisols when wet makes them unsuited for sewage effluent disposal. High plasticity also makes Vertisols unsuited for making bricks.

References

Adams, J. E., and R. J. Hanks, "Evaporation from Soil Shrinkage Cracks," *Soil Science, Soc. Am. Proc., 28:*281–284, 1964.

Agricultural Experiment Stations of the Western States Land-Grant Universities and Colleges and USDA, *Soils of the Western United States,* Washington State University, Pullman, 1964.

Austin, M. E., *Land Resource Regions and Major Land Resource Areas of the United States,* USDA Handbook 296, Washington, D.C., 1965.

Brooks, C. W., C. A. Rogers, J. H. Mayberry, J. D. McSpadden, Jr., W. D. Wayburn and J. W. Huntsinger, "Soil Survey of Ellis County, Texas," USDA and Tex. Agr. Exp. Sta., Washington, D.C., 1964.

Buol, S. W., "Soils of Arizona," *Ariz. Agr. Exp. Sta. Tech. Bull. 171,* 1966.

Buol, S. W., Ed., "Soils of the Southern States and Puerto Rico," Agr. Exp. Sta. of the Southern States and Puerto Rico Land Grant Universities, *Southern Cooperative Series Bulletin* 174, 1973.

Buol, S. W., F. D. Hole, and R. J. McCracken, *Soil Genesis and Classification,* Iowa State Press, Ames, 1973.

Crout, J. D., D. G. Symmank, and G. A. Peterson, "Soil Survey of Jefferson County, Texas," USDA and Tex. Agr. Exp. Sta., Washington, D.C., 1960.

Dixon, J. B., and V. E. Nash, "Chemical, Mineralogical and Engineering Properties of Alabama and Mississippi Black Belt Soils," *Southern Coop. Series No. 130,* Auburn University, Auburn, Alabama, 1968.

Dudal, R., and D. L. Bramao, "Dark Clay Soils of Tropical and Subtropical Regions," FAO Agr. Dev. Paper 83, 1965.

Flack, K. W., and D. F. Slusher, "Soils Used for Rice Culture in the United States," Mimeo, USDA Soil Conservation Service, Washington, D.C., 1977.

Godfrey, C. L., "General Soil Map of Texas," *MP 1034,* Tex. Agr. Exp. Sta., 1973.

Godfrey, C. L., "A Summary of the Soils of the Blackland Prairies of Texas," *MP 698,* Tex. Agr. Exp. Sta., 1964.

Godfrey, C. L., Clarence R. Carter, and Gordon S. McKee, "Resource Areas of Texas," *Bulletin 1070,* Tex. Agr. Ext. Serv. and Tex. Agr. Exp. Sta.

Godfrey, C. L., Harvey Oakes, and Richard M. Smith, "Soils of the Blackland Experiment Station," *MP 419,* Texas Agr. Exp. Sta., 1960.

Gray, F., and H. M. Gallaway, "Soils of Oklahoma," *MP 56,* Okla. Agr. Exp. Sta., 1959.

Gray, F., and M. H. Roozitalab, "Benchmark and Key Soils of Oklahoma," *MP 97,* Okla. Agr. Exp. Sta., 1976.

Johnson, W. M., J. G. Cady, and M. S. James, "Characteristics of Some Brown Grumusols of Arizona," *Soil Sci. Soc. Am. Proc., 26:*389–393, 1962.

Kunze, G. W., H. Oakes, and M. E. Bloodworth, "Grumusols of the Coast Prairie of Texas," *Soil Sci. Soc. of Am. Proc., 27:*412–421, 1963.

Kunze, G. W., and E. H. Templin, "Houston Black Clay, the Type Grumusol: II. Mineralogical and Chemical Characterization," *Soil Sci. Soc. Am. Proc., 20:*91–96, 1956.

McCall, W. W., *Soil Classification in Hawaii,* Hawaii Coop. Ext. Service Circular 476, 1973.

National Cooperative Soil Survey of United States, National Soil Survey Committee of Canada and FAO, *Soil Map of the World,* II: North America, Rome, 1975.

Oakes, H. and J. Throp, "Dark-Clay Soils of Warm Regions Variously Called Rendzina, Black Cotton Soils, Regur and Tirs," *Soil Sci. Soc. Am. Proc., 15:*347–354, 1951.

Parsons, R. B., L. Moncharoan, and E. G. Knox, "Geomorphic Occurrence of Pelloxererts, Willamette Valley Oregon," *Soil Sci. Soc. Am. Proc., 37:*924–927, 1973.

Smith, J. C., C. L. Godfrey, and H. Oakes, "Soils of the Gulf Coast Pasture-Beef Cattle Research Station," *MP735,* Tex. Agr. Exp. Sta., 1964.

Soil Conservation Service, USDA, "Distribution of Principal Kinds of Soils: Orders, Suborders, and Great Groups," *National Atlas Sheet 85 and 86,* Department of the Interior, Washington, D.C., 1969.

Soil Survey Staff, *Soil Taxonomy,* USDA Agriculture Handbook 426, Washington, D.C., 1975.

Templin, E. H., J. C. Mowery, and G. W. Junze, "Houston Black Clay, the Type Grumusol: I. Field Morphology and Geography," *Soil Sci. Soc. Am. Proc., 20:*88–90, 1956.

Uehara, G., H. Ikawa, and H. H. Sato, "Guide to Hawaii Soils," Mesc. Pub., 83 Hawaii Agr. Exp. Sta., 1971.

Westfall, D. G., C. L. Godfrey, N. S. Evatt, and J. Crout, "Soils of the Texas A&M Univ., Agr. Research and Extension Center at Beaumont," *MP 1003,* Texas Agr. Exp. Sta., 1971.

Wheeler, F. F., "Soil Survey of Harris County, Texas," USDA and Tex. Agr. Exp. Sta., Washington, D.C., 1976.

Aridisols of the United States

Central Concept

Aridisols are soils of arid regions. Since desert climate exists on 36 percent of the earth's land, Aridisols are the most extensive soils. Many of the soils of desert regions, such as moving sand dunes, shallow soils on mountain slopes, or soils formed in alluvium along rivers and streams, are not Aridisols. On the other hand, some soils outside of deserts are Aridisols because the soils are coarse and retain little water or because runoff is excessive due to steep slopes or the fine texture of surface horizons. The result is that about 18.8 percent of the world's soils are Aridisols (see Table 1-8). In the United States about 11.5 percent of the soils are Aridisols (see appendix Table 2).

Nature and Genesis of Aridisols

The genesis and nature of Aridisols in deserts is strongly influenced by limited water for weathering, eluviation, and plant growth. Many Aridisols, however, show evidence of considerable pedogenic development caused by soil genesis in an earlier, more pluvial period.

Central Concept of Aridisols

Aridisols are mineral soils that have an aridic soil moisture regime. More specifically, Aridisols:

1. Have one or more pedogenic horizons.
2. Do not have water available for mesophytic plants for long periods of time; for as long as 3 months at a time when the soil is moist and warm enough for plants to grow.

Soil temperature regimes vary from cryic to isohyperthermic.

Genesis and Properties

The features of deserts and Aridisols relate to the limited supply of water. Native vegetation consists of shrubs, cacti, grasses, and forbs that are widely spaced, as shown in Fig. 5-1. Much of the soil surface is bare. Winds are active in moving soil materials and, if parent materials contain gravel, a desert pavement is formed by the removal of fine soil particles. Another explanation for desert pavement is the

Figure 5-1 Widely spaced shrubs and cacti are common features of Aridisol landscapes. This area in southern California receives about 8 centimeters of rainfall annually.

movement of gravel upward under the effect of soil shrinking associated with wetting and drying.

Desert varnish on the gravel consists of a shiny dark coating. Studies indicate that the varnish consists of two layers. The inner layer is rich in SiO_2 and Al_2O_3. The outer layer is rich in FeO and MnO. Colonies of algae seem to mobilize iron ions and produce a concentration of oxides on the gravel surfaces that form the varnishlike appearance. Examples of desert varnish are shown in Fig. 5-2.

Water never moves completely through the soil, although some materials are translocated from the upper to the lower part of the solum. The most striking feature of many Aridisols is a zone of carbonate at varying depths below the surface in the B or C horizon if the parent material is calcareous. In time a high carbonate or calcic layer may become plugged with carbonates and cemented into a *petrocalcic* horizon.

Surface horizons are light-colored ochric epipedons low in organic matter that reflect low biomass production. Present climate results in little, if any, clay eluviation. Many Aridisols have well-developed argillic horizons (Bt) that formed under a more humid climate many years ago (see Fig. 5-3).

Salic horizons form where shallow groundwater tables exist and water moves to the soil surface, evaporates, and leaves the salt on the surface. The well-oxidized conditions inhibit movement of free iron oxides within the profile. The high pH favors solution and downward movement of silica; some Aridisols have *duripans,*

Figure 5-2 Desert pavement in the Sonoran Desert. Note coatings on gravel of the enlarged section.

Figure 5-3 Tubac very gravelly loam (Paleargid) located in the desert near Tucson, Arizona. The clay content increases from 9 percent in the A horizon to nearly 60 percent in the B22t.

where the accumulation of silica causes cementation. Duripans form where parent materials are low in calcium (minimal calcic horizon development) and high in volcanic materials such as glass, which release abundant silica on weathering. Other horizons that may exist in Aridisols include cambic and gypsic (accumulation of calcium sulfate).

The Mohave is a common Aridisol in the southwestern United States, and some representative data are presented in Table 5-1. The Mohave has an argillic horizon, low organic content, and low carbon–nitrogen ratio. Contributing to the low ratio could be blue-green algae, which are often present in surface crusts and fix nitro-

Table 5-1   Selected Properties of Mohave Sandy Clay Loam

Horizon	Depth, centimeters	Per-cent Clay	Percent Organic Matter	Carbon/ Nitrogen Ratio	Percent Exchang-eable Sodium	pH	Percent CaCO₃
A1	0–10	11	0.25	6	1.2	7.8	—
B1	10–25	14	0.19	6	2.0	7.4	—
B2t	25–69	25	0.24	7	2.5	8.5	—
B3ca	69–94	21	0.25	8	4.1	8.9	10
IICca	94–137	17	0.08	—	12.7	9.2	22

Adapted from profile 62 of "Soil Classification, A Comprehensive System," USDA, Washington, D.C., 1960.

gen. A significant amount of exchangeable sodium is present. All horizons have a pH near 8 or above and are entirely base saturated. A calcic horizon exists below the depth of 69 centimeters.

Aridisol Suborders

Aridisols are placed into suborders on the basis of presence or absence of argillic horizons. The suborder Orthids includes Aridisols without argillic horizons; by contrast, the suborder Argids includes Aridisols with argillic horizons. As already noted, the Mohave has an argillic horizon and is an Argid. About 8.6 percent of the soils in the United States are Argids and 2.9 percent are Orthids; the general distribution is shown in Fig. 2-1 with Argids as D1 areas and Orthids as D2 areas.

Relationship of Orthids and Argids to Land Surface Age

Aridisols of the United States occur in a generally mountainous area where mountains and basins are interspersed. Orthids occur on younger land surfaces and argids on older land surfaces. It seems that the Orthids have developed largely within the past 25,000 years in an arid climate that is the same or similar to the present climate. Argids are common on the older land surfaces, where there has been more time for argillic horizon development and where they were under the influence of the more humid climate that existed over 25,000 years ago. Paleargids (old Argids) like the Tubac (see Fig. 5-3) occur on old stable land surfaces that date from the middle or early Pleistocene or earlier. Associated soils are Fluvents and Orthents on very recent alluvial sediments and Orthents and soils such as Alfisols and Inceptisols in the mountains, where more water is available for soil genesis. The relationship of soils to landscape position is shown in Fig. 5-4.

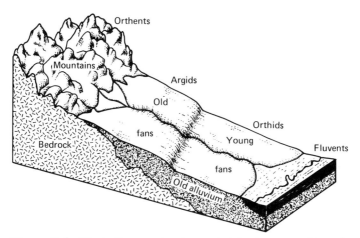

Figure 5-4 Block diagram showing typical landscape positions for soil suborders in the Basin and Range Province of the southwestern United States.

Obliteration of Argillic Horizons in Argids

Argillic horizons typically occur in Aridisols of the southwestern United States that developed under the more humid climate of the Pleistocene. Land dissection and the slow truncation of Argids in some areas after the Pleistocene has resulted in the loss of argillic horizons by erosion and conversion of Argids into Orthids. These landscapes have Argids on undissected broad divides and Orthids on the slopes. In areas of severe dissection Entisols occur. The obliteration of the argillic horizon has also been enhanced by engulfment with calcium carbonate. During the drier post-Pleistocene period, movement of calcium carbonate into argillic horizons and formation of calcic horizons has greatly increased the carbonate content, obliterated clay skins, and caused major color changes. Many of these horizons are no longer able to meet the criteria for argillic horizons.

Some Orthids in southern New Mexico formed in middle Pleistocene sediments of the ancestral Rio Grande. These soils contained evidence that they had argillic horizons at one time. Since the soils occurred on the nearly level surface of a basin, dissection could not account for the loss of argillic horizons. The presence of termite burrows and mounds and rodent tunnels led to the conclusion that animal mixing of soil had destroyed argillic horizons and converted Argids to Orthids.

Orthids with Salic Horizons—Salorthids

Orthids with a *salic* horizon within 75 centimeters of the surface are Salorthids. Salorthids are saturated with water within 1 meter of the soil surface for at least 1 month in most years. Salts are translocated upward from the water table, and salts accumulate within a depth of 75 centimeters and frequently at or near the soil surface. Vegetation consists of salt-tolerant species (halophytes). Although the soil may be moist, the high salt content makes the soils physiologically dry. Salorthids are not extensive in the United States and were formerly called Solonchaks.

Shallow water tables that contribute to the formation of Salorthids are created in basins without a natural outlet when runoff water collects. The extent of saline areas may vary from 1 hectare to hundreds of thousands of hectares. Many of the saline soils of the Great Basin were created by run-on water in basins. In some soils impermeable layers contribute to water tables and salinization.

Argids with Natric Horizons

Salorthids subjected to leaching from improved drainage, natural or artificial, undergo removal of soluble salts. The salts are mainly chlorides and sulfates of calcium, magnesium, and sodium. When leaching results in a concentration of sodium salts of over one-half the total salt, the exchangeable sodium percentage increases to 15 and above, resulting in dispersion of clay and organic matter. Clay migration is favored, and Salorthids may develop argillic horizons that are dark colored because of the simultaneous migration of clay and humus. Natric horizons are argillic horizons with 15 percent or more exchangeable sodium. Soil pH is in the range of 8.5 to 10. A characteristic columnar structure frequently develops in the natric horizon. The soil with these features is a Natrargid (formerly called a Solonetz). These soils are moderately extensive in the western part of the Great Plains.

Orthids in Hawaii

Although Aridisols comprise only 1 to 2 percent of the soils of the Hawaiian Islands, deserts are a characteristic feature in the rain shadows of mountains. The major areas in the state of Hawaii occur in the rain shadow of the Kohala Mountains on the northern part of the island of Hawaii (see Fig. 3-12). The soil, Kawaihae, has developed from volcanic ash and is underlain by cindery lava. The soils occur on the coastal plains at elevations up to 500 meters, with an annual precipitation of 12 to 50 centimeters. The soils lack argillic horizons and are Orthids. Soils are used mainly for pasture, recreation, wildlife, and homesites. Small acreages are irrigated and used for truck crops.

Some Aridisols exist on the other islands of Hawaii and have formed in alluvial materials at low elevation where water tables are close to the surface. The soils have salic horizons and are Salorthids. They are of very minor extent.

Vegetation on Aridisols

Mesophytic plants cannot survive on Aridisols, and irrigation is required to mature agricultural crops. The native vegetation on Aridisols consists of plants that have adapted to a limited water supply.

Drought Tolerance of Desert Plants

One of the most obvious features of deserts is the wide spacing between plants that is caused in some cases by toxins that keep other plants from invading the space close to established plants. The wide spacing permits rooting through larger root volumes for a greater supply of water. Some plants have very extensive lateral root systems that efficiently absorb moisture before it migrates deeply. Other plants, such as mesquite, have very deeply penetrating roots that tap deep underground water supplies.

   Plants that survive and grow where rain is infrequent and/or soil is saline can extract water beyond the 15 bars normally considered to be the wilt point. An Australian shrub has been known to recover from a soil moisture stress as high as 130 bars. On the Great Salt Lake osmotic stress of 220 bars has been observed for some halophytes. Other salt-tolerant species are able to tolerate high internal salt concentrations or to excrete salt from leaves.

Drought Adaptation of Desert Plants

Many desert plants have adapted to desert conditions by means other than those of being able to grow under conditions of high soil moisture stress. Transpiration losses may be reduced by small leaf size, small stomatal openings, efficient photosynthesis or, in cases of severe drought, shedding of leaves. Cacti can fix carbon dioxide at night so they need not have stoma open during the day, when transpiration rates are high. The low photorespiration rates of some plants mean that less photosynthesis is required and less need to have stoma open for carbon dioxide fixation. Chlorophyll in the branches and bark of some species permits photosynthesis with minimal water loss.

   Reproduction by cloning eliminates the need for seed germination and seedling survival—a distinct advantage of desert shrubs. Some annuals have seeds that contain a germination inhibitor. The inhibitor is washed from the seeds by a rain that exceeds 1 or 2 centimeters and is sufficient to insure germination and seedling survival.

   Some desert plants go dormant or greatly reduce activity during a prolonged drought. Rapid regrowth when conditions become favorable is another characteristic that enables shrubs to dominate the arid and semiarid lands. Many plants, including annuals and grasses, grow only during the seasons when water is available. On the Sonoran Desert near Tucson there is a double rainfall maximum; rainfall is highest in midwinter and again in midsummer. This has resulted in two distinct flora—one is adapted to the cool and moist weather and the other to hot and dry weather.

Land Use on Aridisols

The Aridisols of the western United States occur largely in the "western range and irrigated region" (see Fig. 5-5). As the name implies, grazing of sheep and cattle and production of crops by irrigation are the two dominant land uses.

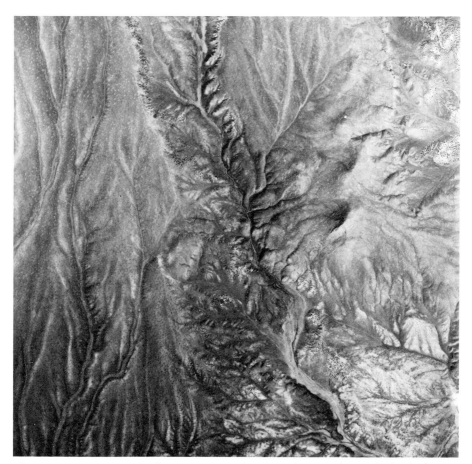

Figure 5-5 Open rangeland in eastern Nevada, where the annual precipitation is about 25 centimeters and sheep and cattle grazing are the major land use. Vegetation is mainly sagebrush, with a sparse admixture of grasses. (USDA photograph.)

Grazing on Aridisols

The use of land for grazing is closely related to precipitation, which largely determines the amount of forage produced. Some areas are too dry for grazing and other areas take advantage of summer grazing on high mountain meadows. As much as 30 hectares or more are needed per head of cattle, thus making large farms and ranches a necessity. Most of the ranchers supplement the range forage by producing some crops on a small acreage that is favorably located for irrigation, as shown in Fig. 5-6. The major hazard in use of land for grazing is overgrazing, which results in invasion of less desirable plant species and increased soil erosion. Most of the land is owned by the federal government and is leased to ranchers.

The extensive presence of shrubs on desert lands has caused some people to question the value of the lands for grazing. Shrubs, however, are the favorite and only browse of some animals. Different animals prefer different kinds of browse. Goats like shrubs better than sheep do, and sheep like shrubs better than cattle do. Obviously, the kinds of plant species on rangelands is of great importance to ranchers.

The feed value of shrubs depends on what part of the plant is consumed. Wood has a low feed value. Animals, however, browse leaves, twigs, buds, flowers, and fruits. It has been reported that mature shrubs have greater protein and phosphorus content than mature grasses. One of the most dramatic means of increasing the productivity of Aridisols is to plant highly productive shrubs such as the *Atriplex*

Figure 5-6   Grazing lands on Aridisols. Note small, irrigated acreage used to produce feed to supplement feed produced on the range. (USDA photograph.)

species, which are high in protein, along with the *Opuntia* species, which are high in carbohydrate.

Giving preference to shrubs as contrasted to grasses as forage plants in deserts is more compatible with existing ecosystems and may result in greatest productivity.

Use of Aridisols for Irrigation Agriculture

Only about 1 or 2 percent of the land of the arid southwest is irrigated. Few Aridisols are irrigated because of lack of water or limiting soil conditions, such as irregular and sloping land surfaces or impermeable argillic horizons. Most of the irrigated land occurs along streams or rivers, and soils are frequently Orthents or Fluvents. These areas tend to have nearly level surfaces that permit gravity flow irrigation. In addition, the rivers are a source of water from natural flow or from reservoirs and make areas adjacent to rivers most desirable for irrigation projects. In Arizona crop production on only 2 percent of the land that is irrigated accounts for 60 percent of the farm income. Grazing, by contrast, utilizes 80 percent of the land and accounts for only 40 percent of the total farm income.

References

Agricultural Experiment Stations of the Western States Land-Grant Universities and Colleges and USDA, *Soils of the Western United States,* Washington State State University, Pullman, 1964.

Austin, M. E., *Land Resource Regions and Major Land Resource Areas of the United States,* USDA Handbook 296, Washington, D.C., 1965.

Buol, S. W., "Present Soil Forming Factors and Processes in Arid and Semiarid Regions," *Soil Sci., 99:*45–49, 1965.

Buol, S. W., "Soils of Arizona," *Ariz. Agr. Exp. Sta. Tech. Bull. 171,* 1966.

Buol, S. W., F. D. Hole, and R. J. McCracken, *Soil Genesis and Classification,* Iowa State University Press, Ames, 1973.

Dregne, H. E., Ed., *Arid Lands in Transition,* American Association for the Advancement of Science, Pub. 190, Washington, D.C., 1970.

Foote, D. E., E. M. Hill, S. Nakamura, and F. Stephens, *Soil Survey of the Islands of Kauai, Oahu, Maui, Molokai, and Lanai,* State of Hawaii, Soil Conservation Service, USDA, Washington, D.C., 1972.

Fuller, W. H., *Soils of the Desert Southwest,* University of Arizona Press, Tucson, 1975.

Gile, L. H., "Causes of Soil Boundaries in an Arid Region: I. Age and Parent Materials," *Soil Sci. Soc. Am. Proc., 39:*316–323, 1975.

Gile, L. H., "Causes of Soil Boundaries in an Arid Region: II. Dissection, Moisture and Faunal Activity," *Soil Sci. Soc. Am. Proc., 39:*324–330, 1975.

Gile, L. H., and R. B. Grossman, "Morphology of the Argillic Horizon in Desert Soils of Southern New Mexico," *Soil Sci., 106:*6–15, 1968.

Hadley, N. F., "Desert Species and Adaption," *Am. Sci., 60:*338–347, 1972.

Hendricks. D. M., and Y. H. Havens, *Desert Soils Tour Guide,* Soil Science Society of America, Tucson, Arizona, 1970.

Marschner, F. J., "Land Use and Its Pattern in the United States," *USDA Handbook 153,* Washington, D.C., 1959.

McCall, W. W., *Soil Classification in Hawaii,* Coop. Ext. Service Circular 476, University of Hawaii, Revised, 1975.

McKell, C. M., "Shrubs—A Neglected Resource of Arid Lands," *Science, 187:*803–807, 1975.

National Cooperative Soil Survey of United States, National Soil Survey Committee of Canada and FAO, *Soil Map of the World,* II: North America, Rome, 1975.

Nettleton, W. D., J. E. Witty, R. E. Nelson, and J. W. Hawley, "Genesis of Argillic Horizons in Soils of Desert Areas of Southwestern United States," *Soil Sci. Soc. Am. Proc., 39:*919–926, 1975.

Sato, H. H., W. Ikeda, R. Paeth, R. Smythe, and M. Takehiro, Jr., Soil Survey of the Island of Hawaii, State of Hawaii, Soil Conservation Service, USDA, Washington, D.C., 1973.

Shantz, H. L., G. F. White, Ed., "History and Problems of Arid Lands Development," *In The Future of Arid Lands,* American Association for the Advancement of Science, *43,* pp. 3–5, 1956.

Scholander, P. F., H. T. Hammel, E. D. Bradstreet, and E. A. Hemmingsen, "Sap Pressure in Vascular Plants," *Science, 148:*339–346, 1965.

Soil Conservation Service, USDA, "Distribution of Principal Kinds of Soils: Orders, Suborders, and Great Groups," *National Atlas Sheet 85 and 86,* Department of the Interior, Washington, D.C., 1969.

Soil Survey Staff, "Soil Classification, a Comprehensive System—7th Approximation," USDA, Washington, D.C., 1960.

Soil Survey Staff, *Soil Taxonomy,* USDA Agriculture Handbook 436, Washington, D.C., 1975.

Solbrig, O. T., and G. H. Orians, "The Adaptive Characteristics of Desert Plants," *Am. Sci., 65:*412–421, 1977.

Springer, M. E., "Desert Pavement and Vesicular Layer of Some Soils in the Desert of the Lahontan Basin," *Soil Sci. Soc. Am. Proc., 22:*63–66, 1958.

Uehara, G., H. Ikawa and H. H. Sato, "Guide to Hawaii Soils," *Mesc. Pub. 83 Hawaii Agr. Exp. Sta.,* 1971.

U.S. Salinity Laboratory Staff, *Diagnosis and Improvement of Saline and Alkaline Soils,* USDA Handbook 60, Washington, D.C., 1954, reprinted 1969.

Mollisols of the United States

Central Concept

Mollisols are the dominant soils of the world's major grasslands. These grassland areas were the last major settlement frontiers because of lack of wood, shortage of water, difficulty of defense against enemies, and difficulty of breaking the prairie sod (Fig. 6-1). Now these are areas of high grain and livestock production and low human density. Four of the top eight cropland countries are associated with large acreages of Mollisols and include the United States, Canada, the Soviet Union, and Argentina.

Mollisols characteristically develop under grass in climates that have a pronounced seasonal moisture deficit. Most Mollisols have a very dark brown to black surface horizon (mollic epipedon) that makes up more than one-third of the combined thickness of the A and B horizons or is over 25 centimeters thick. Organic matter content gradually decreases with increasing soil depth. The mollic epipedon

Figure 6-1   First year homestead on the plains of Alberta, Canada in 1904 in the spring wheat region. (From E. Brown collection, courtesy Provincial Museum and Archives of Alberta.)

has structure or has soft consistence when dry. Base saturation is 50 percent or more to a depth of 1.8 meters, and calcium is the dominant exchangeable cation. Clay minerals are mostly crystalline, with moderate to high exchange capacity. Subsoils are mainly cambic or argillic horizons, depending on degree of development. Horizons with calcium carbonate accumulation are common.

Mollisols are the most extensive soils in the United States, occupying about one-fourth of the land, as shown in Table 6-1. The suborders, in order of decreasing area, are Ustolls, Borolls, Xerolls, Udolls, and Aquolls. Albolls and Rendolls are

Table 6-1   Approximate Area of Mollisol Suborders in the United States

Suborder	Area Square Miles	Hectares	Extent, Percent of United States
Aquolls	46,100	18,670	1.3
Borolls	176,800	71,604	4.9
Udolls	170,450	69,032	4.7
Ustolls	318,400	128,952	8.8
Xerolls	184,250	74,621	4.8
Totals	896,000	362,879	24.6

From Soil Taxonomy, *1975.*

of very limited extent and occupy less than 0.10 percent of the soils of the United States.

Mollisols occur over extensive areas of North America and are bordered mainly by soils of the deserts and mountains on the west and Alfisols and Spodosols on the humid eastern border. The general distribution of the major suborders in North America is shown in Fig. 2-1.

Aquolls

Aquolls are the wet Mollisols that have an aquic soil moisture regime or are artificially drained. They are characterized by black-colored epipedons and gray-colored subsoils, as shown in Fig. 6-2. Most Aquolls developed under a vegetation of

Figure 6-2 Dominant features of Aquolls are black epipedons, gray subsoil, and water table near the soil surface at least part of the year unless artificially drained.

grasses, sedges, and forbs, resulting in high organic matter content. Some Aquolls developed under forest vegetation. Many Aquolls have developed on recent land surfaces from calcareous materials and show evidence of minimal weathering and leaching. Aquolls are typically fertile soils and, when drained, are some of the best agricultural soils. The most extensive areas in North America occur in the Red River Valley between Minnesota and North Dakota and in southern Manitoba, along the major rivers of the central United States and near the gulf coast in Louisiana and Texas (Mla areas of Fig. 2-1).

Aquolls of the Red River Valley

The Aquoll area of the Red River Valley is located between Minnesota and North Dakota and extends northward into Manitoba. The soil is called Humic Gleysol in Canada. The soils formed from sediments deposited in glacial Lake Agassiz about 9000 to 12,000 years ago. The sediments are largely reworked glacial materials with the predominance of montmorillonite in the clay fraction. Principal vegetation was tall prairie grasses mixed with wetland reeds and sedges.

The wet conditions and grass vegetation, along with low soil temperature, have resulted in soils that are high in organic matter. Fargo silty clay is one of the most extensive series and is in the vertic subgroups because of the high content of expanding clay. Fargo has a near neutral black mollic epipedon that overlies a grayish silty clay cambic horizon. The C horizon is calcareous silty clay, and there is a calcic horizon. Average annual precipitation is about 50 centimeters and, combined with a low soil permeability and high water table, has resulted in minimal leaching and clay translocation. The Fargo soils are Haplaquolls.

Hegne soils are also extensive and occur on the slight rises of the microrelief that exists on the lake plain. Hegne soils are calcareous at the surface, have a calcic horizon within 30 centimeters under the soil surface, and are Calciaquolls. Some Hegne soils are saline, and these had a native vegetation that included some salt-tolerant grasses. During the draining of Lake Agassiz there was a period when the lake did not drain south through the Minnesota and Mississippi river. At this time the water was blocked by ice in the north, and salts accumulated as water evaporated. Some of the salty water remained in the area after the ice dam gave way and is the source of salt in some of the soils. The major associated soils are Borolls on the gentle slopes and Ustipsamments on beach ridges and sand dunes.

Nearly all of these naturally fertile fine-textured Aquolls are used for cropland. Drainage is the major soil management problem. Surface drains or ditches are used, but excess water commonly delays planting and temporary ponding of water, as shown in Fig. 6-3. The frigid soil temperature and ustic soil moisture regimes impose limitations on adapted crops. Hard red spring wheat is the major crop and barley is second in area. Sugarbeets and sunflowers are important cash crops. Some flax is grown for oil. The valley was originally a mixed agriculture region but today is a large-scale cash crop farming region.

On relation to engineering concerns, these silty clay soils high in expanding clay

Figure 6-3 Temporary ponding of water on level, silty clay Aquolls in a sugar beet field in the Red River Valley. (USDA-SCS photograph.)

have high shrink-swell potential, low shear strength when wet, and severe limitations for use of septic tank filter fields due to a low soil permeability and a seasonally high water table.

Aquolls of Major River Valleys in the Central United States

Aquolls are extensive on the broad, nearly level floodplains of the Mississippi, Missouri, Ohio and Wabash rivers (see Fig. 2-1). Most of the Aquolls develop some distance from the rivers in fine-textured slack-water sediments in association with coarser-textured materials and Fluvents (and Aquents) near the rivers. Water management is the major soil management problem. Low soil permeability limits effectiveness of tile drains. Surface drainage is commonly impractical because of the low position of the soils on the floodplains. Flooding usually limits farming operations in the spring and summer. The high clay content makes seedbed preparation difficult. Fall plowing is used because soils may be too wet in the spring to be plowed. Soil temperature regime is mesic, and corn and soybeans are the major crops. Some hay is also produced for livestock feed.

Aquolls on Coastal Marshes of Louisiana and Texas

Along the gulf coast in western Louisiana and eastern Texas Aquolls occupy low, marshy, coastal plain areas. Ustipsamments exist on dunes, and Histosols exist in the low-lying areas of the coastal plains. The Aquolls are fine textured and sometimes saline, and they support salt grass. The soil temperature regime is thermic.

Harris soils are the major Aquolls; they are clay textured throughout and are Haplaquolls. They have formed about a meter above sea level. Organic matter content is high, but not high enough to qualify as Histosols. Some areas are flooded with saltwater at high tide and are inundated with saltwater during hurricanes. The dominant clay is montmorillonite, but little shrinking and swelling occur because of the wetness. Land use is mainly for cattle grazing and as a habitat for wildlife, especially waterfowl and muskrats. Grazing cattle are affected by the mosquitoes and other insects. The soil is sometimes mounded up to provide elevated walkways to help cattle cope with the perennial wetness of the range. Some of the soils are being studied for use in shrimp farming in Texas.

Aquolls of Minor Extent

The major factor contributing to development of Aquolls is wetness. As a result, Aquolls occur in landscapes in all or most of the states as small, isolated areas. Some Aquolls form in deserts, where seepage water comes to the surface. Aquolls occur in many small river valleys and are frequently used for pasture.

Aquolls occupy low places on the late Wisconsin glacial plains throughout the midwest. These soils make an important contribution to the total production of the corn belt. Drummer, a Haplaquoll, is one of the most productive soils in Illinois for corn and is the most extensive soil in the state; it occupies 6 percent of the area. Although Aquolls are not considered the dominant soils of the corn belt, their total acreage is considerable, and they make an important contribution to the productivity of the area.

Udolls

Udolls are the more or less freely drained Mollisols of the humid region and have udic soil moisture and mesic or warmer soil temperature regimes. Udolls are the major soils of the corn belt and are the most extensive in Iowa, Illinois, western Indiana, southern Minnesota and Wisconsin, eastern Nebraska, Kansas, and Oklahoma, and northwestern Missouri (see Fig. 2-1). Tall grass was the principal native vegetation, even though the region is humid enough to support forests. A few Udolls supported trees earlier, and some now support trees. Fire seems to have been an important factor in the maintenance of the grasslands in the Udoll region because, in the absence of fire, forests invade these grasslands today. Indians are credited with the maintenance of the grass by their use of fire to drive game. Tall grass was the native vegetation in 75 percent of Iowa and 55 percent of Illinois. Trees were dominant in and along the major river valleys, such as along the Mississippi River, where Alfisols are extensive.

Figure 6-4 Udoll on nearly level till plain in north-central Illinois. Parent material at this site consists of a thin layer of loess over till.

Parent materials for Udolls are mainly loess and till, except for residuum from sedimentary rocks in parts of eastern Kansas and Oklahoma. Many of the slopes are nearly level or only gently sloping on till plains of late Pleistocene age (see Fig. 6-4). The time for soil development has not been excessive, about 15,000 years or less in most cases. Soil solums are acid, and calcic horizons are generally absent. Weathering has been modest, and many weatherable minerals remain. Base saturation of all horizons is over 50 percent, and calcium is the dominant exchangeable cation.

Udoll Great Groups

Four great groups of Udolls are recognized: Hapludolls, Argiudolls, Paleudolls, and Vermudolls. Hapludolls, Argiudolls, and Paleudolls represent soils with increasing development of the argillic horizon, as shown in Fig. 6-5.

Hapludolls do not have an argillic horizon (see Fig. 6-5). Typically, a brownish mollic epipedon rests on a brownish cambic horizon. Some may have a calcic horizon below the cambic horizon. The soils are youthful and tend to occur on recent land surfaces, where slopes are steep, on coarse loess near river valleys, or where precipitation tends to be low for Udolls. Extensive areas occur on thick loess in western Iowa and in Missouri along the Missouri River and on the recent till plain of north-central Iowa and south-central Minnesota (see Fig. 6-6).

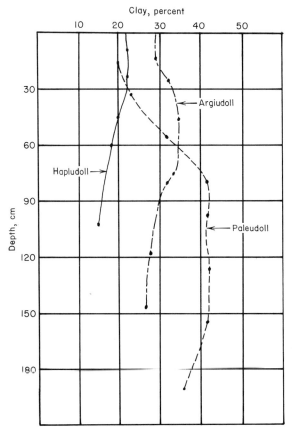

Figure 6-5 Representative clay distribution curves in Hapludolls (Monona), Argiudolls (Tama), and Paleudolls (Dennis).

Argiudolls have a relatively thin argillic horizon or an argillic horizon in which the clay content decreases rapidly with increasing depth (see Fig. 6-5). The epipedon is black to very dark brown and the argillic horizon is usually brown. Many are noncalcareous to a considerable depth, while some have a weakly expressed calcic horizon. Selected properties of an Argiudoll developed from loess in eastern Iowa are given in Table 6-2. Most of the Udolls in the United States are Argiudolls and are most extensive in northern Illinois, southern Wisconsin, southeastern Iowa, eastern Nebraska, Kansas and Oklahoma, as shown in Fig. 6-6.

Paleudolls are reddish Udolls that have a thick argillic horizon in which the clay content decreases slowly with increasing depth (see Fig. 6-5). They are located

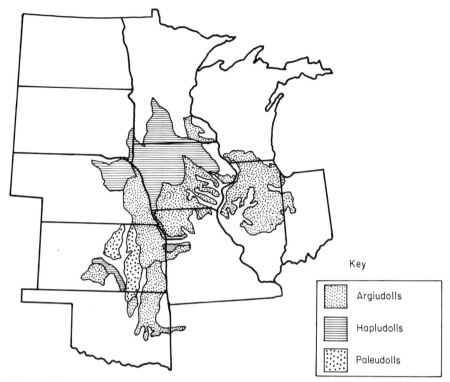

Figure 6-6 Areas where Hapludolls, Agriudolls, and Paleudolls are the dominant soils. (Adapted from National Atlas, Sheet 86, Soils, U.S. Geographic Survey, 1969.)

south of the glaciated region and mainly on land surfaces older than Wisconsin age. They are believed to have formed through at least one glacial and one interglacial stage. They exist only on the southern Great Plains in the United States, as shown in Fig. 6-6.

Paleudolls were formerly called Reddish Prairie soils. They have redder colors and less organic matter than most Argiudolls. The Paleudolls, compared to Argiudolls, have thicker argillic horizons, less granular A horizons, greater exchangeable sodium, and thicker solums (A + B horizons).

Vermudolls are Udolls that have been intensively mixed by earthworms and their predators. The mollic epipedon or the untilled soil below any Ap horizon has 50 percent or more of the soil volume of earthworm casts or animal burrows. The soils consist mostly of wormholes, worm casts, and *krotovinas*. Vermudolls do not have argillic or natric horizons. They are insignificant soils in the United States but are important in some countries.

Table 6-2 Selected Properties of an Argiudoll Developed from Loess in Eastern Iowa

Depth, centimeters	Horizon	Percent			Milliequivalents per 100 grams					Percent Base Saturation	pH
		Organic Carbon	Clay	Silt	Calcium	Magnesium	Sodium	Potassium	CEC		
0–18	Ap	2.4	29	69	14	3.4	0.1	0.5	27	66	5.7
18–28	A12	2.0	32	66	14	4.2	0.1	0.4	30	62	5.8
28–51	AB	1.2	35	63	14	6.1	0.1	0.4	30	70	5.8
51–64	B21t	0.7	35	63	15	6.8	0.1	0.4	30	73	5.7
64–74	B22t	0.5	33	65	15	6.6	0.1	0.3	29	74	5.7
74–89	B23t	0.3	31	68	14	6.6	0.1	0.3	28	76	5.7
89–130	B3	0.2	28	70	15	5.8	0.1	0.3	27	79	6.0
130–155	C	0.1	28	69	16	5.6	0.1	0.4	26	84	6.5

Adapted from Soil Taxonomy, 1975, pedon 89.

Suitability of Udolls for Corn Production

About 50 percent of the world's corn is produced in the United States; most of the corn is produced in the corn belt, where Udolls are extensive soils. Iowa and Illinois rank first and second, respectively, in corn production. The Udolls in the corn belt generally have sufficient moisture and heat (degree days) for corn. Corn production in Canada and the Soviet Union is limited by low temperature and rainfall. The percentage of total production and the distribution of corn production in the Soviet Union and the United States are shown in Fig. 6-7.

The properties of many Udolls as well as the climate are favorable for corn production. Studies by Jenny showed that the organic matter content of grassland soils decreased with increasing annual temperature along a traverse from Canada to Louisiana. Maximum corn yields, however, occured in Iowa. Lower yields north of Iowa were attributed to the effect of low temperature on the development of the corn plant. South of Iowa lower corn yields were attributed to lower soil fertility, especially nitrogen supply. Lower yields with higher temperature is likely also to be related to higher plant respiration.

Most of the Udolls have argillic horizons that are weakly developed, if at all, and do not restrict root penetration. In addition, many Udolls formed in thick loess where rooting depth for corn is about 2 meters and 40 centimeters of available water can be stored in the root zone. This high water storage is associated with the

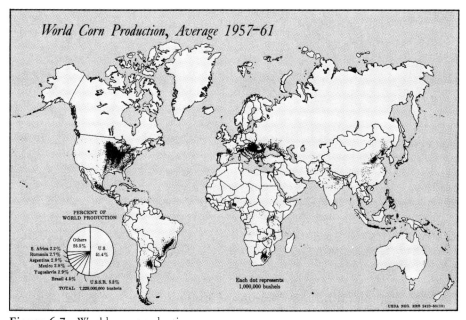

Figure 6-7 World corn production.

high silt content of loessial soils. Summer rainfall is generally well distributed, contributing to drought resistance of soils. The thick mollic epipedons are high in organic matter and nitrogen supplying power. Limited weathering has produced soils with considerable weatherable minerals to supply nutrients. All things considered, it seems that the Udoll region represents an excellent natural combination of soil properties and climate for production of corn as well as some other crops.

Relationship of Surface Soil Thickness to Corn Yields

Many of the Hapludolls of Iowa and Missouri that developed in thick loess occur on steep slopes in an area where corn is the major crop. Water ersosion has significantly reduced the thickness of the surface soil. The weakly developed nature of the subsoil (cambic horizon) results in relatively little change in the properties of the plow layer with loss of soil by water erosion. The most important difference between the mollic epipedon and the cambic horizon for corn production seems to be the difference in organic matter content and, consequently, the difference in nitrogen supplying power.

Results of an experiment conducted on Monona and Marshall silt loam (Hapludolls) provide some insight into the effect of soil loss by erosion on corn yields. In the experiment the surface soil was removed with a bulldozer, and corn yields were compared to untreated soil with the top soil intact. Results in Table 6-3 show that the eroded soil yielded about 30 quintals per hectare less than the normal soil without nitrogen fertilizer. Nitrogen fertilizer use resulted in similar yields; however, more nitrogen fertilizer was required on the subsoil treatment. The nitrogen availability in the normal untreated surface soil was greater than that of the subsoil by 78 kilograms per hectare in 1958 and 84 kilograms per hectare in 1959. The difference in nitrogen availability represents a "permanent" corn production cost

Table 6-3   Average Corn Yields from Subsoil and Surface Soil Treatments of Hapludolls in Southwestern Iowa

	1958			1959	
Nitrogen, kilograms, per hectare	Corn Yield, quintals per hectare		Nitrogen, kilograms per hectare	Corn Yield, quintals per hectare	
	Subsoil	Normal Soil		Subsoil	Normal Soil
0	20	49	0	19	51
75	52	63	67	44	66
150	64	65	134	67	77
225	67	66	202	77	77
300	66	60	381	77	73

Adapted from Soil Science Society of America proceedings, volume 25, page 498, 1961, by permission of the Soil Science Society of America.

on the subsoils exposed by erosion. Although similar maximum yields were obtained with the use of nitrogen fertilizers, cost of production was increased, and erosion on these Hapludolls also creates off-site costs in terms of unwanted flooding and sedimentation from increased water runoff and erosion. The high permeability related to low clay content and high silt content throughout the soil profile of Hapludolls, however, makes level terraces and contour cultivation effective practices for water erosion control.

As the clay content of the subsoil increases, as in Argiudolls and Paleudolls, there is greater reduction in soil productivity as a result of erosion. Corn yields on Argiudolls in Illinois are estimated on severely eroded soils to be 79 percent of that of uneroded soils when the maximum clay content of the argillic horizon is 35 percent and 43 percent of uneroded soil when the argillic horizon contains 49 percent clay.

Responsiveness of Udolls to Management

Today the high natural fertility of the Udoll region is taken for granted. For the early settlers, this was a fact to be discovered. Westward-moving settlers accustomed to converting forest into cropland encountered the tall grass prairies near the Indiana-Illinois border. At first the prairie was avoided, because the sod was difficult to plow (see Fig. 6-1). The high fertility of the soils, however, was discovered about 1830 and, shortly thereafter, the development of the steel plow by John Deere set the stage for rapid prairie conquest.

An experiment at the University of Illinois in 1876 was designed to test whether the dark-colored fertile prairie soils could be depleted. The area now called the Morrow Plots is the oldest soil experiment field in the United States. Yields declined quickly on plots that were continuously planted in corn and not fertilized. Yields for 1920 to 1955 averaged 14 quintals per hectare, and the organic matter content declined 40 percent. Plots that had rotations including legumes and that were fertilized yielded several times more and showed only one-half the loss in organic matter between 1904 and 1955.

In 1955 new questions arose. Had the low-yielding continuous corn plots without fertilizers been permanently damaged, or could the soil be revived with new treatments? Soil tests were used as a basis for applying fertilizer and lime. Plots that had been averaging 14 quintals per hectare yielded 54 quintals per hectare after treatment in 1955 and 71 quintals per hectare in 1956. By comparison, plots that had crop rotations and fertilizers for many years yielded only slightly more. From this came the conclusion that the affects of continuous corn without fertilizers for 80 years on nearly level Udolls was not permanent and was essentially depletion of nutrients that could be easily restored by fertilizers. The durability of Udolls is seen in the fact that farmers continue to increase crops yields after 150 years of farming.

Figure 6-8 Aerial photograph showing land use in north-central Illinois. Nearly all the land is in farms and is used as cropland. Note the absence of woodlots and trees mainly at farmsteads. (USDA photograph.)

Even though Udolls have high native fertility relative to most other soils, the fertility is rapidly depleted when high yields are produced. Price and cost factors are conducive to the use of considerable amounts of fertilizer to utilize fully the fairly long growing season and high amount of available water when corn is grown. Corn responds especially to nitrogen fertilizer so that 1 metric ton of fertilizer can increase yields equivalent to 6 to 8 hectares of unfertilized land. A large part of the nitrogen fertilizer is usually applied as a preplanting operation using anhydrous ammonia (NH_3). The ammonia is under pressure and is released within the soil, where it is converted to ammonium (NH_4^+) and retained on cation-exchange sites for use later in the growing season. Liming is a common practice to reduce soil acidity.

Land Use on Udolls

The Udoll region is the central feed grains and livestock region. It is the world's greatest corn- and hog-producing region. Nearly all of the land is in farms and is used as cropland where slopes are favorable, as shown in Fig. 6-8. Corn and soybeans are the two major crops (see Fig. 6-9). Most of the corn is fed to hogs and beef cattle in the region, resulting in high numbers of livestock. Hay crops and pasture are important on the more sloping areas and where the climate is less favorable for corn. Winter wheat and pasture are important on Udolls in Kansas and Oklahoma. A small amount of cotton is grown on the most southerly Udolls in Oklahoma.

Ustolls

Ustolls are Mollisols with an ustic soil moisture regime and soil temperature regimes warmer than frigid. Annual precipitation ranges from about 350 to 900 millimeters and increases from west to east and from north to south. The effectiveness of the precipitation increases from south to north and from west to east. The precipitation comes mainly in the spring and summer and is erratic (see Fig. 6-10). Droughts are common. For an 81-year period in Pierre, South Dakota, 5 years were arid, 33 were semiarid, 31 were dry subhumid, and 12 were moist subhumid. None of the years were humid, although records of longer periods do show an occasional humid year. Ustolls are extensive soils on the Great Plains south of Borolls, west of Udolls, and east of Aridisols (see Fig. 2-1).

Figure 6-9 Corn and soybeans are the two major crops grown on Udolls in the corn belt.

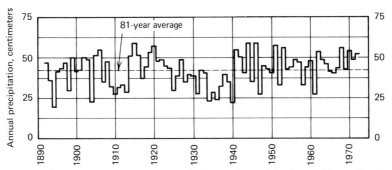

Figure 6-10 Eighty-one-year record of annual precipitation at Pierre, South Dakota. (Adapted from Schumaker, 1974, and used by permission of the *Journal of Soil and Water Conservation.*)

In addition to a mollic epipedon, Ustolls usually have a layer or spheres of powdery lime, as shown in Fig. 6-11. A few Ustolls have formed in acid parent material and do not have secondary lime. Some Ustolls contain appreciable exchangeable sodium and have natric horizons. If a natric horizon is present there is also likely to be an albic horizon overlying it. Most Ustolls had a vegetation of short and midgrasses before settlement. The generally favorable level to gently sloping topography has made many Ustolls adaptable to large-scale farming.

Ustoll Great Groups

Five great groups of Ustolls are recognized in the United States. Haplustolls, Argiustolls, and Paleustolls comprise a sequence of soils with increasing development of the argillic horizon, as in the case of Udolls. Haplustolls occur on the younger surfaces and parent materials and are of late Pleistocene or Holocene age. Argiustolls have thin argillic horizons and, compared to Haplustolls, develop in older parent materials, on more stable land surfaces, and with higher precipitation. Most of the Ustolls in the United States are Haplustolls or Argiustolls, and parent materials are mainly glacial sediments and loess.

Paleustolls are found on old stable land surfaces; as a consequence, they are mostly outside the areas dominated by glacial sediments and loess in the southern Great Plains. Many of these soils have been called claypan soils. Significant acreages of Paleustolls occur in central Oklahoma, eastern Colorado, western Texas, and eastern New Mexico.

Calciustolls have caliche horizons or calcic horizons, and most horizons above the zones of lime accumulation are calcareous. Their formation is enhanced by a high lime content of parent material, limited precipitation, and steep slopes.

Ustolls of limited extent in the United States are Natrustolls. Natrustolls have natric horizons and are associated with restricted drainage and accumulation of exchangeable sodium. They occur mainly as small areas surrounded by larger areas of Ustolls.

Land Use on Ustolls

Ustolls are generally quite fertile and have good physical properties; many have fair precipitation. Management on Ustolls, however, is dominated by the variability of precipitation and not by the total amount. Wet and dry cycles occur irregularly, causing droughts that last longer than 5 years (see Fig. 6-10). Early settlers met disaster unless they had a source of water and could irrigate some of their land. Farming required more land than in the humid east, and the homestead was

Figure 6-11   Profile of an Ustoll showing dark-colored mollic epipedon and a calcic layer near the trowel.

Figure 6-12   Large-scale wheat farming on Ustolls in Kansas. (USDA photograph.)

increased from one quarter section to one section. Today, wheat and cattle grazing are the dominant enterprises on farms and ranches that are much larger than one section and with operations on a large scale (see Fig. 6-12).

Winter wheat occupies more land of Ustolls than any other crop. Kansas, Nebraska, and Oklahoma rank first, fourth, and fifth in wheat hectarage in the United States, and Kansas produced the most wheat of any state. The winter wheat is planted in the fall and makes significant growth during the low potential evapotranspiration period of the winter. Rains come mainly in the spring and early summer, and the wheat matures before the long, dry summer occurs. The amount of land planted in wheat is importantly related to the amount of soil moisture at the time of planting. The effect of stored soil moisture and the rainfall during the growing season on yields is shown in Table 6-4.

Grain sorghum is the second major grain crop on Ustolls; it is more drought resistant than wheat and is thus better suited for the sandier and drier soils. Sorghum also fits well into winter wheat production programs. If wheat has not been planted in the fall due to low soil moisture storage, sorghum can be planted the following spring if winter and spring rains are sufficient. Sorghum is more heat tolerant than wheat and is better adapted to soils with thermic temperature regime. Sorghum and cotton are important crops on Ustolls with a thermic temperature regime, such as the High Plains of Texas near Plainwell. A wide variety of crops including alfalfa, corn, and sugarbeets, is grown where irrigation water is available. In recent years the use of central pivot sprinkler systems has greatly expanded the irrigated acreage of Ustolls onto land where slopes were too irregular or steep and soils too sandy for surface flow irrigation methods.

Table 6-4   Relation of Wheat Yields to Depth of Moistened
Soil at Seeding Time in the Fall and Weather After Planting

Depth of Wetted Soil, centimeters	Percent Relative Yield	
	Unfavorable Weather	Favorable Weather
0–30	5	35
30–60	12	41
60–90	32	75
over 90	48	100[a]

[a]Equivalent to 17.6 quintals per hectare.

Summer Fallowing on Ustolls for Water Storage

The annual precipitation decreases from east to west on the Great Plains. The use of summer fallowing to increase stored soil moisture increases from about 1 year in every 4 or 5 years in the east to about every other year in the west. In many cases a 3-year sequence of winter wheat, sorghum, and summer fallow is used because it provides considerable water storage before each crop and allows two crops every 3 years.

To increase the amount of water stored in the soil by fallowing, the land is left bare after the wheat is harvested, and it is occasionally cultivated. The cultivation kills weeds and other vegetation to eliminate the loss of water by transpiration. After each rain during the fallow period, some water enters the soil and is stored and some water is lost from the soil surface by evaporation. During the fallow period, however, the depth of soil brought up to field capacity is increased. This results in the creation of alternating strips of fallowed and nonfallowed land, with strips perpendicular to the wind to reduce the wind erosion hazard, as shown in Fig. 6-13.

The fallowing system is far less than 100 percent efficient because some runoff occurs and some water is lost by evaporation. A good estimate is that in western Kansas about 20 percent of the rainfall during the fallow period will be stored in the soil. This extra quantity of water, however, is important for increasing wheat yields (see Table 6-4). Mineralization of nitrogen during the fallow years also contributes to increased yields. Disadvantages of fallowing are the decline in organic matter content of soils and exposure of soil to wind erosion during the fallow period.

Wind Erosion on Ustolls

Wind erosion is a perennial hazard on Ustolls exposed by summer fallowing and overgrazing. The erosion is most severe in dry years, particularly during periods of prolonged drought. A 7-year drought began on the southern Great Plains in 1931

Figure 6-13 Summer fallowing on Ustolls in eastern Colorado, where annual precipitation is about 40 centimeters and about one-half of the wheatland is fallowed each year. Some irrigation of crops occurs in the distance, where trees can be seen growing along a river.

and created the Dust Bowl. The personal physical discomfort and mental distress of people caused by dust storms in Colorado has been vividly depicted by James Michener in *Centennial*. Many fields lost 5 to 30 centimeters of topsoil and thousands of farmers abandoned their farms. It has been found that wheat yields decline 2 to 4 percent for each 2½ centimeters of topsoil lost by erosion in Kansas and Colorado. Many hectares of cropland were converted into grasslands.

Rains returned to the region in the fall of 1938; yields of wheat quickly increased and returned to nearly the same level that existed before the Dust Bowl. Measures to control wind erosion include planting crops in strips perpendicular to wind, leaving crop residues on the surface of the soil, and plowing deep enough to bring argillic horizon material to the surface to increase the clay content. Deep plowing, however, is only effective if future erosion is controlled so that the newly brought up clay is not removed from the surface soil. An equally dry period occurred again in the 1950s, but Dust Bowl conditions were not produced because of the improved management of the land.

Cattle Grazing on Ustolls

Extensive areas of Ustolls are in native grass and shrub vegetation and are used for cattle grazing in areas where soil moisture regime borders on the aridic, soils are sandy, slopes are steep, or soils are shallow to rock. Five of the six states that have the most beef cattle (Texas, Nebraska, Kansas, Oklahoma, and South Dakota) are located in the Great Plains and have large hectarage of Ustolls. Forage production is related to rainfall, and ranchers have the same general problem as wheat farmers—erratic rainfall. Prevention of overgrazing during prolonged droughts is the key to the maintenance of a permanent and healthy rangeland.

Borolls

Borolls are the cool to cold well-drained Mollisols with a continental climate. Mean annual soil temperature is lower than 8°C. Soil temperature regimes are mainly frigid or cryic, but some have a pergelic regime. The frigid regime has a higher summer soil temperature and greater temperature fluctuation between winter and summer than the cryic regime. Borolls are extensive on the northern Great Plains of North America, as shown in Fig. 2-1. About 4.9 percent of the soils in the United States are Borolls. The soils developed mainly from late Pleistocene and Holocene sediments, and native vegetation was mainly grasses. Many Borolls occur in the western mountains and support trees as well as grass (see Fig. 6-14). Some Borolls in Alaska support spruce, birch, aspen, and alpine tundra. Soil moisture regimes are udic, and ustic. Borolls in Canada are mainly Chernozemic and occupy about 5 percent of the land area.

Figure 6-14 Cool summer temperature makes Borolls in the foothills of the Rocky Mountains ideal recreational sites. Natural vegetation is grass, sagebrush and ponderosa pine.

Boroll Great Groups

There are seven great groups in the Boroll suborder. The Haploborolls, Argiborolls, and Paleborolls comprise an age or development sequence similar to that for the Udolls and Ustolls. The coldest Borolls are Cryoborolls, which have frigid or pergelic temperature regimes. Summers are cool or short. Most Cryoborolls occur in the mountains of the western states, and some occur in Alaska. Calciborolls are Borolls with calcic horizons and are calcareous in all horizons above the calcic horizon. In most of them a mollic horizon rests on a cambic horizon. Their distribution is closely related to recent parent materials in areas where limestone is common. Natriborolls have a natric horizon with or without an albic or A2 horizon. Natriborolls typically occur as small inclusions in larger areas of Borolls. Zones of carbonate or salt accumulation below the natric horizon are common. Vermiborolls are Borolls that have been extensively mixed by earthworms and their predators. The mollic epipedon is usually the only genetic horizon, and 50 percent or more by volume is composed of wormholes, wormcasts, or filled animal burrows. Vermiborolls are rare in North America. Most of the Borolls in the United States are Haploborolls, Argiborolls, or Cryoborolls and were formerly called Chernozems.

Land Use on Borolls

Low temperature and precipitation are the two major limitations for agriculture on Borolls. Across the northern plains the annual precipitation ranges from over 50 centimeters in western Minnesota to only 30 centimeters in the most arid parts of Montana. This decrease in precipitation is associated with thinner mollic epipedons with lower content of organic matter and less intensive leaching. Spring wheat by dry farming methods is the principal crop. The area of spring wheat production in North America is the approximate area of Borolls, as shown in Fig. 6-15. From Fig. 6-15 it can also be seen that most of the wheat production in the Soviet Union is also quite far north and is mainly spring wheat. Roughly one-third of the wheat in the United States is spring wheat compared to two-thirds in the Soviet Union. This difference reflects the dominance of winter wheat in the United States on Ustolls and of spring wheat in the Soviet Union on Borolls. Other important crops include oats, barley, and flax. Grazing becomes a more important land use as the precipitation and temperature decrease and slopes become less desirable for mechanized agriculture.

Saline Seep on Borolls

The lower temperature of Borolls compared to Ustolls results in reduced evapotranspiration. Increased water storage from fallowing with reduced evaporation in Borolls as compared to Ustolls has created excess soil water storage and saturated subsoil zones in some Borolls. The water that accumulates as a saturated zone

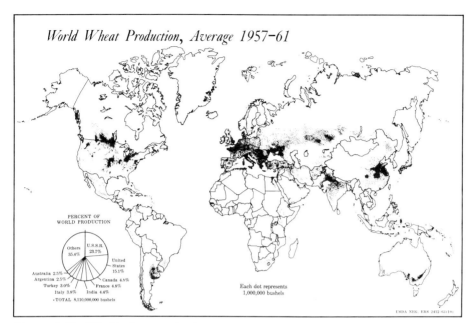

Figure 6-15 World wheat production.

above an impermeable layer moves laterally and creates a seep. As the water evaporates, salt accumulates and creates the phenomenon called *saline seep.* Factors that favor saline seep development in Borolls are frequent fallowing, sandy soils with low water-holding capacity, low temperature, high precipitation, and cropping with "medium- or shallow-" rooted grain crops. The areas of saline seep are increasing in recent years; where salt crusts form, the area may be devoid of vegetation.

Control of saline seep depends on management of the area where water recharge occurs. Management practices for control include reduced frequency of fallowing, more intensive annual cropping, use of fertilizers to increase plant growth and water consumption, and production of alfalfa. Alfalfa is a deep-rooted perennial that can effectively dry the soil to depths much greater than that normally penetrated by the roots of wheat and sorghum. When the alfalfa has dried out the soil, it will stop growing or grow very slowly, indicating that it is time to begin wheat production.

Xerolls

Xerolls are Mollisols with a Mediterranean climate or xeric moisture regime. Xerolls are dry for extended periods in the summer and become almost completely dry every summer. Winter precipitation, however, may completely recharge the water storage capacity, and there may be some surplus water. Xerolls are most extensive in Oregon and Washington and less extensive in Idaho, Nevada, Utah, and California (see Fig. 2-1).About 5 percent of the soils in the United States are Xerolls (see Table 6-1).

Most Xerolls developed under a vegetation of grass and shrubs and characteristically have a mollic epipedon that overlies a cambic or argillic horizon. An accumulation of calcium carbonate exists in the lower part of many of them, and most horizons are neutral.

Xerolls of the Northwestern United States

The Xerolls of Washington, Oregon, and Idaho dominate one of the most important wheat regions of North America (see Fig. 6-16). The soils are mainly Haploxerolls and have cambic horizons. Parent material is dominantly loess that varies from about 1 meter to over 75 meters in thickness and overlies basalt. Several dust storms occur in the area each year, and loess is still accumulating at a rate of about 0.2 millimeter per year at Pullman, in eastern Washington. Increasing amounts of volcanic ash exist in the loess with increasing distance toward the soil surface. The ash likely originated mainly from the eruption of Mount Mazama (which formed Crater Lake) in Oregon about 6600 years ago and from Glacier Peak in Washing-

Figure 6-16 Xerolls on gentle slopes of the Columbian Plateau are used mainly for winter wheat production.

Table 6-5   Selected Properties of the Ritzville Silt Loam, Haploxeroll

Depth, centimeters	Horizon	Percent Silt	Percent Clay	Cation-Exchange Capacity milliequivalents per 100 grams	pH	Percent CaCO₃ Equivalent	Percent Organic Carbon
0–15	A1	67	6	14	7.0	0.0	0.85
15–30	A12	74	6	14	7.4	0.0	0.80
30–45	A13	72	7	14	7.7	0.0	0.63
45–65	B11	69	6	14	8.1	0.3	0.51
65–84	Bca1	71	8	13	8.3	2.7	0.38
84–94	Bca2	68	7	12	8.3	2.7	0.26
94–105	Bca3	66	8	14	8.4	3.3	0.18
105–123	Bca4	69	9	16	8.4	3.3	0.13
123–150	Bca5	62	9	15	8.5	3.4	0.12
150–170	C1	67	9	13	8.6	6.6	0.14

Adapted from Gilkeson, 1965.

ton about 12,000 years ago. The ash content is generally too low to dominate soil properties by virtue of the amorphous nature of the ash. Some properties of one of the most extensive soils, Ritzville, are given in Table 6-5.

The data show that all horizons are high in silt and low in clay, resulting in textures of silt loam. The B horizon is a cambic horizon. All horizons have a pH of 7 or more, and the pH increases with increasing depth. The lower soil horizons are influenced by sodium. The upper horizons have been leached of any origianl carbonate, but the lower horizons are calcareous. Organic carbon decreases slowly with increasing soil depth. The great depth of loess provides for high water storage capacity. Many Xerolls have properties very similar to Udolls, which developed from thick loess in western Iowa. The difference in climate, however, results in Xerolls being well adapted to winter wheat production and Udolls to corn production.

Land Use on Xerolls

Xerolls of the Pacific Northwest have a mesic temperature regime and are well suited for winter wheat production, which is the dominant crop. The wheat is planted in the fall, when rains begin, and grows through the cool, moist winter season and spring. By the time hot, dry weather occurs in the summer, the wheat is harvested. In areas of greatest precipitation, as in the eastern part of the area, there is sufficient moisture to recharge almost completely soil water storage so that soils contain a lot of water when the winter rains stop. In fact, there is sufficient water for very high yields when high rates of nitrogen fertilizer are used unless an exceptionally dry year occurs, as in the winter of 1976 to 1977. The short-stemmed wheat varieties were developed in the region and are able to produce high yields without lodging. The high available moisture and high rate of nitrogen fertilization has enabled Whitman County in eastern Washington to produce more wheat than any other county in the United States.

The precipitation decreases in a westerly direction, and fallowing for water conservation becomes very important on the Columbia Plateau. Winter rainfall during periods of low temperature results in much more efficient storage of moisture by fallowing than on Ustolls, where most of the rainfall occurs in the hot summer period. In fact, wheat is grown on soils with annual precipitation as low as 20 centimeters; on the Great Plains these soils would not only be unable to produce a crop of wheat with fallowing but would be more like a desert.

The Palouse is a unique region of the northwest wheat region in eastern Washington, western Idaho, and northeastern Oregon. Thick loess overlies steep hills of basalt (see Fig. 6-17). The gentle nature of the winter rains and permeable soil allows continuous wheat production on slopes up to 40 percent without excessive erosion where measures are taken to leave crop residues on the soil surface. Soil erosion, however, is a serious problem on the steep, northeast-facing slopes, where snowdrifts accumulate and warm winds in the spring cause rapid melting on frozen soils. Nearly all the land is cultivated and used for winter wheat. Dried field peas are the second most important crop in this, the highest rainfall area of the Pacific

Figure 6-17 Fallowing scene in the Palouse. Nearly all the land is cultivated utilizing slopes of up to 40 percent.

Northeast wheat region. The Palouse has some similarity with the Sand Hills of Nebraska. Sand instead of loess and summer rainfall instead of winter rainfall in the Sand Hills, however, have resulted in distinctly different soils and land use patterns.

The Spokane Flood and Scablands

About 18,000 to 20,000 years ago glacial ice moving southward blocked the Clark Fork River near the Idaho-Montana border and formed glacial Lake Missoula. The water at the ice dam reached a depth of about 600 meters, and the lake contained one-half the volume of present-day Lake Michigan. The ice dam gave way quite abruptly, creating the Spokane Flood, which allowed the water from Lake Missoula to cascade across the countryside. Entire loess hills that had over 30 meters of loess were destroyed, and a large quantity of the underlying basalt was stripped away and removed. The new landscape that was created is called the "channeled scablands." The scablands cover about 40,000 square kilometers in eastern Washington and are suited more for grazing than for wheat production.

Xerolls of Other Regions

Xerolls have developed on gentle or moderate slopes from residuum weathered from underlying bedrock or in unconsolidated materials on the higher alluvial fans in southeastern Oregon, northeastern Nevada, and southwestern Idaho. Most of these Xerolls have argillic horizons and are Argixerolls. Argixerolls also occur on mountains slopes and dissected plateaus in Idaho and Utah and are used mainly for grazing. About 11 percent of the rice produced in California is produced on Haploxerolls that contain more than 35 percent of montmorillonitic clay. Even though the soils have a water table that does not come as close as 75 centimeters of the soil surface, the soils have sufficiently low permeability to be suited for flooded rice production.

Albolls

Albolls typically have an albic (A2) horizon sandwiched between mollic and argillic horizons. Very few have a natric horizon. They have significant clay eluviation and are wet some seasons of the year because of the fluctuating water table. Albolls generally develop on nearly level or depressional areas where leaching has been more intense than on surrounding areas. Thus Albolls tend to occur as inclusions in larger areas of Udolls, Ustolls, and Borolls. Albolls are mostly cultivated and are used similarily, as are the dominant landscape soils that surround them.

mollic epipedon

Figure 6-18 Rendoll with a mollic horizon that rests on soft lime
stone in Puerto Rico.

Rendolls

Rendolls are Mollisols that formed under grass or trees from highly calcareous
parent material such as chalk, marl, or soft limestone. Most Rendolls were formerly
called Rendzina. The mollic epipedon rests on calcareous material or a cambic
horizon rich in carbonates (see Fig. 6-18).

Rendolls are not extensive soils in North America. Only 13 series are recognized
by the USDA, and 4 of them exist in Puerto Rico.

References

Aandahl, A. R., *Soils of the Great Plains,* Map, P.O. Box 81242, Lincoln, Nebraska,
68518, 1972.

Agricultural Experiment Stations of the Western States Land-Grant Universities and Colleges and USDA, *Soils of the Western United States,* Washington State University, Pullman, 1964.

Arneman, H. F., "Soils of Minnesota," *Minn. Agr. Ext. Bull. 278,* 1963.

Austin, M. E., *Land Resource Regions and Major Resource Areas of the United States,* USDA Handbook 296, Washington, D.C., 1965.

Baver, L. D., "How Serious is Soil Erosion?", *Soil Sci. Soc. Am. Proc., 14:*1–5, 1950.

Bidwell, O. W., "Soils of Kansas," *Kan. Agr. Exp. Sta. Dept. of Agron. Contrib. 1359, 1973.*

Brun, L. J., and B. K. Worcester, "The Role of Alfalfa in Saline Seep Prevention," *Farm Research, 31:*9–14, N. Dak. Agr. Exp. Sta., 1974.

Brun, L. J., and B. K. Worcester, "Soil Water Extraction by Alfalfa," *Agron. Jour., 67:*586–588, 1975.

Carter, W. T., "Soils of Texas," *Tex. Agr. Exp. Sta. Bull. 431,* 1931.

Cheney, H. B., W. H. Foote, and E. G. Knox, "Field Crop Production and Soil Management in the Pacific Northwest," in *Advances in Agronomy,* Vol. 8, pp. 2–61, Academic, New York, 1956.

Chepil, W. S., "Wind Erosion Problems," in *Advances in Agronomy,* Vol. 10, pp. 56–62, New York, Academic, 1958.

Clayton, J. S., W. A. Ehrlich, D. B. Cann, J. H. Day, and I. B. Marshall, *Soils of Canada,* Canada Department of Agriculture, Ottawa, 1977.

Cooper, C. F., "The Ecology of Fire," *Sci. Am., 204:*150–160, 1961.

Crout, J. D., D. G. Symmank, and G. A. Peterson, *Soil Survey of Jefferson County, Texas,* USDA and Tex. Agr. Exp. Sta., Washington, D.C., 1965.

Department of Agronomy, "The Morrow Plots," *College of Agr. Circular 777,* University of Illinois, 1960.

Engelstad, O. P., and W. D. Shrader, "The Effect of Surface Soil Thickness on Corn Yields; II. As Determined by an Experiment Using Normal Surface Soil and Artificially-Exposed Subsoil," *Soil Sci. Soc. Am. Proc. 25:*497–499, 1961.

Evans, C. E., and E. R. Lemon, "Conserving Soil Moisture," in *Soil,* USDA Yearbook, pp. 340–359, Washington, D.C., 1957.

Fehrenbacher, J. B., G. O. Walker, and H. L. Wascher, "Soils of Illinois," *Ill. Agr. Exp. Sta. Bull. 725,* 1967.

Flack, K. W., and D. F. Slusher, "Soils Used for Rice Culture in the United States," Mimeo, *USDA Soil Conservation Service,* Washington, D.C., 1977.

Foth, H. D., "Properties of the Galva and Moody Series of Northwestern Iowa," *Soil Sci. Soc. Am. Proc., 18:*206–211, 1954.

Gilkeson, R. A., "Ritzville Series, Benchmark Soils of Washington," *Wash. Agr. Exp. Sta. Bull., 655,*1965.

Gray, F. and M. H. Reezitalab, "Benchmark and Key Soils of Oklahoma," *MP 97 Okla. Agr. Exp. Sta.,* 1976.

Guidry, N. P., *A Graphic Summary of World Agriculture,* Miscellaneous Publication 705, USDA, Washington, D.C., 1964.

Hetzler, R. L., R. H. Dahl, K. W. Thompson, K. E. Larson, B. C. Baker, and C. J. Erickson, *Soil Survey of Walsh County, North Dakota,* USDA and N. Dak. Agr. Exp. Sta., Washington, D.C., 1972.

Heyne, E. G., "Field Crop, Trends and Problems in the Great Plains," in *Advances in Agronomy,* Vol. 10, pp. 8–15, New York, Academic, 1958.

Higbee, E., *American Agriculture,* Wiley, New York, 1958.

Hobbs, J. A., "The Winter Wheat and Grazing Region," in *Soil,* USDA Yearbook, pp. 505–515, Washington, D.C., 1957.

Horner, G. M., W. A. Starr, and J. K. Patterson, "The Pacific Northwest Wheat Region," in *Soil,* USDA Yearbook, pp. 475–481, Washington, D.C., 1957.

Jacobson, M. U., *Soil Survey of Norman County, Minnesota,* USDA and Minn. Agr. Exp. Sta., Washington, D.C., 1974.

Klages, K. H. W., *Ecological Crop Geography,* Macmillan, New York, 1942.

Leo, M. W. M., "Effects of Cropping and Fallowing on Soil Salinization," *Soil Sci.,* 96:422–427, 1963.

Lotspeich, F. B., and H. W. Smith, "Soils of the Palouse Loess: I. The Palouse Catena," *Soil Sci.,* 76:467–480, 1953.

Lyles, L., "Possible Effects of Wind Erosion on Soil Productivity," *Jour. Soil Water Cons.,* 30:279–283, 1975.

Mathews, O. R., "The Place of Summer Fallow in the Agriculture of Western United States," *USDA Circular 886,* Washington, D.C., 1951.

Mayer, J., "Food and Population: The Wrong Problem?", *Daedalus, 93*:830–844, 1964.

Mitchener, J. A., *Centennial,* Random House, New York, 1974.

Moreland, D. C., and R. E. Moreland, *Soil Survey of Boulder County Area, Colorado,* USDA and Col. Agr. Exp. Sta., Washington, D.C., 1975.

Morgan, M. F., J. H. Gourley, and J. K. Albeiter, "The Soil Requirements of Economic Plants," in *Soils and Men,* USDA Yearbook, pp. 753–776, Washington, D. C., 1938.

National Cooperative Soil Survey of United States, National Soil Survey Committee of Canada and FAO, *Soil Map of the World,* II: North America, Rome, 1975.

North Central Regional Committee on Soil Survey, *Soils of the North Central Region of the United States,* North Central Regional Pub. 76, Bulletin 544, Univ. of Wis. Agr. Exp. Sta., Madison, 1960.

North Central Regional Technical Committee 3, "Productivity of Soils in the

North Central Region," *Univ. of Ill. Agr. Exp. Sta. Bull. 710,* North Central Regional Research Publication 166, 1965.

Olson, R. V., "The Great Plains Area," in *Advances in Agronomy,* Vol. 10, pp. 3–8, New York, Academic, 1958.

Omodt, H. W., G. A. Johnsgard, D. D. Patterson, and O. P. Olson, *The Major Soils of North Dakota,* N. Dak. Agr. Exp. Sta. Bull. 472, 1968.

Oschwald, W. R., F. F. Riecken, R. I. Dideriksen, W. H. Scholtes, and F. W. Schaller, "Principal Soils of Iowa," *Iowa State University Special Report 42,* Department of Agronomy, 1965.

Russell, M. B. "All the Way Back in One Year?" *Plant Food Review, 2* 1956.

Schumaker, C. M., "The Great Plains—Wet or Dry?", *Jour. Soil Water Cons., 29:* 157–159, 1974.

Scrivner, C. L., J. C. Baker, and B. J. Miller, "Soils of Missouri," *C823,* Missouri Agricultural Extension Division, 1966.

Simonson, R. W., F. F. Riecken, and G. D. Smith, *Understanding Iowa Soils,* W. C. Brown, Dubuque, 1952.

Smith, G. D., W. H. Allaway, and F. F. Riecken, "Prairie Soils of the Upper Mississippi Valley," in *Advances in Agronomy,* Vol. 2, pp. 157–205, New York, Academic, 1950.

Soil Conservation Service, "Facts About Wind Erosion and Dust Storms on the Great Plains," *USDA Leaflet 394,* Washington, D.C., 1966.

Soil Survey Division of Bureau of Soils, "Soils of the United States," in *Soils and Men,* USDA Yearbook, Washington, D.C., 1938.

Soil Survey Staff, *Soil Taxonomy,* USDA Agriculture Handbook 426, 1975.

Stewart, O. C., "Fire as the First Great Force in Changing the Face of the Earth," in *Changing the Face of the Earth,* William L. Thomas, Ed., University of Chicago Press, 1956.

Thorp, J., B. H. Williams, and W. I. Watkins, "Soil Zones of the Great Plains States–Kansas to Canada," *Soil Sci. Soc. Am. Proc., 13:*438–445, 1948.

U.S. Department of Interior-Geological Survey, *The Channeled Scablands of Eastern Washington,* Washington, D.C., 1973.

Weaver, J. E., *North American Prairie,* Johnson, Lincoln, Neb., 1954.

Westfall, D. G., C. L. Godfrey, N. S. Evatt, and J. Crout, "Soils of the Texas A and M University Agricultural Research and Extension Center at Beaumont in Relation to Soils of the Coast Prairie and Marsh," *MP-1003,* Texas A & M University, 1971.

Westin, F. C. and D. D. Malo, "Soils of South Dakota," *Bulletin 656,* S. Dak. Agr. Exp. Sta., 1978.

Worcester, B. K., L. J. Brun, and E. J. Doering, "Classification and Management of

Saline Seeps in Western North Dakota," *Farm Research, 33:3–7*, N. Dak. Agr. Exp. Sta., 1975.

Worcester, B. K., and B. D. Seeling, "Plant Indicators of Saline Seep," *Farm Research, 34:18–20*, N. Dak. Agr. Exp. Sta., 1976.

World Book Encyclopedia, Field Enterprises Educational Corp., Chicago, 1968.

Alfisols of the United States

Central Concept

Alfisols are extensive soils in widely separated landscapes. Most of the Alfisols developed under forest in a humid-temperate or cool region. The central concept of Alfisols is soils that have: (1) argillic horizons, (2) ochric epipedons, (3) medium to high base saturation, and (4) water available to mesophytic plants for a considerable part of the time in most years. Since Alfisols are required to have an argillic horizon, they develop on land surfaces that are sufficiently stable to result in significant eluviation of silicate clays. Most Alfisols are characterized by a base saturation of 35 percent or more at a depth of 1.25 meters below the upper boundary of the argillic horizon or 1.8 meters below the soil surface. Alfisols contain significant amounts of weatherable minerals that supply bases through weathering; they are generally considered fertile soils. Increased weathering and leaching, producing a nearly complete loss of weatherable minerals and base saturation less than 35 percent, cause Alfisols to evolve into Ultisols.

Degradation of Argillic Horizons

Evidence of the destruction of argillic horizons and conversion of argillic horizon material into albic material has been observed in several soil orders but is perhaps best expressed in the Alfisol order. The degradation begins in the upper part of the argillic horizon, progresses downward, and is interpreted to mean that the argillic horizon is moving deeper into the profile. The destruction begins on ped surfaces and progressively destroys the argillic nature of the argillic horizon material, leaving a grayish-colored albic material. When the albic coatings on peds become greater than 5 to over 15 millimeters thick (depending on texture) and have a vertical dimension greater than 5 centimeters, the soil is considered to have tonguing of albic material into the argillic horizon (see Fig. 7-1). In some soils the tongues penetrate the entire depth of the argillic horizon.

The formative element, *gloss,* is used to indicate tonguing (Gr. *glossa,* meaning tongue) at the great group level. Glossaqualfs, Glossoboralfs, and Glossudalfs are the only three great groups recognized (see *Soil Taxonomy*). Tonguing at a lower level of classification is recognized by the adjective glossic in several soil orders, including Alfisols.

Theories to explain argillic horizon degradation include (1) weathering and destruction of clay mineral lattices, (2) clay eluviation, and (3) a combination of the first two theories. Studies of Alfisols in northwestern New York support clay translocation as the primary cause of argillic horizon degradation, as shown in Fig. 7-2. Although weathering is believed to be the most intense in the zone of degradation, clay translocation is the primary cause of the development of albic material. The clay mineral suite was found to be similar throughout the argillic horizon.

Suborders

Alfisols occur under a wide range of soil moisture and temperature regimes. The suborders include Aqualfs, Boralfs, Udalfs, Ustalfs, and Xeralfs. Alfisols occur on about 13.4 percent of the land in the United States. The extent of the suborders is: Udalfs, 5.9 percent, Boralfs 3 percent, Ustalfs 2.6 percent, Aqualfs 1 percent, and Xeralfs 0.9 percent (see appendix Table 2).

Boralfs

Boralfs are the generally well-drained Alfisols of cool places. They have frigid or cryic temperature regimes, and most of them have a udic moisture regime. They occur in two very different environments in the United States. One environment is the Great Plains along the grassland-forest border (see A2a areas of Fig. 2-1). The second environment is mountainous and is in the Rocky Mountains and Black Hills (A2S areas of Fig. 2-1).

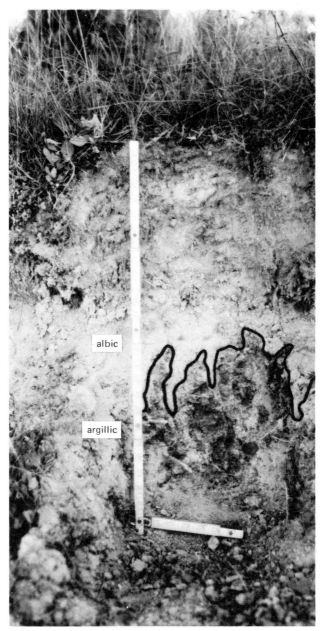

Figure 7-1 Tonguing of albic material into the argillic horizon of a Boralf.

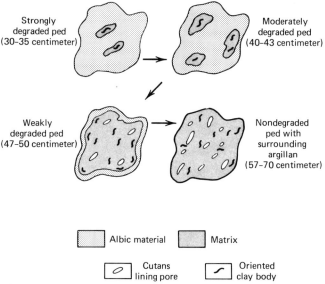

Figure 7-2   Changes in ped morphology with increasing soil depth that support clay translocation as the major process in argillic horizon degradation. (Reproduced from *Soil Sci. Soc. Am Proc.,* 38:623, 1974, by permission of the Soil Science Society of America.)

Boralfs typically have an 0 surface horizon underlain by an A2 horizon, and an argillic horizon (see Fig. 7-1). Although they may support forest, most Boralfs have quite a high pH and base saturation. Apparently the low temperature and relatively low precipitation can support trees, but surplus water for leaching is minimal. Many Boralfs are frozen much of the year when leaching cannot occur. Most Boralfs were previously called Gray-Wooded in the United States and are called Gray Luvisols in Canada.

Cryoboralfs of the Rocky Mountains

Most of the Boralfs of the Rocky Mountains have a frigid or cryic soil temperature regime and are Cryoboralfs. They are mostly forested and occur in the montane zone, where coniferous forest is the dominant vegetation. At lower elevations in drier environments, the Boralfs share a common boundary with Borolls. At higher elevations the Boralfs merge into Inceptisols with B horizons that have the appearance of spodic horizons (weak Spodosols) of the spruce-fir forests of the subalpine zone, as shown in Fig. 7-3. Above the subalpine zone is the tundra zone, which is dominated by Cryaquepts in the Rocky Mountains.

Figure 7-3 Boralfs are common in the montane zone of the Rocky Mountains and Inceptisols that are weak Spodosols are common in the subalpine zone. In northern Colorado the altitude of the montane zone is about 2100 to 2800 meters.

Many Boralfs in the Rocky Mountains occur on thick sediments of glacial origin. Steep slopes are common, and there are many rock outcrops. Summers are warm but short (frigid regime). Nearly all the land is in forest, and virtually none of it is cultivated. Some of the gentler slopes and more open areas are grazed and are used for pasture. Deer and elk use the areas for summer browse. Skiing is a popular winter sport.

Boralfs of the Black Hills

The Black Hills are an island of forested hills in the grasslands of western South Dakota and eastern Wyoming. The hills consist of a core of granite and slate surrounded by sedimentary rocks, mostly limestone. The limestone forms a dissected plateau. Most of the Boralfs have developed from residuum and alluvium. Elevations range from 1200 to 2100 meters. Slopes are mainly steep and hilly; there are gentler slopes on the plateau divides. Annual precipitation ranges from 40 to 60 centimeters. Wetter areas have an udic moisture regime, and vegetation is mainly coniferous forest. Much of the parent material is calcareous; the soils are high in bases and are Eutroboralfs.

Eutroboralfs have 60 percent or more base saturation in most horizons, and some have an accumulation of calcium carbonate in the lower B or upper C horizon. At lower elevations on the plains, the soils are surrounded by Ustolls and Aridisols. Land uses are mainly recreation, forestry, grazing, and mining.

Boralfs of the North-Central States

An extensive area of Boralfs exists along the eastern side of the grassland soils in northern Minnesota and Wisconsin (see Fig. 2-1). Less extensive areas occur in Michigan. Parent materials were mainly glacial deposits of Wisconsin age. They developed under a forest cover of deciduous, coniferous, or mixed deciduous and coniferous forest. Most of the soils are Eutroboralfs. Although they developed under forest and have very pronounced whitish A2 horizons, soil pH and base saturation are surprising high, as shown in Table 7-1.

The Nebish soil developed from calcareous till in northern Minnesota and has a well-developed argillic horizon, high base saturation, and a ca layer beginning at 58 centimeters. These properties stand in marked contrast to those of Udalfs to the south and east that are quite acid and have no ca layers.

Some of the land in the northern part of the region is used to produce forage for livestock; grain crops are limited because of cool summers and short growing seasons. These areas are mostly in second-growth aspen-pine-birch forests. The use of land for agriculture increases in a southerly direction; the cleared land is used mainly for small grains and forage for dairy cattle. Potatoes are an important crop locally.

Many of the Boralfs in Wisconsin show tonguing of the A2 horizon into the upper argillic horizon, and they are Glossoboralfs. Base saturation and pH are lower than in Eutroboralfs.

Boralfs occur in lower Michigan and are located near the mesic-cryic soil temperature boundary; they have a greater crop potential than Boralfs, which are fur-

Table 7-1   Selected Properties of Nebish Silt Loam, Typic Eutroboralf

Depth, centimeters	Horizon	Percent Clay	pH	Percent Base Saturation
5–0	0	—	6.6	—
0–13	A2	9	7.4	88
13–25	B21t	29	7.1	87
25–38	B22t	30	7.0	87
38–58	B3	13	7.0	—
58–74	C1ca	9	7.8	100

Pedon 49 from Soil Taxonomy, *1975.*

ther north. Many of these soils are cleared and used for pasture and crops. They represent a soil that is transitional to the Udalfs, which developed under a warmer and more humid environment.

Udalfs

Udalfs are well-drained Alfisols with udic moisture regime and mesic or warmer temperature regimes. There is sufficient moisture for leaching in most years. As a result, Udalfs are mainly on land surfaces no older than late Pleistocene, and most have developed in glacial materials (till, outwash, loess, etc.). Udalfs on older surfaces develop from highly calcareous parent material or limestone. Udalfs generally have high agricultural potential because they combine favorable climate with moderate to high soil fertility. Udalfs are very extensive in the middle United States (see A3a areas in Fig. 2-1), and most have been intensively used for agriculture in both the United States and western Europe.

Most Udalfs have or once had a deciduous forest cover. Normally the undisturbed soil has a thin A1 horizon darkened by humus that is underlain by an A2 and argillic horizon. The profile of a Udalf in a cultivated field is shown in Fig. 7-4. On sloping land some of the soils have lost their A horizons because of erosion, and argillic horizons are exposed at the soil surface. The soils were formerly called Gray-Brown Podzolic in the United States.

Udalfs of the Midwest

Udalfs of the midwest have a mesic soil temperature regime. The soils do not show extreme development of the argillic horizons. The base of the argillic horizon is less than 1.6 meters below the soil surface. Minimal development of the argillic horizon in most of them results in their classification as *Hapludalfs*. The soils generally occur north of the confluence of the Mississippi and Ohio rivers. Many of the soils developed on nearly level to rolling till plains and are intensively used for agriculture.

In Iowa and Illinois Udalfs and Udolls occur in the same landscapes. Both Udalfs and Udolls are acid and need lime and fertilizers for high yields. The Udolls, however, are slightly less weathered and contain more organic matter. Corn yield potentials between Udalfs and Udolls are not large. The corn and wheat yield potential of Udalfs is 85 percent of that for Udolls with high-level management and similar slopes and parent materials. Yield potential for soybeans is similar for the two soils. The most extensive soils of Indiana are Hapludalfs, and Indiana ranks third in corn production after Iowa and Illinois. In fact, much of the corn belt is located where Hapludalfs are the dominant upland soils in Indiana, Ohio, and southern Michigan. (Ohio ranks sixth and Michigan tenth in corn production.) On

Figure 7-4 Udalf in a cultivated field showing well-developed argillic horizon that makes this soil marginal for use for septic filter field in developing subdivision due to slow permeability.

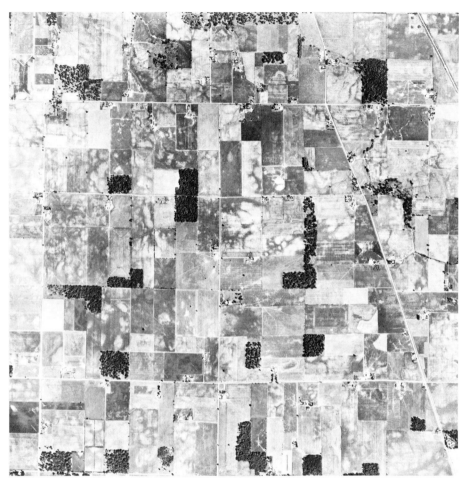

Figure 7-5 Hapludalf-Argiaquoll landscape in north-central Indiana, where most of the land is used for cropping. Udalfs, like Miami, occur on the well-drained sites, and Aquolls, like Brookston, occur on the poorly drained sites (dark-colored areas).

nearly level till plains such as those that are common in central Indiana, 80 percent of the land is used for cropping (see Fig. 7-5). Corn, soybeans, and other feed grains are the dominant crops. Soils were originally forested, and now only occasional small farm woodlots remain. Fence rows commonly have rows of trees and shrubs so that one gets a "closed-in" feeling when traveling through the area. Trees are always on the horizon. Aquolls are quite extensive and make an important contribution to the agricultural potential here as they do in the Udoll-Aquoll landscapes of northern Illinois and Iowa.

Udalfs in the more northerly locations, such as New York and Wisconsin, are less suited for corn production, and these areas became the focus of the dairy industry. The cooler weather near the mesic-cryic soil temperature border is more suited for forage and small grain production. The production of legume forage crops, clover, and alfalfa in rotation with grain crops produced a complementary relationship with potential for both greater grain and forage production. Legume forage crops increased yields of grain because of an increased soil nitrogen supply, and the rotations avoided low forage yields associated with old stands. Manure was carefully conserved and used. Forage crops on the more rolling lands provided good erosion control. The result was the establishment of a permanent and prosperous agriculture based on the production of dairy products.

A fairly large area of Udalfs occurs in a 25- to 70-kilometer-long band bordering both sides of the Mississippi River floodplain in southern Wisconsin and Minnesota and northern Iowa and Illinois. Austin (1965) refers to the area as the Northern Mississippi Valley Loess Hills. The area is highly dissected, and much of it is in the driftless area. The uplands are covered with a blanket of loess that varies in thickness from less than 2 meters to more than 15 meters. The loess thickness decreases with increasing distance from the river valley, resulting in a series of soils (Seaton, Fayette, Clinton) with increasing clay content of argillic horizons and increasing loess fineness with increasing distance from the river. Fayette silt loam, Typic Hapludalf, has about 27 to 35 percent clay and 55 to 85 percent base saturation in the argillic horizon. The Udalfs are very similar to soils formed from similar age loess on much less sloping landscapes and where the loess is primarily overlying till in the upper midwest.

Many Scandinavian people settled in the area, which has dairying as the major agricultural product because of the steep slopes and limited amount of land suited to row crops in the narrow stream valleys. The steepest land is still forested, and many farms have more forested lands than croplands. The cropland is concentrated on the less steep ridge tops and the less steep slopes and bottomlands of narrow valleys. Soil erosion is a major problem, and one of the first erosion control experiment stations was established near La Crosse, Wisconsin, to study erosion control and reclamation of severely eroded Fayette soils.

The area of Udalfs in western New York is located on the Erie-Ontario plain. Most of the area is a nearly level to rolling glacial drift plain. The underlying rocks are mainly limestones and calcareous shales that resulted in till that is high in lime and, in general, soils that are the most naturally fertile in New York. Some of the important Udalf series include Hilton, Honeoye, Lansing, Lima, Marlette, and Ontario; all of these soils are Glossoboric Hapludalfs.

The Erie-Ontario plain slopes gradually to the south from about 80 meters near the lake to about 300 meters at the Allegheny escarpment, which forms the approximate southern boundary of the Udalfs near Ithaca. South of the Udalfs the soils on the Allegheny plateau are mainly Inceptisols because of the steep slopes and glacial materials, which are derived mainly from acid or low lime content sandstones and shales.

The largest acreage of these Udalfs in New York is used for production of feed and forage crops for dairy cattle. Other important crops, however, include many canning and truck crops, fruits, and winter wheat. About 20 percent of the land in farms is in permanent pasture, and a nearly equal amount in forest.

Udalfs of the Lower Mississippi Valley

A band of Udalfs occurs along the eastern side of the Mississippi River Valley from Kentucky to Louisiana (see Fig. 2-1). The soils are similar to many Udalfs in the north-central states that have also formed in loess parent material, as shown in Fig. 7-6. The loess is about 20 meters thick near the Mississippi River and thins in an easterly direction. Although the band of Udalfs is rather narrow, the influence of a thin layer of loess occurs into central Kentucky and Tennessee.

Figure 7-6 Loess bluff 20 meters thick at the edge of the Mississippi River floodplain in western Tennessee. The loess overlies coastal plain sand and gravel, exposed where the man at the left is walking. Udalfs developed here are similar to Udalfs developed in thick loess in Iowa and Illinois.

At least two different loess deposition periods occurred. An early loess, Rox-anna, was deposited about 30,000 to 70,000 years ago. Weathering of the Roxanna loess before the Peoria loess was deposited produced a soil with a fragipan. Because the loess thins with increasing distance east of the Mississippi River, soils near the river formed in thick Peoria loess are unaffected by the fragipan. These soils are Hapludalfs. Further from the river the loess thins and soils on interfuves as well as eroding slopes have fragipans close enough to the surface (within 1 meter) to be a factor in plant growth and soil classification. These soils are Fragiudalfs. The catenary relationships are shown in Fig. 7-7.

Fragipans are brittle when wet, hard when dry, and have a high bulk density. Fragipans occur in a wide variety of soils, and several theories have been advanced for their genesis. It seems that fragipans occur below an eluvial horizon that is in a position to receive eluviated material from above and to have been alternatively water saturated and dried. Iron reduction occurs during saturation in a soil layer that produces a perched water table. Silicate clays decompose during the saturation period, leaving a matrix of gray-colored sand and silt grains (albic material). Cracks develop during the drying cycle, and albic materials are moved downward into the cracks. Rewetting causes expansion in the fragipan layer, and compression increases bulk density. The result is a layer of large prisms that are coated with gray albic material and have a high bulk density. The layer is slowly permeable or impermeable to roots and water (see Fig. 7-8).

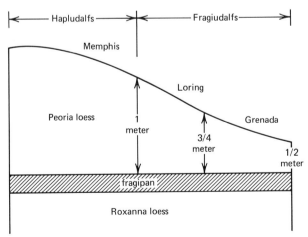

Figure 7-7 Catenary relationship of Memphis, Loring, and Grenada soils. Memphis is a Hapludalf because the fragipan is more than 1 meter below the soil surface. Loring and Grenada are Fragiudalfs.

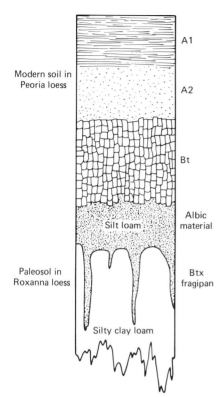

Modern soil in
Peoria loess

A1

A2

Bt

Silt loam

Albic
material

Paleosol in
Roxanna loess

Btx
fragipan

Silty clay loam

Figure 7-8 Diagrammatic representation of Fragiudalf (Grenada) in the lower Mississippi Valley. The fragipan has large prisms with gray-coated surfaces and high bulk density. (After Lytle, 1968.)

The mesic-thermic line is just north of the Tennessee border near the Mississippi River; this brings about a thermic soil temperature regime for most of the soils. The Hapludalfs are very similar to the Hapludalfs that developed from loess in the midwest and are among the best agricultural soils of the states along the lower Mississippi River. The Fragiudalfs have shallower rooting depth and less available water, so they are slightly less productive. Memphis soils (Hapludalfs) on gentle slopes with good management (without irrigation) produce about 20 to 25 percent more corn and lint cotton per hectare than the Grenada soils (Fragiudalfs) under similar conditions. The Udalfs have high agricultural potential but require good management to achieve that potential.

The loess was deposited over a highly dissected part of the coastal plain. The many steep slopes in the region make erosion an important management consideration. About one-third of the land is used for crops; cotton, corn, and soybeans are the major crops. Farms are quite small; many have less than 50 hectares. One-fourth of the land is forested with timber, and pulpwood is the major forest product.

Udalfs of the Nashville Basin

The Nashville Basin is located in central Tennessee; Nashville is in the northwestern part of the basin. The area occupies about 15,000 square kilometers stretching from northern Alabama almost to the Kentucky border (see Fig. 2-1). The soils developed mainly from limestones (Ordovician). Most of the soils supported hardwood deciduous forest before settlement. Soil temperature regime is thermic. Many of the flatter and broader ridge tops are capped with 45 to 60 centimeters of loess. The loess is thickest on the flattest surfaces and in areas closest to the Mississippi River. The basin is surrounded by the higher Highland Rim, which is composed of cherty limestones and shales that weather more slowly than the limestones of the basin; the dominant soils are Ultisols.

The basin consists of an inner and outer part. The inner basin is at a lower elevation and has gentler slopes than the outer basin. Seventy-five percent of the slopes are less than 5 percent compared to 40 percent in the outer rim. Rocks of the inner basin are predominately argillaceous limestones that are low in phosphorus. Although slopes are gentle, the soils tend to be shallower and lower in phosphorus than soils in the outer rim. Much of the inner rim has been cleared of forest and is in cropland and pasture (see Fig. 7-9).

Figure 7-9 Landscape typical of the inner Nashville Basin where slopes are gentle and much of the land is used as cropland. Soils are generally shallow to limestone rock fragments, reddish-brown colored, and without the albic horizons that are so typical of Udalfs developed from loess and glacial sediments.

The outer rim area is larger than the inner rim area and contains about 11,000 square kilometers; it is underlain by high-phosphorus limestones. The outer rim is deeply weathered and has a strongly dissected topography. Most of the soils are about 1 meter or more thick over bedrock, are high in phosphorus, and are productive for agriculture (see Fig. 7-10). The steep slopes, however, result in much more forest and permanent pasture and less cropland than in the inner part of the basin.

One of the striking features of the Nashville Basin is the reddish soil color and absence of albic horizons in Alfisols developed under forest. Soils are generally fine-textured and represent some of the most productive soils in Tennessee. Early settlers were attracted to this area. About 35 percent of the soils are Alfisols and about 25 percent are Ultisols (considering the Maury soils as Paleudalfs and not Paleudults). A significant amount of the soils are Paleudalfs and represent about 18 percent of the Alfisols, while most of the Alfisols are Hapludalfs.

For the basin as a whole, 40 percent of the land is cropped. Corn, small grains, and hay are the principal crops that support a livestock-based agriculture. Beef cattle are the major livestock; dairying is important near the large cities. Tobacco is the main cash crop. About one-third of the land is in pasture, and one-fourth is still forested.

Figure 7-10 Highly dissected landscape of the outer part of the Nashville Basin. Forest and permanent pasture occupy the steeper slopes, and cropland is restricted to the more gently slopes. Soils are thick, high in phosphorus, and productive.

Udalfs of the Kentucky Bluegrass Region

Soils similar to those of the Nashville Basin have developed from high-phosphate limestones in the Bluegrass region in north-central Kentucky. The region is similar to the Nashville Basin in that the inner part has the most gentle slopes; it differs in that the inner Bluegrass area has soils with a higher phosphate content than soils in the outer Bluegrass area.

Fayette County, where Lexington is located, is in the heart of the Bluegrass region of Kentucky. Maury is the most extensive series and occurs on 42 percent of the land in Fayette County. Maury soils occur on nearly level uplands, mostly with slopes of 6 percent or less and developed mainly in material weathered from phosphatic limestone and influenced by a thin mantle of loess. The soils are deep to limestone and well suited for agriculture. The Maury soil is classified as a Paleudalf (in early literature the classification was Paleudult) and has a mesic temperature regime. Surface soil texture is silt loam, and the thick argillic horizon has a texture of silty clay and/or silty clay loam.

Maury soils occur in close association with McAfee soils on more sloping areas, as shown in Fig. 7-11. McAfee soils are younger than Maury soils and are Hapludalfs. Armour soils developed in old alluvium associated with gentle slopes of Maury areas are Ultic Hapludalfs, and the Huntington soils developed in recent alluvium in areas dominated by McAfee soils are Fluventic Hapludolls (see Fig. 7-11).

Agriculture has been important in the Bluegrass region since the area was first settled. About two-thirds of the farmland in Fayette County is in pasture. The high phosphate content of the soils and the forage produced has resulted in excellent bluegrass and clover pastures for thoroughbred racehorses. White buildings and

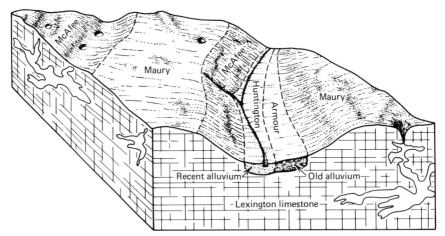

Figure 7-11   Relationship of Maury, Paleudalf, and McAfee, Hapludalf, to landscape position in the Bluegrass region of Kentucky. (From Sims, 1968.)

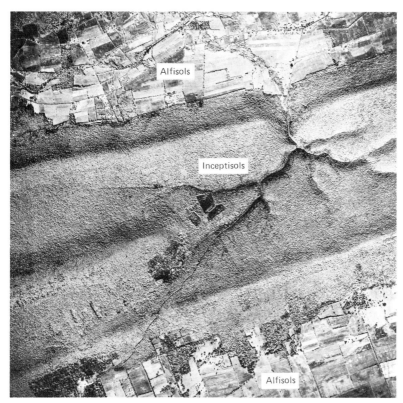

Figure 7-12 Landscape in central Pennsylvania, where nearly all the land in limestone valleys is cropland and soils are Alfisols and the sandstone-shale ridges dominated by Inceptisols are nearly all forested. (USDA photograph.)

fences are a characteristic feature of the region. Important agricultural crops include burley tobacco, which is the main cash crop, corn, barley, and wheat. Alfalfa, red clover, grass, and lespedeza are the main hay crops. Beef, dairy, and sheep are also important in the livestock industry of the region.

Udalfs of the Limestone Valleys of the Ridge and Valley Province

Throughout the Appalachian region are limestone valleys where the most extensive soils are Hapludalfs and Paleudalfs. These valleys were caused by the solution of limestone in the midst of more resistant strata of sandstones and shales that now exist as topographic highs. As one travels through the region, one observes a dramatic change from the limestone valleys used intensively for agriculture in the midst of forested lands; there is much abandoned land and meager agriculture (see Fig. 7-12). The soils of these limestone valleys are similar to those of the Nashville

Basin and Bluegrass region in Tennessee and Kentucky. Hagerstown (Hapludalf) is an extensive series of the limestone valleys of the Appalachian region.

Aqualfs

Aqualfs are Alfisols that have aquic soil moisture regimes or are artificially drained. Subsoil colors are gray and mottled. Aqualfs in the United States are predominantly of two types. In Michigan and Ohio the Aqualfs have developed on nearly level lake and glacial plains with naturally high water tables. The other major Aqualf areas in Illinois, Missouri, and Louisiana have developed because water tables were perched above impermeable argillic horizons (see A1a areas of Fig. 2-1).

Grass and forest were native vegetation before settlement, although most seem to have supported forest at some time in the past.

Aqualfs of the Lake Plain of Ohio and Michigan

The Wisconsin ice remained stationary for a time near the north Ohio border and impounded water to create a glacial lake. Some of the previously deposited till was reworked by water, and lacustrine sediments were produced. This created a landscape in which Aqualfs developed from fine-textured calcareous till and Aquepts developed from lacustrine sediments, as shown in Fig. 7-13. The Nappanee and

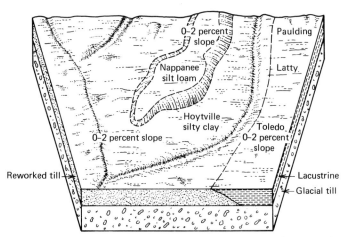

Figure 7-13 Parent material and relief relationships for Aqualfs developed from till and Aquepts developed from lacustrine sediments on the lake plain of northwestern Ohio and southeastern Michigan. (After Blevins and Wilding, 1968.)

Hoytville soils are Ochraqualfs and have an ochric epipedon that rests on an argillic horizon without an abrupt textural change. The slightly better drainage of the Nappanee results in classification as Aeric Ochraqualf and Mollic Ochraqualf for the Hoytville.

The lake plain was a forest swamp before settlement and elm, ash, and soft maple were the dominant trees and soils developed under the influence of a naturally high water table. The soils are high in organic matter, about 6 percent in Ap horizon, and are only slightly leached. The mean pH of the surface soil is near 7 and gradually increases with increasing soil depth. Leaching depth of calcium carbonate in the Hoytville ranges from 84 to 137 centimeters. The B/A clay ratio of the Hoytville is minimal for the soil to qualify as having an argillic horizon, and all horizons have about 40 percent or more clay. The clay is dominantly illite; montmorillonite is second in abundance.

Natural fertility is high, and corn and soybeans are the major crops in this cash grain area in the easternmost part of the corn belt. Nearly all the farmland is cropland; only 5 percent of this land is in forest. Drainage is a severe problem and, in some cases, both surface and tile drainage are used to minimize the danger of low yields caused by poor soil aeration from high rainfall in March, April, and May. The soils are severely limited for use as highway foundations, building sites, or septic tank filter fields because of a seasonally high water table, high plasticity, high compressibility and low load-bearing capacity when wet, slow runoff, and low percolation rate.

Characteristics of Albaqualfs

The Aqualfs of southern Illinois, northeastern and western Missouri, and southwestern Louisiana (see Fig. 2-1) are Albaqualfs; they were formerly called Planosols or claypan soils. The soils developed under trees or grass on nearly level uplands where erosion was minimal. The soils that are forested likely were Hapludalfs at an earlier age, and many of the soils that developed under grass were Mollisols at an earlier age. Continued soil evolution, however, resulted in intensive translocation of clay and development of slowly permeable argillic horizons that created seasonally perched water tables. The saturated zone above the argillic horizons resulted in limited grass root growth and limited the addition of organic matter to the zone immediately above the argillic horizon. Lateral movement of water above the argillic horizon produced lateral leaching that caused the formation of albic horizons in soils that previously were Mollisols without albic horizons. Albaqualfs that developed under grass, therefore, now have properties very similar to Albaqualfs that developed under trees. The Albaqualfs generally are quite acid, have medium to low base saturation, low to moderate amount of organic matter, and an abrupt textural change from the albic (A2) horizon to the argillic horizon (see Fig. 7-14).

Figure 7-14   Albaqualf showing albic horizon that rests abruptly on a slowly permeable argillic horizon. Considerable cracking has occurred with drying of the argillic horizon of this Crowley silt loam. (USDA photograph.)

Some of the more extensive Albaqualf series in the United States include the Cisne and Cowden in Illinois, Putnam in Missouri, Lufkin in Texas, Dayton in Oregon, and Crowley in Louisiana. They all have montmorillonite as the dominate clay; however, some Albaqualfs have clays dominated by illite or have mixed or siliceous mineralogy.

Albaqualfs of Southern Illinois

The area of Albaqualfs in southern Illinois, which comprises about one-fourth of the state, is the largest area in the United States where Aqualfs are the dominant soils. The soils occur mainly on nearly level uplands in association with better drained and less well-developed soil on the more sloping sites. Parent material

Figure 7-15 Albaqualf landscapes in southern Illinois showing lighter-colored soil developed under trees and darker-colored soil developed under grass on the right. Note level land surface and the surface drainage ditch in landscape on the left.

consisted mainly of about 0.50 to 1.50 meters of fine loess over Illinoian till. Some soils have B horizons that developed mainly in till. Silt loam is the typical surface soil texture, and argillic horizons are silty clay loam or silty clay and create a perched water sometime during the year, especially in the spring. The albic horizon, however, is dry for short periods in most years, so the soils can be droughty in dry years.

The southern two-thirds of the Albaqualf area was originally forested, and the northern one-third was tall grasslands. Although the soils of both areas have similar properties, the soils developed under grass have darker-colored A horizons and contain more organic matter than the lighter-colored forested soils, as shown in Fig. 7-15.

The nearly level surface and claypan make drainage a major management problem in the use of soils in the claypan area of southern Illinois. Tile drains do not

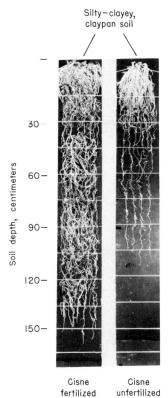

Figure 7-16 Use of fertilizer is the most effective means of increasing rooting depth, available water supply, and crop yields on Albaqualfs. Note in fertilized Cisne that there were more roots both above and below a depth of about 30 centimeters in the argillic horizon. (Photograph courtesy Dr. J. B. Fehrenbacher, University of Illinois.)

function properly, and surface drainage is used where necessary. The pH of the surface soils in natural conditions is as low as 4, and large applications of lime are needed for cropping. Soils are naturally low in fertility. Fertilizers are used to increase soil fertility and are also the most practical means of increasing the available water supply by increasing rooting depth, as shown in Fig. 7-16. Studies in Missouri on the Putnam soil showed that only subsoil shattering with no fertilizer was ineffective. Yields on Albaqualfs vary greatly from year to year, depending on rainfall. Yields on Albaqualfs nearly approach yields of some of the best soils in Illinois in years when rainfall is sufficient and well distributed.

The soils are responsive to good management, as shown in Table 7-2. Although the productivity index of the Albaqualfs is considerably lower than that of the best soils in Illinois (Udolls and Aquolls), the Albaqualfs are nearly as responsive to good soil management, especially those that developed under grass vegetation. Note also that the forested Albaqualfs have a significantly lower productivity index than grassland Albaqualfs and are less responsive to high management.

Albaqualfs of Missouri

The Albaqualfs of northeastern Missouri are located only a short distance to the west of the claypan area of southern Illinois. Soils developed in thin loess over till under conditions similar to those in Illinois, with grass as the native vegetation. The soils have similar properties, management problems, and uses as the Albaqualfs in Illinois. Smaller areas of Albaqualfs occur in southeastern Iowa.

The Aqualfs of southwestern Missouri and eastern Kansas are Albaqualfs that developed in the unglaciated parts of the states. Native vegetation was grass, and parent materials consist mainly of acid micaeous shales (Pennsylvanian). A thin layer of loess covers the area. Soils are located on nearly level upland plains and have well-developed argillic horizons. These claypan soils are used for general farming.

Some Albaqualfs occur in northeastern Oklahoma. On these and other soils with strongly developed argillic horizons gophers build mounds to create nests at higher elevations in order to escape the seasonal periods of soil wetness.

Albaqualfs of Louisiana

Soils in the Aqualf area of southern Louisiana (see Fig. 2-1) are mainly Albaqualfs. The soils are claypan soils that developed on nearly level marine terraces over the past 30,000 years in silty and clayey alluvium. Native vegetation was tall grass prairie. The dominant clay mineral is montmorillonite, and the soils are less acid

Table 7-2   Productivity Indexes and Response to Management of Selected Soil
Associations in Illinois

| Soil Association | Dominant Soil | Productivity Indexes | | Difference of Reponse |
		High Management	Basic Management	
Sidell, Catlin, Flanagan, Drummer	Aquolls and Udolls	151	97	54
Hoyleton, Cisne, Huey	Albaqualfs, grassland	112	59	53
Ava, Bluford, Wynoose	Albaqualfs, forested	86	43	43

Adapted from Fehrenbacher, Henrickson, Alexander, and Oschwald.

and more fertile than the Albaqualfs in Illinois and Missouri. Soil temperature
regime is thermic.

The dominant series is Crowley silt loam, which is well suited to rice production
because of an impermeable argillic horizon (see Fig. 7-14) and nearly level surface.
A rice experiment station is located near Crowley, the parish seat. The dominant
use of the soil is for rice production; Louisiana ranks second of the states in land
area used for rice production, and ranks third in total rice production (as shown in
Table 7-3).

Rice requires only moderate soil fertility, an acid or slightly acid soil, a high
temperature during the growing season, and a dependable water supply. A water
level of about 12 centimeters is maintained for most of the growing season. The
water prevents drought and helps control weeds. About 120 centimeters of water
are used per year in Louisiana; 50 centimeters are supplied by natural precipitation.

Table 7-3   Rice Production in the United States, 1976

State	Area Harvested, 1000 hectares	Quantity, 1000 metric tons
California	163	886
Southern States		
Arkansas	341	1815
Louisiana	244	968
Mississippi	58	255
Texas	217	1051
Other	7	25
Total	1030	5000

From Flack and Slusher, 1977.

Land used for rice production is leveled and tilled for seedbed preparation; contour levees or ridges are constructed to hold the irrigation water (see Fig. 7-17).

The rice is commonly seeded into the water by airplane, and supplemental nitrogen fertilizer is also applied by air after plants are well established. About 2 weeks before harvest, the levees are broken and the surplus water drains off the field to permit soil drying and use of large machinery for harvesting.

Flooding the soil causes reducing conditions in the root zone; the availability of phosphorus is increased and enhances the growth of rice plants. Reducing conditions favor the reduction of nitrates to gaseous forms of nitrogen and the loss of nitrogen from the soil by volatilization. Ammonium (reduced) forms of nitrogen fertilizer are preferred. Undesireable toxic materials, such as sulfides, build up in water-saturated soils and create a need for the occasional growth of nonflooded crops to permit improved aeration and dissipation of toxic incomplete decomposition products. For this reason rice is commonly grown in rotation with other crops.

Figure 7-17 Well water used for rice irrigation on Albaqualfs on nearly level marine terraces in Louisiana. Note tip of contour ridge used to hold irrigation water on leveled areas. (USDA photograph.)

Other important crops grown on the Albaqualfs in Louisiana include cotton, corn, sweet potatoes, and hay.

Ustalfs

Ustalfs are the reddish and reddish-brown Alfisols of the warm subhumid and semiarid regions. Most of these soils have an ustic soil moisture regime and a thermic or hyperthermic soil temperature regime. Surplus water moves through some Ustalfs in an occasional year. Ustalfs tend to occur between the drier Aridisols and the humid region Ultisols in the southern United States; they occur on about 2.6 percent of the land in the United States. The soils are most extensive in Texas, Oklahoma, and New Mexico (see A4a areas of Fig. 2-1). In this region many soils developed under grasslands and were formerly called Brown, Chestnut, and Reddish Prairie soils. Many of them developed under grasses and shrubs and have a high base saturation, but the high temperature and sandy nature of many parent materials prevented the development of mollic epipedons, so the soils do not qualify as Mollisols. On the other hand, some of the Ustalfs near the Ultisol border in east Texas developed under a more humid climate, were most likely Ultisols at an earlier age, and now have a base saturation too high for Ultisols because of the inflow of bases in dust and precipitation. In some of these Ustalfs there are a few weatherable minerals, and they contain some kaolinitic clay.

Current vegetation on most Ustalfs is savannah or grassland. The dry season results in trees that are either deciduous or xerophytic.

Ustalf Great Groups

Six Ustalf great groups are recognized. **Durustalfs** have a duripan within 1 meter of the surface and are not known to occur in the United States. **Natrustalfs** have natric horizons and are not extensive in the United States. **Plintustalfs** have continuous plinthite in the subsoil, or a significant part of the argillic horizon is plinthite. There are no series recognized in the United States. **Rhodustalfs** are the dark red (color hue redder than 5YR) Ustalfs with thin argillic horizons. Rhodustalfs develop on erosional or young land surfaces from basic rocks and are rare in the United States. Most of the Ustalfs in the United States are either **Haplustalfs** or **Paleustalfs**; Paleustalfs are the most extensive. In Texas, where most of the Ustalfs are found, the Paleustalfs occur on 16 percent and Haplustalfs on 2.35 percent of the land. Haplustalfs and Paleustalfs were scattered among Chestnut, Reddish Chestnut, Reddish Prairie, Reddish-Brown, and Red-Yellow Podzolic soils in the 1938 and 1949 classifications.

Haplustalfs are the reddish to brown young Ustalfs that have relatively thin argillic horizons and do not have an abrupt boundary at the top of the argillic horizon. The boundary between the A horizon and the argillic horizon is gradual or clear. Haplustalfs occur on relatively recent erosion surfaces or deposits mostly of late Pleistocene age.

Paleustalfs are Ustalfs that are older than Haplustalfs and have thick argillic horizons. The upper argillic horizon is clayey, and the boundary with the A horizon is abrupt. They occur on old surfaces, and many of them contain some plinthite. Landscapes are relatively stable and have gentle slopes; most Paleustalfs began to evolve before the late Pleistocene. In the United States they characteristically have a ca horizon in or below the argillic horizon as a result of the addition of atmospheric carbonates. Some Paleustalfs near the Aridisol border have petrocalcic horizons. Near the Udult and Udalf borders, they do not have ca horizons. The vegetation before they were cultivated included a mixture of grasses and woody plants.

Land Use on Ustalfs

The use of Ustalfs is importantly related to their ustic moisture regime and their thermic or hyperthermic temperature regime. In eastern and northern Texas and in Oklahoma peanuts, winter wheat, grain sorghum, and cotton are the main cash crops. In southern Texas the Ustalfs are mainly in association with Ustolls and Usterts and are used mainly for range and beef cattle production. In western Texas and in New Mexico the smooth upland areas are mainly dry farmed to cotton and grain sorghum, and the steeper slopes along the major rivers are in a range of native grasses and shrubs and are used for grazing. Cattle, sheep, and goat range, however, are the most important overall land use of Ustalfs. Sandy Ustalfs in the panhandle of Texas are intensively used in irrigation agriculture.

Xeralfs

Xeralfs are the mostly reddish Alfisols of regions with a Mediterranean climate. Soil moisture regime is mainly xeric. Although soils are dry for extended periods in summer, winter precipitation may be sufficient to produce surplus water, and some Xeralfs are intensely weathered and leached.

Extent and Distribution of Xeralfs

Xeralfs occur on about 0.9 percent of the land in the United States. In North America Xeralfs occur almost exclusively in the interior valleys and foothill regions of northern California and in the coastal plain and valleys of southern California,

Figure 7-18 Xeralf landscape in foothills of Sierra Nevada in eastern California showing typical grass and shrub vegetation and used mainly for grazing. At higher elevations, the Xeralfs merge with Xerults.

excluding the desert regions. There are about 1.8 million hectares of Xeralfs in California, and small hectarages occur in Oregon and Washington (see areas A5S1 and A5S2, Fig. 2-1). Soil temperature regimes are mainly mesic and thermic.

Natural vegetation was a mixture of annual grasses, forbs, and woody shrubs. Most of the land is in similar vegetation today and is used for grazing (see Fig. 7-18). Some Xeralfs are irrigated and produce mainly horticultural crops.

Dominant Properties of Xeralfs

Except for a few Xeralfs that have developed from basic rocks, Xeralfs are characterized by epipedons that are hard and massive when dry (unless sandy) and that rest abruptly on argillic horizons. Most Xeralfs were formerly called Noncalcic Brown, and Harradine (1963) has summarized the distinguishing features of Noncalcic Brown soils as follows.

1. Very weak surface soil structures when moist, or essentially massive and hard to very hard when dry.
2. Low organic carbon content with maximum values of 1.5 percent in sur-

face horizons, decreasing rapidly to amounts less than 0.5 percent in subsoils.

3. The color of soils will vary in accordance with the mafic mineral content of their soil material. As a general range, surface soils (moist and dry) are between 7.5YR and 10YR in hue, 4 to 6 in value, and 3 to 6 in chroma, with little change in the subsoils of young soils. In older soils, the subsoils become browner, yellowish-red, or redder with age.

4. At all stages of profile development, the proportion of clay relative to silt plus clay increases with depth, and the ratio becomes progressively wider the older the soils.

5. The pH values increase with profile depth. Surface soil reactions are near neutral to slightly acid. Deep subsoils, with some exceptions, are near neutral to slightly basic in reaction.

6. The percent of base saturation is consistently higher than 50 in surface horizons, and it increases in subsoils to values that often are higher than 80 or 90.

7. Calcium and magnesium are by far the dominant exchange cations, with the exchange-calcium content ranging from 2 me per 100 g or less in surface horizons to about 15 in lower subsoils. Ca/Mg ratios are greater than one, but variable in surface horizons and narrow, or with magnesium in excess of calcium in the subsoils of older profiles.

8. All stages of profile development, including the maximum with an iron-silica cemented hardpan, are attained in California.

Xeralf Great Groups

Most of the Xeralfs in the United States have argillic horizons that are only modestly developed, contain less than 35 percent clay, have a gradual or clear boundary between the A horizon and argillic horizon, and are Haploxeralfs. They have formed mostly in late Pleistocene deposits or on erosional surfaces of that age. Palexeralfs have thick argillic horizons, an abrupt A horizon and argillic horizon boundary, and represent older soils than Haploxeralfs. The Palexeralfs exist on stable land surfaces, and their genesis began before the late Pleistocene. Carbonates seem to have been almost completely removed from the argillic horizons during the Pleistocene pluvial periods, but recalcification has apparently occurred.

Durixeralfs are Xeralfs with a duripan whose upper boundary is within 1 meter of the soil surface and below an argillic or natric horizon. Some of these soils exist on very old surfaces and have complex genesis that indicate polygenesis. The duripan consists of an iron-silica cemented hardpan. The San Joaquin soils are Abruptic Durixeralfs and occur on surfaces about 140,000 years old. Ripping the duripan improves plant rooting; some orange groves east of Fresno are found on San Joaquin soils. Redding soils, which are also Abruptic Durixeralfs, occur on surfaces older than 650,000 years.

Plinthoxeralfs have plinthite, which forms a continuous phase or comprises more than one-half the matrix of some subhorizon within 1.25 meters of the soil surface. Few of these soils exist in the United States.

Rhodoxeralfs are the more or less dark red Xeralfs that form in limestone, basalt, and other highly basic parent materials. These soils are rare in the United States.

Natrixeralfs have natric horizons and are not extensive. They support sparse vegetation that usually consists of salt-tolerant grasses and forbs. These soils were considered Solonetz or Solod in the 1938 classification.

References

Agricultural Experiment Stations of the Western States Land-Grand Universities and Colleges and USDA, *Soils of The Western United States,* Washington State University, Pullman, 1964.

Aldrich, S. R., "Illinois Field Crops and Soils," *Circular 901,* University of Illinois Cooperative Extension Service, 1965.

Allgood, F. P., and F. Gray, "Genesis, Morphology and Classification of Mounded Soils in Eastern Oklahoma," *Soil Sci. Soc. Am. Proc., 37:746–753,* 1973.

Anyononomus, *Facts About American Rice,* The Rice Council, Houston, 1967.

Arkley, R. J., *Soil Survey of Merced Area, California,* USDA and Calif. Agr. Exp. Sta., Washington, D.C., 1962.

Austin, M. E., *Land Resource Regions and Major Resource Areas of the United States,* USDA Handbook 296, Washington, D.C., 1965.

Bartelli, L. J., "Soil Development in Loess in the Southern Mississippi Valley," *Soil Sci., 115:254–260,* 1973.

Blevins, R. L., and L. P. Wilding, "Hoytville Soils: Their Properties, Distribution, Management and Use," *Res. Bull. 1006,* Ohio Agricultural Research and Development Center, Wooster, Ohio, 1968.

Brock, A. R., and D. R. Urban, *Soil Survey of Putnam County, Ohio,* USDA, Ohio Department of Natural Resources and Ohio Agricultural Research and Development Center, Washington, D.C., 1974.

Bullock, P., M. H. Milford and M. G. Cline, "Degradation of Argillic Horizons in Udalf Soils of New York State," *Soil Sci. Soc. Am. Proc., 38:621–628,* 1974.

Buntley, G. J., and F. F. Bell, "Yield Estimates for the Major Crops Grown on the Soils of West Tennessee," *Bulletin 561,* University of Tennessee Ag. Exp. Sta., 1976.

Buntley, G. J., R. B. Daniels, E. E. Gamble, and W. T. Brown, *The Relation*

Between Loess Deposits and Fragipan Horizons in West Tennessee, (mimeo), American Society of Agronomy Field Trip, August 23, 1975.

Buol, S. W., Ed., "Soils of the Southern States and Puerto Rico," *Southern Cooperative Series Bulletin 174,* Agr. Exp. Sta. of Southern States and Puerto Rico Land Grant Universities and USDA, 1973.

Clark, H. L., G. J. Haley, E. J. Hebert, R. M. Hollier, Jr., and A. J. Roy, *Soil Survey of Acadia Parish, Louisiana,* USDA and La. Agr. Exp. Sta., Washington, D.C., 1962.

Clayton, J. S., W. A. Ehrlich, D. B. Cann, J. H. Day, and I. B. Marshall, *Soils of Canada,* Canada Department of Agriculture, Ottawa, 1977.

Cline, M. G., *Soil Association Map of New York State,* Department of Agronomy, Cornell University, Ithaca, N.Y., 1961.

Cline, M. G., "Soils and Soil Association of New York," *Extension Bulletin 930,* Cornell University, Ithaca, N.Y., 1970.

Edwards, M. J., J. A. Elder, and M. E. Springer, "The Soils of The Nashville Basin," *Bulletin 499,* USDA and Tenn. Agr. Exp. Sta., 1974.

Ehrlich, W. A., and W. Odansky, "Soils Developed Under Forest in the Great Plains Region," *Agr. Inst. Rev.,* 29–32, Agriculture Institute of Canada, Ottawa, March–April, 1960.

Fehrenbacher, J. B., R. T. Hendrickson, J. D. Alexander, and W. R. Oschwald, "Productivity Indexes of Soil Associations in Illinois," *Extension Service Circular 1041,* University of Illinois, Urbana, 1971.

Fehrenbacher, J. B., B. W. Ray, and J. D. Alexander, "How Soils Affect Root Growth," *Crops and Soils Magazine, 21:*14–18, January 1969.

Fehrenbacher, J. B., B. W. Ray, and J. D. Alexander, "Root Development of Corn, Soybeans, Wheat and Meadow in Some Contrasting Illinois Soils," *Ill. Res.,* 9(2):3–5, Ill. Agr. Exp. Sta., Spring 1967.

Fehrenbacher, J. B., G. O. Walker, and Wascher, H. L., "Soils of Illinois," *Ill. Agr. Exp. Sta. Bull. 725,* 1967.

Flack, K. W., and D. F. Slusher, "Soils Used for Rice Culture in the United States," (mimeo) *USDA Soil Conservation Service,* Washington, D.C., 1977.

Godfrey, C. L., "General Soil Map of Texas," *MP 1034,* Tex. Agr. Exp. Sta., 1973.

Grossman, R. B., and F. J. Carlisle, "Fragipan Soils of the Eastern United States," *Adv. Agron., 21:*237–279, 1969.

Harmon, A. B., E. Lusk, J. Overton, J. H. Elder, Jr., and L. D. Lewis, *Soil Survey of Maury County, Tennessee,* USDA and Tenn. Agr. Exp. Sta., Washington, D.C., 1959.

Harradine, F., "Morphology and Genesis of Noncalcic Brown Soils in California," *Soil Sci., 96:*277–287, 1963.

Hays, O. E., A. G. McCall, and F. G. Bell, "Investigations in Erosion Control and the Reclamation of Eroded Land at the Upper Mississippi Valley Conservation Experiment Station Near La Crosse, Wisconsin, 1933–43," *USDA Tech. Bull. 973,* 1949.

Heady, Earl O., "Economics of Cropping Systems," in *Soil,* USDA Yearbook Washington, D.C., 1957.

Hole, F. D., "Photo-Mosaic Soil Map of Wisconsin," *A2822-1,* University Extension and Geological and Natural History Survey, Madison, Wis., 1977.

Hole, F. D., *Soils of Wisconsin,* University of Wisconsin Press, Madison, 1976.

Lytle, S. A., "The Morphological Characteristics and Relief Relationships of Representative Soils in Louisiana," *Bulletin 631,* Department of Agronomy, Louisiana State University, 1968.

Marschner, F. J., "Land Use and Its Patterns in the United States," *USDA Handbook 153,* Washington, D.C., 1959.

National Cooperative Soil Survey of United States and National Soil Survey Canada and FAO, *Soil Map of the World,* II: North America, Rome, 1975.

North Central Regional Technical Committee on Soil Survey, "Soils of the North Central Region of the United States," *North Central Regional Pub. 76, Bulletin 544,* Univ. Wis. Agr. Exp. Sta., Madison, 1960.

North Central Regional Committee of Soil Survey, "Soil Yield Potential," *Crops and Soils,* 1970.

North Central Regional Technical Committee 3, "Productivity of Soils in the North Central Region," *Univ. Ill. Agr. Exp. Sta. Bull. 710, North Central Regional Research Pub. 166,* 1965.

Odell, R. T., and W. R. Oschwald, "Productivity of Illinois Soils," *Circular 1016,* University of Illinois Cooperative Extension Service, 1970.

Oschwald, W. R., F. F. Riecken, R. I. Dideriksen, W. H. Scholtes, and F. W. Schaller, "Principal Soils of Iowa," *Iowa State University Special Report 42,* Department of Agronomy, 1965.

Pearson, R. W., and L. E. Ensminger, "Southeastern Uplands," in *Soil,* USDA Yearbook, Washington, D.C., 1957.

Petersen, G. W., R. W. Ranney, R. L. Cunningham, and R. P. Matelski, "Fragipans in Pennsylvania Soils: A Statistical Study of Laboratory Data," *Soil Sci. Soc. Am. Proc., 34:*719–722, 1970.

Radeke, R. E., and F. C. Westin, "Gray Wooded Soils of the Black Hills of South Dakota," *Soil Sci. Soc. Am. Proc., 27:*573–576, 1963.

Ranney, R. W., E. J. Ciolkosz, R. L. Cunningham, G. W. Petersen, and R. P. Matelski, "Fragipans in Pennsylvania Soils: Properties of Bleached Prism Face Materials," *Soil Sci. Soc. Am. Proc., 39:*695–698, 1975.

Regional Research Committee on Soils of the North Central Region, "Field Descriptions and Analytical Data of Certain Loess-Derived Gray-Brown Podzolic Soils in the Upper Mississippi River Valley," *Ill. Agr. Exp. Sta. Bull. 587,* 1955.

Scrivner, C. L., J. C. Baker, and E. J. Miller, "Soils of Missouri," *Circular 823,* Missouri Agricultural Extension Division, 1966.

Shaprio, R. E., "Effect of Flooding on Availability of Phosphorus and Nitrogen," *Soil Sci.,* 96:190–197, 1963.

Simonson, R. W., F. F. Riecken, and G. D. Smith, *Understanding Iowa Soils,* W. C. Brown, Dubuque, Iowa, 1952.

Sims, R., D. G. Preston, A. J. Richardson, J. H. Newton, and D. Isgrig, *Soil Survey of Fayette County, Kentucky,* USDA and Ky. Agr. Exp. Sta., Washington, D.C., 1968.

Smith, G. D., W. H. Allaway, and F. F. Riecken, "Prairie Soils of The Upper Mississippi Valley," in *Advances in Agronomy,* Vol. 2, pp. 157–205, Academic, New York, 1950.

Soil Conservation Service, *Classification of Soil Series of the United States,* USDA, Washington, D.C., 1977.

Soil Survey Staff, *Soil Taxonomy,* USDA Agriculture Handbook 426, 1975.

Storie, R. E., and F. F. Harradine, "Soils of California," *Soil Sci.,* 85:207–227, 1958.

Sturgis, M. B., "Managing Soils for Rice," in *Soil*, USDA Yearbook, pp. 658–663, Washington, D.C., 1957.

Thorp, J., J. G. Cady, and E. E. Gamble, "Genesis of Miami Loam," *Soil Sci. Soc. Am. Proc.* 23:156–161, 1959.

Veatch, J. O., *Soils and Land of Michigan,* Michigan State College Press, East Lansing, 1953.

Westin, F. C., L. F. Puhr, and G. J. Buntley, "Soils of South Dakota," *Soil Survey Series Number 3,* S. Dak. Agr. Exp. Sta. and USDA, 1967.

Whiteside, E. P., I. F. Schneider, and R. L. Cook, "Soils of Michigan," *Michigan State University Extension Bulletin E-630,* 1968.

Woodruff, C. M., and D. D. Smith, "Subsoil Shattering and Subsoil Liming for Crop Production on Claypan Soils," *Soil Sci. Soc. Am. Proc.,* 11:539–542, 1946.

Ultisols of the United States

Central Concept

Ultisols are extensive soils that occur on 40 percent of the land area of the 13 southeastern states and on 12.9 percent of the land in the United States. Ultisols are soils of the middle to low latitudes; they have argillic horizons and low base statuses. Soil temperature regimes are mesic or warmer, and soil moisture regimes are udic, aquic, ustic, or xeric. Native vegetation is forest. More specifically, Ultisols have:

1. Argillic horizons.
2. Base saturation less than 35 percent 1.8 meters below the soil surface or 1.25 meters below the upper boundary of the argillic horizon.
3. Mesic or warmer soil temperature regimes.

If plinthite is present it does not form a continuous phase within 30 centimeters of the soil surface.

Ultisols are related to and found in landscapes with Alfisols where Alfisols represent younger soils. Ultisols are related to Oxisols and are found in landscapes where Oxisols represent older soils. Ultisols were mostly considered Red-Yellow Podzolic in earlier classification systems in the United States.

Dominant Properties of Ultisols

Ultisols are intensively weathered and leached (L. *ultimus,* last). The depth distribution of clay, base saturation, and cation-exchange capacity of a representative Ultisol are given in Fig. 8-1. The argillic horizon is well developed and thick. The low cation-exchange capacity reflects the low organic matter content and the low

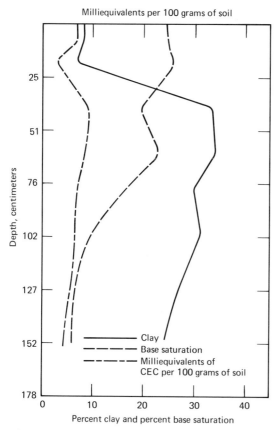

Figure 8-1   Selected properties of an Ultisol.

Figure 8-2 Gross morphology of an Ultisol. Note similarity to many Alfisols in terms of A1, A2, Bt, and C horizon sequence and a blocky Bt with clay skins.

cation-exchange capacity of kaolinitic and oxide clays in the clay fraction. Kaolinite, gibbsite, aluminum interlayered clay minerals, and oxide clays are common in the clay fraction. Base saturation is low and decreases with increasing soil depth. The higher base saturation in the upper soil layers reflects the cycling of bases by the vegetation. Extractable aluminum is high in many Ultisols, and calcium-deficient argillic horizons are common. Ultisols are generally well supplied with water for agriculture, but the main deficiency is plant nutrients. The release of bases by weathering is usually equal to or less than the removal by leaching, and normally most of the bases are held in the vegetation and the upper few centimeters of soil. The agriculture is shifting cultivation where fertilizers are not used. The gross morphology of an Ultisol is shown in Fig. 8-2; it is similar to that of many Alfisols. This is to be expected, since the major difference in the soils of the two orders is degree of weathering and base saturation.

Early Settlement on Ultisols in the United States

The plight of the Pilgrims at Plymouth in farming infertile Spodosols is paralleled by the early settlers on Ultisols in Maryland and Virginia. The Ultisols developed in sandy marine sediments that had few if any weatherable minerals in the sand and silt fractions. The native soil fertility was very low, and tobacco was a prized crop grown for export to Europe. Many early farmers were shifting cultivators who grew tobacco for a few years, abandoned the land, and then moved west. Early in the eighteenth century Edmund Ruffin inherited a tidewater farm in Virginia from his father; he was determined to succeed there and not move west. Ruffin was also an agitator for secession of the South; he secured a permanent place in history for supposedly firing at Fort Sumter the first shot that started the Civil War. Less well known, however, is the enormous contribution Ruffin made to the agricultural revival of the South beginning about 1820.

Ruffin spent a short time at William and Mary College and was intelligent and well read. After returning to the farm, Ruffin described a familiar pattern of farming. It was a shifting cultivation system that included growing crops for a few years until yields of grain were very low, abandonment of the land, a return to pine trees after 20 to 30 years, and clearing of the land for another brief period of cropping. At this time there was a great exodus to the west; the growth rate in Virginia dropped from 38 percent in 1820 to 2 percent in 1840. Ruffin said "All wished to sell, none to buy." Craven later described George Washington's estate at Mount Vernon as a wide spread and perfect agricultural ruin beyond imagination. One of the soils at Mount Vernon is Keyport, an Ultisol.

Many good agricultural practices dating back to the Greeks and Romans that were successful in Europe produced little results on the Ultisols. Manure produced little effect. Ruffin noted in a book by Davy (1813) that if a soil contained salts of iron, such as iron sulfate, or any acid matter, it may be ameliorated by the use of quick lime. Ruffin had his workers dig marl and put it on some land as an experiment. Effects of the marl on corn growth were immediate, and yields increased 40 percent. This ushered in a new era of increased agricultural productivity for the abandoned ultisolic lands—an era that has continued to this day. It is now known that the manures and other practices brought over from Europe were ineffective because of the low soil pH, which had an inhibitory effect on nutrient availability and caused aluminum toxicity. Ruffin's book, *An Essay on Calcareous Manures,* is one of the most important pieces of American literature; it was reprinted in 1961 by Belknap Press of Harvard University. Ruffin, who never took a course in either soil science or chemistry, has been called the "father of soil chemistry in America" for his discovery that liming was the most important first step in the productive management of Ultisols. The European settlers had farming experience based mainly on Alfisols, which responded to different management practices.

Suborders and Their Distribution

Ultisols are placed into suborders mainly on the basis of soil moisture regime. Ultisols with aquic moisture regime are Aquults and have gray-olive-colored subsoils. The water table is close to the soil surface during part of the year, usually in winter and spring, and deep in other times. Aquults are extensive on the lower Atlantic coastal plains as shown in Fig. 2-1 (U1a areas).

Humults are Ultisols that are more or less freely drained, humus rich and of the middle and low latitudes. They are mainly in the mountainous areas that have high rainfall and are common in the coastal range of Oregon and Washington, as shown in Fig. 2-1 (U2S areas).

Udults are the more or less freely drained, humus-poor Ultisols in humid climates that have well-distributed rainfall or udic soil moisture regimes. Most of the Ultisols of the United States are Udults, as seen in Fig. 2-1 (U3a and U3S areas).

Ustults are Ultisols that are well drained but have pronounced dry seasons; that is, they have ustic moisture regimes. Ustults have low organic matter content and are not common in the United States. In other parts of the world they have a native vegetation of deciduous forest or savanna that is probably anthropic.

Xerults are Ultisols with xeric soil moisture regimes. Winter rains and winter soil temperatures are adequate for effective weathering and leaching. Xerults have ochric epipedons that rest on brownish or reddish argillic horizons. The soils are not extensive in the United States but are important locally in the Pacific states.

Ultisols of the Piedmont Province

Most of the Ultisols of North America occur in the southeastern United States on the coastal plains and piedmont. The Piedmont province consists of a dissected plain sandwiched between the Appalachian Mountains and the Atlantic coastal plain, as shown in Fig. 8-3.

Topography and Parent Material of the Piedmont

The Piedmont province consists of a dissected plain created by erosion; it slopes gently to the east. Near the mountains monadnocks, mountains, and hills are surrounded by and rise above an erosional plain. The interior of the province typically has a gently rolling landscape (see Fig. 8-4). Many interstream divides have gently sloping, rounded summits and shallow valley heads. Large broad flats are rare. These interstream-divide areas are more or less sharply set off by steeper valley slopes that grade to low terraces or to the modern streams. The amount of dissection generally increases as the coastal plain is approached where the rocks of the piedmont pass beneath the sediments of the coastal plain.

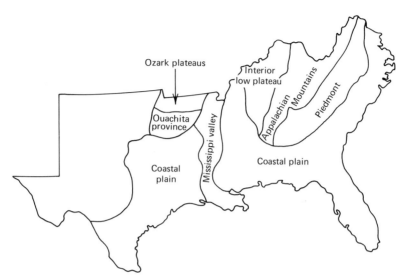

Figure 8-3 Physiographic divisions important in the southeastern United States. (Adapted from Buol, 1973.)

Figure 8-4 Rolling landscape of the piedmont.

Figure 8-5 Exposure of deeply weathered saprolite in the piedmont with a resistant dike of pegmatite feldspar. Soil on the dike is a Dystrochrept. Soil on the deeply weathered saprolite is a Hapludult.

The variety of rocks in the piedmont is large and includes almost all types of metamorphoric and igneous rocks. The rocks are deformed and generally strong. The piedmont plain, however, seems to cut across all rock types without appreciable change in form. There is little structural control of the drainage except for a few local rivers that follow faults or soft beds. The diverse rocks have been deeply weathered to a soft saprolite that is several meters thick, and the soils have formed in this weathered material, as shown in Fig. 8-5.

Hapludults of the Piedmont

Most of the soils of the piedmont are Hapludults, or "youthful" Ultisols. The effect of erosion on sloping land has resulted in Hapludults, even though soils have developed in the residuum of ancient rocks. The Hapludults are characterized by thin or only moderately thick argillic horizons. The percentage of clay in the argillic horizon decreases from its maximum by over 20 percent of that maximum within 1.5 meters of the soil surface. Hapludults have some primary minerals in the sand and silt fractions of argillic horizons that release cations for plant growth. Hapludults have 10 percent or more weatherable minerals in the 20 to 200 millimeter fraction in the upper 50 centimeters of the argillic horizon.

The Cecil soil (Typic Hapludult) is one of the most extensive soils of the piedmont and formed in parent material derived from granite gneiss, mica gneiss, granite, and mica schist. Cecil soils have light-colored sandy loam A horizons and red-colored argillic horizons. Argillic horizons have thick and continuous clay skins, which may restrict nutrient ion diffusion in and out of the soil matrix. The dominant clay mineral is kaolinite. Vermiculite is the next most important constituent of the clay; it contains lesser amounts of gibbisite, goethite, and quartz. Some selected properties of a Cecil soil from western North Carolina are given in Table 8-1. The Cecil has a well-developed argillic horizon that qualifies as a Hapludult. Even though the clay content of the argillic horizon is sufficient for the clay textural class, the argillic horizon has moderate shrink swell potential, permeability to water, and water retention capacity because of the low activity of kaolinite. The low cation-exchange capacity is related to generally low organic matter content and low activity clays. The most striking features are the near absence of exchangeable nutrient cations, the high extractable or exchangeable acidity, and the extremely low base saturation. Nutrients such as calcium, magnesium, and potassium for plant uptake come mainly from organic matter decomposition and weathering of primary minerals in the sand and silt. A study of five Cecil soils showed that exchangeable aluminum is present in all horizons and ranges from 0.3 to 5 milliequivalents per 100 grams of soil. In general, exchangeable aluminum increases with soil depth. The pH of soil horizons ranged from 4.2 to 5.8, with a pH less than 5 for many horizons. The soils have good physical condition but low natural fertility, and crops grown on the soils are very responsive to lime and fertilizers.

Rhodudults of the Piedmont

Rhodudults are similar to Hapludults in that both are well drained. Rhodudults of the piedmont have developed in the residuum of basic rocks. Most of them have dark color values in all horizons; colors range from dark reddish-brown to dark red. Rhodudults show less evidence of eluviation, and light-colored A2 horizons are typically absent (see Fig. 8-6). Compared to Hapludults, Rhodudults have more oxides, usually contain more primary minerals, are better supplied with bases, and contain more phosphorus. All in all, both kinds of soils are similar in physical and chemical properties. Rhodudults were considered Reddish-Brown Lateritic in earlier classification systems in the United States.

Rhodudults are most extensive in South Carolina and Georgia and are conspicuous by their bright red color. Only about 2 percent of the soils in the Ultisol area of the southeastern United States are Rhodudults but, in landscapes where Rhodudults occur, they comprise about 40 percent of the Ultisols. The next most extensive soils of the piedmont are Dystrochrepts and Paleudults.

Land Use in the Piedmont

Most of the land in the piedmont was cleared and cotton was the major crop before the Civil War. Today about 60 percent of the land in the piedmont is in forest, and only about 20 percent is in cultivated crops. Principal forest products are pulp-

Table 8-1   Selected Properties of a Cecil Soil, Typic Hapludult

Depth, centimeters	Horizon	Percent Clay	Percent Organic Carbon	CEC	Milliequivalents per 100 grams				Extractable Acidity	Percent Base Saturation
					Calcium	Magnesium	Sodium	Potassium		
0–3	A1	11	1.63	10.3	Tr[a]	0.1	Tr	0.1	10.1	2
3–8	A2	16	0.64	6.0	Tr	0.1	Tr	0.1	5.8	3
8–13	B21	33	0.32	7.7	Tr	0.1	Tr	0.1	7.5	3
13–23	B22	44	0.22	8.8	Tr	0.2	Tr	0.1	8.5	3
12–28	B23	46	0.17	9.1	Tr	0.3	Tr	0.1	8.7	4
28–38	B31	48	0.11	9.6	Tr	0.1	Tr	0.1	9.4	2
38–48	B32	32	0.07	8.0	Tr	0.1	Tr	0.1	7.8	3
48–70	C1	12	0.04	6.1	Tr	Tr	Tr	0.2	5.9	3
70–84	C2	5	0.02	3.8	Tr	Tr	Tr	0.1	3.7	3
84–108	C3	2	0.02	4.6	Tr	Tr	Tr	0.2	4.4	4

Adapted from Soil Survey Investigations Report 16, 1967, for soil number S60NC-45-2 from Henderson County, North Carolina.

[a]Tr = trace.

Figure 8-6 Dark red and reddish-brown colors throughout and absence of A2 horizons are characteristic of Rhodudults of the piedmont that develop in basic rock residuum.

wood, hardwood, and pine lumber. Soil erosion, along with changes brought about by the Civil War, have been responsible for the reduction in cropland acreage. On many slopes the A horizons have been eroded away, and argillic horizons are exposed. Gullying is common. Erosion control practices are widely used; erosion is a major soil management problem. The use of fertilizers and lime is essential for cropping. An aerial view of land use in the piedmont of Virginia is shown in Fig. 8-7.

From Maryland south the soil temperature regime for most Udults of the piedmont is thermic, and a wide variety of crops are grown. Forage and grain for livestock production dominate the agricultural use of land. Other important crops include tobacco, corn, and wheat. Georgia is the major broiler chicken producer; the industry is located in the piedmont in the northern part of the state.

North of Maryland the soils of the piedmont have mesic temperature regime. More of the land is cultivated and less is in forest. Dominant agricultural use is the same as farther south: forage and grain crops for livestock. Truck crops and orchards, however, are more important.

Ultisols of the Coastal Plain of the Southeastern United States

The coastal plain of the southeastern United States is the largest area where Ultisols are the most extensive soils. It extends from New England to Texas and varies in width from about 240 kilometers in the Atlantic section to 640 kilometers in the gulf section (see Fig. 8-3).

Topograpy and Parent Materials of the Atlantic Coastal Plain

The coastal plain slopes upward from the sea, and local relief generally increases with distance inland. Many streams have produced dissection and flow to the Atlantic Ocean or Gulf of Mexico. Compared to the piedmont, slopes are generally much more gradual, and the topography well suited to large-scale farming. Parent materials orginated from the weathering and erosion of the Appalachian highlands

Figure 8-7 Typical land use pattern in piedmont. Most of the land is forested. Contour farming reflects the sloping topography and need for erosion control practices. (USDA-SCS photograph.)

and consist mainly of a mixture of quartz sand and aluminum-saturated kaolinite. These materials represent the end products of intense weathering, and about the only soils that could develop in the parent materials are Ultisols. A more recent mantle of loess was deposited in a strip along the eastern side of the Mississippi River Valley, and soils here are mainly Alfisols. In areas where the sediments are thick sands, Quartzipsamments have developed.

Ochraquults of the Lower Coastal Plain

Aquults are Ultisols with a water table close to the soil surface at sometime during the year; they have gray-colored or gleyed argillic horizons. They occur mainly in the Atlantic and gulf coast flatwoods from Delaware to Georgia and in southern Louisiana and Texas (see Fig. 2-1). The soils are usually water saturated in winter and spring, and the water table declines, leaving the soils relatively dry in the autumn of most years. Organic matter accumulates in the upper few centimeters, forming a thin A1 horizon with gray, dark gray, or black color. The soils are Ochraquults, and a typical horizon sequence is Ap, A2g, B2tg, B3tg, and IICg. Some of the Aquults have umbric A1 horizons and are Umbraquults.

The Ochraquults are the most extensive Aquults on the coastal plain and have developed in alluvium and marine sediments that are of Pleistocene age or older. Argillic horizons are thin or moderately thick, and there is no abrupt textural change between the A and Bt horizons. In Georgia and the Carolinas there are quite a few Paleudults in association with Ochraquults. Many of the Ochraquults of the coastal plain have a significant amount of weatherable minerals in the subsoil. Slopes are gentle, and most are forested, as shown in Fig. 8-8. The major product is pulpwood. As pressure on the better-drained sites increases, there is expansion onto less well-drained soils such as Aquults. Extensive areas have been drained and cultivated to produce truck and fruit crops, corn, soybeans, wheat and barley.

Paleudults of the Middle and Upper Coastal Plain

Sediments on the middle and upper coastal plain were largely unaffected by Pleistocene events and are older than the parent materials from which Aquults formed on the lower coastal plain. Soils have been exposed to weathering for 2 to 3 million years on the middle coastal plain and for 10 million years or more on the upper coastal plain. Highly weathered sediments on gently sloping topography and the long period of soil evolution has produced soils that are mainly Paleudults. Quartzipsamments have developed on thick quartz sand sediments.

Paleudults are characterized by intense weathering to produce sand and silt fractions with less than 10 percent weatherable minerals and thick argillic horizons. The argillic horizons of Paleudults of the coastal plain are usually 150 to 200 cen-

Figure 8-8 Pine forest and nearly level land surface typical of the lower coastal plain.

timeters thick and 2 to over 3 meters to the bottom of the argillic horizon. Paleudults of the coastal plain typically have thick sandy A1 and A2 horizons that overlie argillic horizons that range in color from yellow to red to brown, depending on the state of oxidation of iron and the presence of organic matter (see Fig. 8-9). Primary minerals that weather to release cations are few or absent, and natural fertility is uniformily low. Basic cations are mainly in organic matter of the surface soil and upper B horizons and are biologically recycled. Low activity of the clay fraction makes the soils generally have good physical properties.

The Paleudults of the coastal plain have differences resulting from differences in age. With increasing age the soils have thicker solums, greater gibbsite content, and reduced kaolinite content. Plinthite, a reticulate mottled material that hardens irreversibily upon drying, is more common in soils as they become older. Plinthite occurs in the lower solum, and its formation appears related to water table influence (see Fig. 8-10). The plinthite has little or no significance on the use of soil for agriculture, but it may have some effect in restricting tree roots. The reduced permeability associated with the plinthite reduces the suitability of the soil for septic tank filter fields. Plinthite is of fairly common occurrence on the coastal plain.

Ruston and Norfolk are two well-known series of Typic Paleudults with red and

Figure 8-9 Paleudult on the middle coastal plain showing thick, sandy A1 and A2 horizons overlying thick argillic horizon. (Depth in centimeters.)

yellowish argillic horizons, respectively. The soils were formerly called Red-Yellow Podzolic.

Hapludults are associated with Paleudults on the middle and upper coastal plain and are found on younger surfaces and where erosion has been more severe on slopes and delayed soil evolution. Paleudults tend to occupy the broad interstream uplands with Hapludults on slopes along streams.

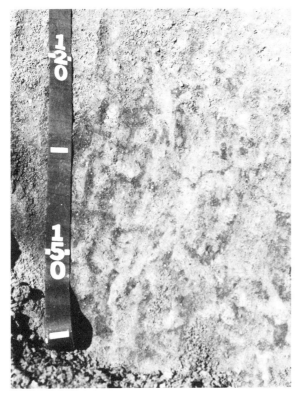

Figure 8-10 Reticulate mottling characteristic of plinthite layer with plinthite as the darker or red areas. Closeup of a lower part of soil in Fig. 8-9, a Plinthic Paleudult.

Land Use on Paleudults

Although the Paleudults of the middle and upper coastal plains have fewer weatherable minerals and more exchangeable aluminum than Ultisols of the lower coastal plain and piedmont, the agriculture of the southeastern United States is concentrated on the Paleudults. The Paleudults are generally well drained, and slopes are conducive to the use of large equipment (see Fig. 8-11). In Georgia, for example, 81 percent of the corn, 92 percent of the tobacco, and 99 percent of the peanuts are grown on the middle and upper coastal plains. The concentration of harvested cropland in Georgia on Paleudults is shown in Fig. 8-12. Paleudults of the coastal plain are well suited for peanut production because the thick, sandy A horizons are ideal for pegging, and the harvested peanuts have little soil clinging to them. In Sumter county, Georgia, where the Jimmy Carter farm is located, over 86 percent of the soils are Paleudults.

Figure 8-11 Agriculture of the southeastern United States is concentrated on the middle and upper coastal plain, where slopes are gentle and water tables are generally deep enough that drainage is no problem. The sandy surface soil is, however, susceptible to wind erosion.

The sandy A horizons, however, are subject to forming a compact state by the gradual movement of silt and small sand particles into spaces between the large sand particles. Furthermore, compaction by tractor tires when soils are at certain moisture contents creates compaction that is not obliterated by subsequent wetting and drying because of the low content of clay, which is nonexpanding. Plow pans or plow soles are common and restrict rooting depth of crops.

Paleudults are well suited for tobacco production; the major area of production is in North Carolina, where flue-cured tobacco is grown for cigarettes. The soils are low in organic matter and, consequently, in nitrogen supplying power. The tobacco plants are fertilized to produce a large, leafy plant early in the season, but just enough nitrogen is used to make sure the plants become nitrogen deficient before maturity. This produces low-nitrogen-content leaves with low nicotine content. As the lower leaves become nitrogen deficient and turn yellow, the leaves are harvested and cured in a heated shed (flue cured). The curing process results in the conversion of starch to sugar and increases the sweetness; the color turns from yellow to a golden brown.

Land Forming on the Coastal Plain

Land forming to smooth the land surface has become more popular in recent years on nearly level land of the coastal plain. Land forming allows excess water to move slowly off fields; this permits time for infiltration but is fast enough to prevent

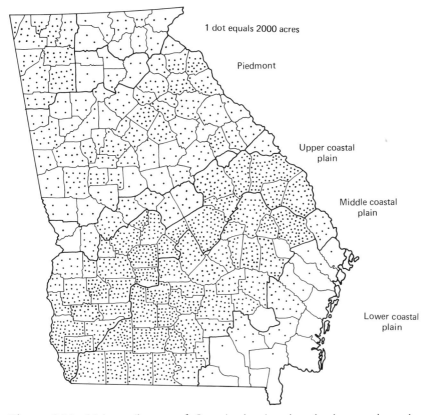

Figure 8-12 Major soil areas of Georgia showing that the harvested cropland acreage is concentrated on the middle and upper coastal plain. (Adapted from Cary, 1968, and used by permission of the Soil Conservation Society of America.)

flooding in depressions. The result is more efficient use of farm machinery on larger fields, more uniform planting surface for improved seed germination and seedling survival, and more efficient use of rainfall and irrigation water. Cutting or removal of surface soil from "high areas," however, exposes clayey argillic horizons with poor physical properties, and low pH, and high exchangeable aluminum, as shown in Table 8-2.

At the low pH found in most horizons of the Ultisols of the coastal plain, silicate clay lattices (including kaolinite) slowly disintegrate along edges and release trivalent aluminum that is strongly adsorbed on the exchange so that as much as 90 percent of the exchange complex becomes aluminum saturated (see Table 8-2). Aluminum saturation over 60 percent has been found to reduce seriously the yields of many crops. Experimental work in North Carolina has shown that growth of corn and wheat and rooting depth have been decreased when grown on cut materials. Successful land forming seems to be related to minimization of exposure of argillic horizons and the use of large amounts of lime and phosphorus where argillic horizons are exposed or are close to the soil surface. The major problems associated with liming to increase pH and reduce aluminum toxicity of argillic horizons are the limited solubility of calcium carbonate and the high cost of applying lime deep in the soil.

Land Use Changes in the Old Cotton Belt

Cotton was king in the early years of settlement of the Atlantic coastal plain and piedmont of the old cotton belt (including North and South Carolina, Georgia, Alabama, Tennessee, Mississippi, Arkansas, and Louisiana). As late as 1930, 90 percent of the farms in Alabama and Mississippi produced some cotton. By 1910, however, some counties in South Carolina and Georgia had already passed their peak cotton hectarage and, since 1929, the hectarage of cotton in southeastern United States has declined. Cash receipts for cotton as a percentage of all farm income declined from 51 percent in 1935 to only 4 percent in 1944. Cotton production has shifted west, and much of it is now produced with irrigation in west Texas, Arizona, and California.

Table 8-2   Exchangeable Aluminum, Aluminum Saturation, pH, and Available Phosphorus of Three Soils of the Norfolk Catena

Depth, centimeters	Horizon	Aluminum		pH	Phosphorus, parts per million
		Milliequivalents per 100 Grams	Percent Saturation		
		Norfolk, Typic Paleudult			
0–15	Ap	0.2	7	5.8	120+
15–30	A21	1.0	47	4.9	120+
30–45	A22	0.6	35	5.2	—
45–65	B1	2.8	79	4.7	—
		Lynchburg, Aeric Paleaquult			
0–15	Ap	0.6	25	5.3	70
15–30	A2	0.5	39	4.9	12
30–42	B1	2.3	67	4.6	—
42–75	B21t	3.5	82	4.7	—
		Portsmouth, Typic Umbraquult			
0–20	Ap	2.2	53	4.6	14
20–26	A2	4.2	80	4.2	5
26–45	B21tg	3.1	82	4.6	—
45–75	B22tg	3.3	89	4.4	—
75–85	B3tg	4.0	90	4.1	—
85–100	C1	6.5	90	3.9	—

Adapted from Phillips and Kamprath, 1973.

Table 8-3   Land Use Changes in the
Old Cotton Belt, 1930–1950

Land Use	Percent Change
Cotton	−40
Corn	−19
Cropland harvested	−11
Soybeans	+1402
Oats (grain)	+349
Hay	+115
Peanuts	+84
Pasture	+54
Woodland	+35

Data from Pearson and Yeager, 1957.

The decline of cotton in the old cotton belt has been offset by large increases in soybean production. Increases have occurred in forage and pasture crops. Much of the retired land from cotton production has been returned to woodland. These and other changes are presented in Table 8-3.

Livestock enterprises have also shown large increases, particularly in broilers and cattle production. The Ultisols of the southeastern United States that proved to be such a problem for early settlers comprise one of the most productive agricultural regions of the United States because of the favorable climate and use of technology.

Other Areas of Ultisols in the United States

The Ultisols of the piedmont and coastal plain are the most extensive areas of Ultisols in the United States and have been intensively studied. Other important areas of Ultisols, however, include the Ultisols of the interior dissected plateaus and of the mountainous areas of western United States, Hawaii, and Puerto Rico. Ultisols are not recognized in Canada.

Ultisols of the Dissected Plateaus

Hapludults and Paleudults are important soils of dissected plateaus in southern Indiana, central Kentucky and Tennessee, northeastern Alabama, northern Arkansas, southern Missouri, northeastern Oklahoma, and western Pennsylvania. The soils formed from shale, sandstone, and limestone and occur mainly on level to rolling ridge tops, benches, and foot slopes. The soils in Indiana, Kentucky, Tennessee, Mississippi, and Alabama have a thin to moderately thick mantle of loess. Within this region Alfisols tend to occur in valleys that have a limestone floor and

Figure 8-13   A landscape in the dissected plateau region where Ultisols are extensive soils.

in the Nashville Basin and bluegrass region of Kentucky. Temperature regimes are mesic or thermic.

These Ultisols occur in landscapes where many slopes are not conducive to agriculture and much of the land is in forest, as shown in Fig. 8-13. About one-third to two-thirds of the land is in forest. Principal crops include corn, feed grains, and hay for dairy cattle and other livestock. In some places soybeans, cotton, and tobacco are important cash crops.

Humults of the Pacific Northwest and Hawaii

Humults are Ultisols that are more or less freely drained and develop in the middle to low latitudes under high rainfall, typically in mountainous areas. Humults have (1) 0.9 percent or more organic carbon in the upper 15 centimeters of the argillic horizon, (2) 12 kilograms of organic carbon in a unit volume of 1 square meter to a depth of 1 meter, exclusive of an 0 horizon, or (3) Humults meet both criteria. Humults have developed mostly from basic rocks and are dark colored and occur on surfaces that are late Pleistocene or older. Their slopes are commonly strong and, if cultivated, erosion is likely to have exposed the argillic horizon. The native vegetation in midlatitudes is mostly coniferous forest and rain forest at low latitudes (see Fig. 8-14).

Humults are the dominant soils in landscapes of minor extent in the continental United States in California, Oregon, and Washington (see Fig. 2-1). Their moisture

Figure 8-14 High rainfall, steep slopes, and coniferous forest characterize the Humult landscapes of Oregon and Washington.

regime is udic or xeric, and the temperature regime is dominantly mesic. These Humults are mostly Haplohumults with some weatherable minerals in thin argillic horizons. The associated soils are principally Xerolls on water-deposited materials and terraces. The soils are mainly covered with Douglas fir forests, and lumber is the major product.

Humults in Hawaii are mostly Tropohumults; their temperature regime is iso-mesic, isothermic, or isohyperthermic. The soils occur on gentle to moderate slopes of terraces, alluvial fans, and foot slopes. The soils formed from basaltic rock materials and are used mainly for the production of pineapples and sugarcane.

Xerults of the Pacific Northwest

Xerults are freely drained Ultisols of Mediterranean climates that have a moderate or small amount of organic matter. Xerults have an ochric epipedon that rests on a brownish or reddish argillic horizon. The soils are not extensive in the United States but are important locally. No area of Xerults is shown on the map in Fig. 2-1. The soils occur in the Sierra Nevada and Cascade mountains under coniferous forest, mainly in California and Oregon. Temperature regime is mesic or thermic, and moisture regime is xeric.

The Xerults of the United States are mainly Typic Haploxerults that have a thin or moderately thick argillic horizon, have an appreciable amount of weatherable minerals in the argillic horizon, or both. These soils occur in the mountains, where slopes are steep and support coniferous forests (see Fig. 7-18). Xerults were mostly considered Red-Yellow Podzolic or Yellowish-Brown Lateritic soils in earlier classification systems in the United States.

References

Agricultural Experiments Stations of the Western States Land-Grant Universities and Colleges and USDA, *Soils of The Western United States,* Washington State University, Pullman, 1964.

Austin, M. E., *Land Resource Regions and Major Land Resource Areas of the United States,* USDA Handbook 296, Washington, D.C., 1965.

Barnhisel, R. I., and C. I. Rich, "Clay Mineral Formation in Different Rock Types of a Weathering Boulder Conglomerate," *Soil Sci. Soc. Am. Proc., 31:*627–631, 1967.

Bonner, J. C. *A History of Georgia Agriculture 1732–1860,* University of Georgia Press, Athens, 1964.

Bradford, W., *Of Plymouth Plantation, The Pilgrims in America,* Selected and Edited with an Introduction by Harvey Wish, Capricorn, New York, 1962.

Brewer, E. O., *Soil Survey of Catawba County, North Carolina,* USDA, Washington, D.C., 1975.

Buol, S. W., Ed., "Soils of the Southern States and Puerto Rico," *Southern Cooperative Series Bulletin 174,* Agr. Exp. Sta. of Southern States and Puerto Rico Land Grant Universities and USDA, 1973.

Cady, J. G., "Rock Weathering and Soil Formation in the North Carolina Piedmont Region," *Soil Sci. Soc. Am. Proc., 15:*337–342, 1950.

Campbell, R. B., D. C. Reicosky, and C. W. Doty, "Physical Properties and Tillage of Paleudults in the Southeastern Coastal Plains," *Jour. Soil Water Con., 29:*220–226, 1974.

Carson, C. D., and G. W. Kunze, "Red Soils of East Texas Developed in Glauconitic Sediments," *Soil Sci., 104:*181–190, 1967.

Cary, H. C., "Cropping and Land Use Trends in Georgia," *Jour. Soil Water Con., 23:*100–102, 1968.

Corliss, J. F., *Soil Survey of Alsea Area, Oregon,* USDA and Ore. Agr. Exp. Sta., Washington, D.C., 1968.

Craven, A. O., *Edmund Ruffin, Southerner,* Appleton, New York, 1932.

Craven, A. O., "Soil Exhaustion as a Factor in the Agricultural History of Virginia and Maryland, 1606–1860," *University of Illinois Studies in Social Sciences,* Vol. 13, 1925.

Daniels, R. B., E. E. Gamble, and L. A. Belson, "Relation Between A2 Horizon Characteristics and Drainage in Some Fine Loamy Ultisols," *Soil Sci., 104:*364–369, 1967.

Daniels, R. B., E. E. Gamble, and J. G. Cady. "Some Relations Among Coastal Plain Soils and Geomorphic Surfaces in North Carolina," *Soil Sci. Soc. Am. Proc., 34*:648–653, 1970.

Daniels, R. B., E. E. Gamble, and L. J. Bartelli, "Eluvial Bodies in B Horizons of Some Ultisols," *Soil Sci., 106*:200–206, 1968.

Daniels, R. B., H. F. Perkins, B. F. Hajik, and E. E. Gamble, "Morphology of Discontinuous Phase Plinthite and Criteria for Its Field Identification in the Southeastern United States," *Soil Sci. Soc. Am. Jour., 42*:944–949, 1978.

Davy, H., *Elements of Agricultural Chemistry,* Longman, London, 1813.

Devereux, R. E., G. H. Robinson, and S. S. Obenshain, "Genesis and Morphology of Virginia Soils," *Bulletin 540,* Va. Agr. Exp. Sta., 1962.

Edwards, M. J., J. A. Elder, and M. E. Springer, "The Soils of The Nashville Basin," *Bulletin 499,* USDA and Tenn. Agr. Exp. Sta., 1974.

Fiskell, J. G. A., and J. F. Perkins, Eds., "Selected Coastal Plain Soil Properties," *Southern Coop. Bull. 148,* University of Florida, 1970.

Foote, D. E., E. Hill, S. Nakamura, and F. Stephens, *Soil Survey of Islands of Kauai, Oahu, Maui, Molokai, and Lanai, State of Hawaii,* USDA and Univ. Hawaii Agr. Exp. Sta., Washington, D.C., 1972.

Francis, J. K., and N. S. Loftus, "Chemical and Physical Properties of Cumberland Plateau and Highland Rim Forest Soils," USDA Forest Res. Paper SO-138, Southern Forest Exp. Sta., New Orleans, 1977.

Gamble, E. E., and R. B. Daniels, "Parent Material of Upper and Middle-Coastal Plain Soils in North Carolina," *Soil Sci. Soc. Am. Proc., 38*:633–637, 1974.

Giddens, J., H. F. Perkins, and R. L. Carter, "Soils of Georgia," *Soil Sci., 89*:229–238, 1960.

Godfrey, C. L., "General Soil Map of Texas," *MP 1034,* Tex. Agr. Exp. Sta., 1973.

Hart, J. F., "The Demise of King Cotton," *Ann. Assoc. Am. Geog., 67*:307–322, 1977.

Kellogg, C. E., "Shifting Cultivation," *Soil Sci., 95*:221–230, 1963.

Khalifa, E., "Properties of Clay Skins in a Cecil (Typic Hapludult) Soil," unpublished Ph.D. Thesis, North Carolina State University, 1968.

Lee William D., "The Soils of North Carolina," *Tech. Bull. 115,* N.C. Agr. Exp. Sta., 1955.

Lytle, S. A., "The Morphological Characteristics and Relief Relationships of Representative Soils in Louisiana," *Bulletin 631,* Department of Agronomy, Louisiana State University, 1968.

Marschner, F. J., "Land Use and Its Patterns in the United States," *USDA Handbook 153,* Washington, D.C., 1959.

McCaleb, S. B., "The Genesis of the Red-Yellow Podzolic Soils," *Soil Sci. Soc. Am. Proc., 11*:164–168, 1959.

McCaleb, S. B., and W. D. Lee, "Soils of North Carolina: I. Factors of Soil Formation and Distribution of Great Soil Groups," *Soil Sci. 82:*419–431, 1956.

McCracken, R. J., E. J. Petersen, L. E. Aull, C. I. Rich, and T. C. Peele, "Soils of the Hayesville, Cecil and Pacolet Series in the Southern Appalachian and Piedmont Regions of the United States," *Southern Coop. Series Bull. 157,* North Carolina State University, 1971.

National Cooperative Soil Survey of the United States, National Soil Survey of Canada and FAO, *Soil Map of the World, Volume II: North America,* Unesco, Paris, 1975.

Pearson, R. W., and L. E. Ensminger, "Southeastern Uplands," in *Soil,* USDA Yearbook, Washington, D.C., 1957.

Pearson, R. W., and J. H. Yeager, "Agricultural Trends in the Old Cotton Belt," in *Advances in Agronomy,* Vol. 9, pp. 1–29, Academic, New York, 1957.

Perkins, H. F., and F. T. Ritchie, Jr., "Physical Features of Georgia," *Jour. Soil Water Con., 23:*97–100, 1968.

Perkins, H. F., and M. E. Shaffer, *Soil Associations and Land Use Potential of Georgia Soils,* Map Sheet, University of Georgia, 1977.

Phillips, J. A., and E. J. Kamprath, "Soil Fertility Problems Associated with Land Forming in the Coastal Plain," *Jour. Soil Water Con. 28:*69–73, 1973.

Pilkinton, J. A., *Soil Survey of Schley and Sumter Counties, Georgia,* USDA and Ga. Agr. Exp. Sta., Washington, D.C., 1974.

Robinson, G. H., R. E. Deverux, and S. S. Obenshain, "Soils of Virginia," *Soil Sci., 92:*129–142, 1961.

Ruffin, E. *An Essay on Calcareous Manures,* Belknap Press, Harvard University Press, Cambridge, Mass., 1961. (Originally published in 1832 by J. W. Campbell in Petersburg, Va.)

Scrivner, C. L., J. C. Baker, and B. J. Miller, "Soils of Missouri," *Circular 823,* Missouri Agricultural Extension Division, 1966.

Simonson, G. H., "Soils of Oregon" (a map), *Oregon State University,* 1975.

Smith, B. R., M. A. Granger, and S. W. Buol, "Sand and Course Silt Mineralogy of Selected Soils on the Lower Coastal Plain of North Carolina," *Soil Sci. Soc. Am. Jour., 40:*928–932, 1976.

Smith, H. N., *Virgin Land,* Vintage, New York, 1957.

Soil Conservation Service, *Classification of Soil Series of the United States,* USDA, Washington, D.C., 1977.

Soil Conservation Service, "Soil Survey Laboratory Data and Descriptions for Some Soils of Georgia, North Carolina and South Carolina," *Soil Survey Investigations Report 16,* USDA, Washington, D.C., 1967.

Soil Survey Division of Bureau of Chemistry and Soils, "Soils of the United States" in *Soils and Men,* USDA Yearbook, pp. 1019–1161, Washington, D.C., 1938.

Soil Survey Staff, *Soil Taxonomy,* USDA Agriculture Handbook 436, Washington, D.C., 1975.

Southern Regional Project S-14, "Certain Properties of Selected Southeastern United States Soils and Mineralogical Procedures for Their Study," *Southern Coop. Series Bull. 61,* Virginia Agricultural Experiment Station, Blacksburg, 1959.

Strahler, A. N., and A. H. Strahler, *Elements of Physical Geography,* Wiley, New York, 1976.

Truog, E., "Putting Soil Science to Work," *Jour. Am. Soc. Agron.,* 30:973–985, 1938.

Spodosols of the United States

Central Concept

Spodosols are the most extensive in cool, humid, or perhumid regions but are also found in the humid tropics. Conditions favoring Spodosol formation are sands rich in quartz and humid climates with intensive leaching. Most Spodosols developed under a conifer forest vegetation. The feature most characteristic of Spodosols is the presence of a spodic horizon—an illuvial horizon enriched with an *active* amorphous mixture of organic matter and aluminum with or without iron. Spodosols were called Podzols in earlier soil classification systems in the United States.

Genesis and Properties of Spodic Horizons

The genesis of spodic horizons is believed to be related to the precipitation or complexing action of sequioxides (used here to mean compounds of iron and aluminum). Organic particles moving downward in water are precipitated in the spodic horizon by mobile sequioxides released in the weathering of primary minerals or from plant cycling. Spodic horizons can develop in only a few hundred years under ideal conditions; many have formed in a few thousands of years. Holzhey et al. (1975) calculated that there was enough organic carbon available on the coastal plain in North Carolina to develop a spodic horizon 7 meters thick in less than 30,000 years or about 15 centimeters thick in 600 years. Biological destruction of spodic horizons can also be rapid in some soils if lime and fertilizers are used.

At some sites where shallow water tables exist, upward migration of organic matter and sequioxides from capillarity enhances spodic horizon formation. The spodic horizon, however, seems not to develop in soils that are permanently water saturated.

In the field spodic horizons can be recognized by their sandy nature, weakly developed structure, and color, which varies from brown to yellowish and reddish-brown to black. The upper boundary is usually abrupt and commonly underlies an albic (A2) horizon composed mainly of uncoated quartz sand grains (see Fig. 9-1).

Chemical Criteria for Spodic Horizons

Reliable identification of a spodic horizon rests on laboratory measurements. The first requirement concerns the extent of development and presence of *active* amorphous materials. The product of milliequivalents of cation-exchange capacity minus one-half the clay percentage and the thickness of the spodic horizon in centimeters must be 65 or greater. An illuvial horizon with a cation-exchange capacity of 10, containing 4 percent clay and having a thickness of 12 centimeters, would have a thickness criterion of 96 and would thus qualify as a spodic horizon. On the other hand, a horizon may have a value of 65 or more and not qualify as spodic.

The second chemical criterion seeks to establish that the amorphous materials are active; that is, they have a high cation-exchange capacity, high surface area, and high water retention. The active amorphous fraction is extractable in sodium pyrophosphate solution at pH 10. Dithionite-citrate extraction removes less active forms of iron and aluminum complexes. The percentage of pyrophosphate-extractable iron and aluminum divided by the percentage of dithionite-citrate-extractable iron and aluminum must be equal to or less than 0.5.

Third, spodic horizons must meet criteria based on iron content as follows.

1. If extractable iron is 0.1 percent or more:

$$\frac{\text{percent extractable iron} + \text{percent extractable aluminum}}{\text{percent clay}} = 0.2 \text{ or more}$$

2. If extractable iron is less than 0.1 percent:

$$\frac{\text{percent extractable aluminum} + \text{percent extractable carbon}}{\text{percent clay}} = 0.2 \text{ or more}$$

Figure 9-1 Spodosol (Orthod) showing typical horizons.

The distribution with depth of pyrophosphate-extractable iron and aluminum in a Spodosol is shown in Fig. 9-2. From this last criterion it is obvious that the more clay in the horizon, the less likely it is that the horizon will qualify as spodic.

Dominant Properties of Spodosols

Spodosols are typically sandy soils (low in clay content) with low natural fertility. Selected properties of a Spodosol in an idle field in New Hampshire are given in Table 9-1. The soil has high sand and low clay content in all horizons. There is considerable silt in this particular soil and, in many Spodosols, sand content is 90 percent or more in all horizons. There is a secondary maximum of organic carbon

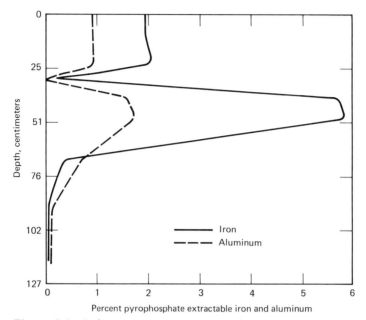

Figure 9-2   Sodium pyrophosphate extractable iron and aluminum of a representative Spodosol with the upper boundary of the spodic horizon at about 30 centimeters depth. (Adapted from *Soil Taxonomy,* 1975.)

or organic matter in the B or spodic horizon that has more organic matter in the upper part than in the lower part. The cation-exchange capacity reflects the differences in organic matter and particularly the accumulation of active amorphous materials in the spodic horizon (B21h, B22ir). Exchangeable nutrient cations are low and the exchange is dominated by exchangeable hydrogen and aluminum (extractable acidity). The low saturation with bases and high extractable acidity is reflected in the low base saturation percentage and pH.

Cation exchange in Spodosols is due mainly to carboxyl and hydroxyl sites of the organic matter and the active amorphous organic-iron-aluminum complexes. This cation-exchange capacity is pH dependent, as illustrated in Fig. 9-3. The high cation-exchange capacity shown in Table 9-1 is caused partly by determination at pH 8.2. The significance of the pH-dependent charge is that as lime is used to increase soil pH, the cation-exchange capacity increases, bringing about an increase in the amount of lime needed to achieve increases in soil pH. In some cases the use of lime could result in little or no increase in soil pH and base saturation but would result in more basic cations in the exchange.

Spodosols have low native fertility. The soils are high in water infiltration but generally low in water retention for the soil as a whole. Increased development and

Table 9-1 Selected Properties of a Typic Haplorthod

| Depth, centimeters | Horizon | Percent Sand | Percent Clay | Percent Organic Carbon | CEC | Milliequivalents per 100 Grams | | | | Extractable Acidity | Percent Base Saturation | pH |
						Calcium	Magnesium	Sodium	Potassium			
0–20	Ap1	56	3	3.5	25.7	2.9	0.2	Tr[a]	Tr	22.6	12	4.9
20–25	Ap2	55	3	3.3	26.3	3.6	0.2	0.1	Tr	22.4	11	5.1
25–36	A2	57	3	0.8	7.2	1.9	0.2	Tr	Tr	5.1	29	5.1
36–43	B21h	53	3	3.9	44.7	2.4	0.3	Tr	Tr	42.0	6	5.0
43–53	B22ir	50	2	2.9	44.8	1.7	0.1	Tr	Tr	43.0	4	4.9
53–84	B3	56	2	1.2	18.3	0.5	0.1	Tr	Tr	17.8	3	4.9
84–105	C1	65	2	0.3	6.0	0.5	0.1	Tr	Tr	5.5	8	4.9
105–120	C2	77	1	0.1	2.2	0.1	0.1	Tr	Tr	2.1	4	4.9

Adapted from Soil Taxonomy, 1975, p. 534, pedon 25.

[a]Tr = trace.

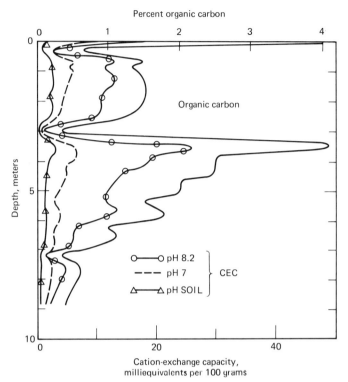

Figure 9-3 Relation of cation-exchange capacity to soil pH and organic carbon in a Spodosol on the coastal plain in North Carolina. (Reprinted from the *Soil Sci. Soc. Am. Proc.*, *39:*1182–1187, 1975, by permission of the Soil Science Society of America.)

accumulation of amorphous material in the spodic horizon, however, increases both water and nutrient retention and makes the soil capable of supporting more demanding species. The soils generally are droughty in the summer unless rains are frequent. The soils are, however, responsive to good management. Extreme development of the spodic horizon may result in cementation to produce *ortstein,* which limits permeability and root growth.

Early Agriculture on Spodosols in New England

The Pilgrims landed in an area dominated by Spodosols in Plymouth county, Massachusetts, in 1620. Thick, quartz sands on moraines and outwash plains were the dominant parent materials. Soils are mainly Entic Haplorthods and/or Psamments interspersed with areas of wet soils, as shown in Fig. 9-4.

By the spring of 1621 the Pilgrims had met Squanto, an Indian who spoke

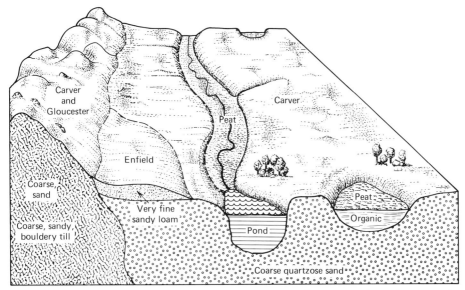

Figure 9-4 Soils formed in deep sands on moraines and outwash plains dominate the landscape where the Pilgrims landed. (From Upham, 1969.)

English and showed them where and how to grow corn. Years before Squanto had been captured and taken to Spain; from there he made his way to England, where he learned English. Sickness and death had decimated Squanto's people, and their cornfields were idle. Squanto told the Pilgrims that "the corn would come to nothing in these old grounds unless fertilized." He showed them how to fertilize the corn with fish, which were plentiful in local streams. There is some question as to whether fertilizing corn with fish was an indigenous Indian practice of the area or whether it was a practice that Squanto had learned in Europe. The first year's harvest was small, but there was a lot of wild game, which the Pilgrims gathered to get them through their second winter in America. William Bradford, governor of Plymouth, referred to Squanto as a "special instrument sent by God for their good beyond their expectation."

The incoming settlers slowly cleared the Spodosol lands and established farms. Soil fertility depletion was evident within 10 years. Even the pastures declined markedly in productivity. The limited number of livestock meant a scarcity of manure for fertilizing fields. By 1800 soil depletion was characteristic in all cultivated regions along the Atlantic coast. The unsuitability of Spodosols for agriculture was described by Swanson (1954) as follows.

When the settlers cleared New England forests 300 years ago, the topsoil they found was only 2 to 3 inches thick. Below this was sterile subsoil and when the plow mixed the two together, the blend was low in nearly everything a good soil should have. It was not the lavish virgin soil of popular fancy. Such soil could not support extractive agriculture which takes nutrients out of the soil and does not replace them. Many New England lands that were treated in this way soon went back into forest.

Today nearly all the Carver and Gloucester soils of Plymouth County are covered with pitch pine and scrub oak. The clear, sandy-bottomed lakes in the area provide recreation.

Some progressive farmers were able to establish a permanent agriculture by building and maintaining soil fertility in parts of New York and Massachusetts. Most of the farmers were unable to make a living, and abandonment was widespread. Many of these farmers, however, moved west, only to repeat their unfortunate experiences.

Spodosol Suborders and Their Distribution in North America

Suborders of Spodosols are Aquods, Ferrods, Humods, and Orthods. Aquods have aquic soil moisture regime. Ferrods have spodic horizons rich in iron relative to those of carbon. The ratio of free iron to carbon is 6 or more. Humods have spodic horizons that are rich in humus relative to those of iron. The ratio of free iron to carbon is less than 0.2. Orthods have ratios of free iron to carbon that are 0.2 or more but less than 6. Orthods are considered the "normal" Spodosols and are about six times more extensive than Aquods in the United States. Ferrods do not occur in the United States, and Humods are not extensive in North America. The general distribution of the suborders is shown in Fig. 2-1.

Orthods

Orthods are the more or less freely drained Spodosols that have an accumulation of aluminum, iron, and organic carbon in the spodic horizon. Orthods are the most extensive Spodosols in North America and northern Europe. Most have formed in coarse-textured, acid Pleistocene and Holocene deposits. Native vegetation is mainly conifers, but some Orthods also support hardwood forest. Normally Orthods have thin solums, and cultivation mixes the albic horizon with the spodic horizon, as shown in Fig. 9-5. Moisture regimes are mainly udic and temperature regimes are cryic or frigid except for a few mesic ones that border on the cryic or frigid.

Most of the Orthods in North America are Cryorthods and Haplorthods and do not have placic horizons or fragipans. The Cryorthods have cryic or pergelic temperature regimes; the Haplorthods have frigid or warmer temperature regimes.

Figure 9-5   Recently plowed field showing mixing of soil horizons.

Cryorthods of the Cook Inlet-Susitna Lowland, Alaska

The Cook Inlet-Susitna lowland is an intermountain basin that contains the largest area of agricultural land in Alaska. The basin is located in southern Alaska (see area S2S3 of Fig. 2-1) near Anchorage and includes the Matanuska Valley and the city of Palmer, where an agricultural experiment station is located. The valley is about 320 kilometers long and averages nearly 100 kilometers wide.

The climate is moderated by the Gulf of Alaska and has high precipitation, cool summers, and mild winters. Nearly all the land was covered by mountain glaciers, and the entire lowland has been covered by loess ranging from a few centimeters to 100 centimeters thick. Much of the loess was derived from glacial materials, but a large porportion is volcanic, particulary in the southern part of the valley. Forests originally covered most of the land; the principal species included white spruce (Pices glauca), paper birch (Betula papyrifera), and quaking aspen (Populus tremuloides).

Dominant soils of the Cook Inlet-Susitna lowland are Cryorthods. They have high silt content, about 60 percent, and low clay content, generally less than 10 percent. The very fine sand fraction is about 10 to 20 percent in most horizons of four soils studied in detail by Rieger and Dement (1965). Base saturation is low, and most horizons have a pH less than 5. 0 horizons 20 centimeters or more thick exist at the soil surface. Soils in the southern part of the lowland have been influenced by volcanic ash. Andepts orginally developed here under grass. More recently, forests have replaced the grasslands, and the Andepts have been converted into Spodosols. These soils also have a significant amount of allophane in the clay fraction.

Over one-half of the agricultural land in Alaska is located in the lowland, and much of the land orginally cleared to meet homestead requirements is no longer cropped. About 25 percent of the lowland has been inventoried by the Soil Conservation Service. Of this inventoried land, less than 2 percent is in cropland, 12 percent is pasture, and 56 percent is in forests. Of the inventoried land in Alaska, the Cook Inland-Susitna lowland contains 71 percent of the cropland, 99 percent of the pastureland, and 63 percent of the forests. Soils are very acid, and heavy fertilization is required. With good management good yields of grass, small grains, potatoes, and other vegetables are obtained. About 30 percent of the lowland is muskeg and is not suited for agriculture.

Cryorthods in Southeastern Alaska

In southeastern Alaska in the vicinity of Juneau the dominant soils are Humic Cryorthods. Annual precipitation is about 2500 millimeters per year, and the vegetation is a dense forest composed mainly of spruce and hemlock. Slopes are moderate or steep, and parent materials are till and colluvium. The moderating effect of the Pacific Ocean results in cool summers and winters; the soil is seldom frozen. Principal land uses are timber production, watershed protection, and wildlife.

Haplorthods of the Northeastern United States

Spodosols of the northern states are mainly Haplorthods and are extensive in all of New England, northern New York, and northern Michigan and Wisconsin (see Fig. 2-1). The soils occur in a small area in northern Minnesota along the Canadian border. The soils developed on nearly level to gently sloping plains from sandy glacial materials. The boundary between Alfisols and Spodosols in southern Michigan and northern Wisconsin is essentially a boundary between loamy glacial materials in the south and sandy glacial materials in the north. Parent materials are generally thicker than in Canada, and the more moderate climate results in less well-developed organic horizons at the soil surface. Orthods with fragipan, called Fragiorthods, are an important component of the Spodosol region of the northeastern United States.

A considerable area in New England has Spodosols that, without cobalt fertilization, produces forage that is low in cobalt for cattle and sheep (see Fig. 9-6). Cobalt

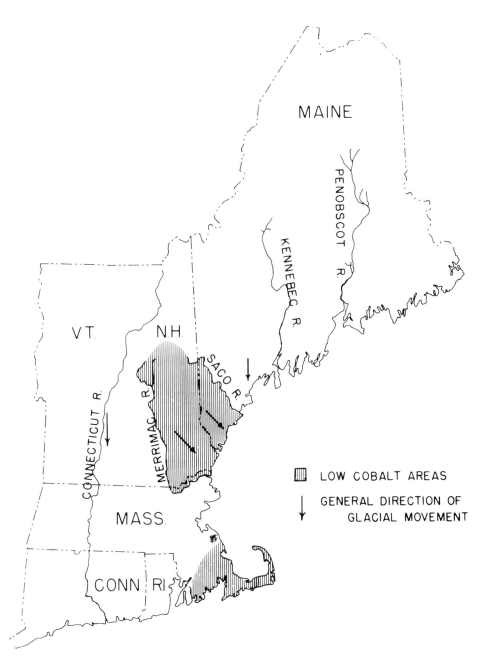

MAINE

PENOBSCOT R.

KENNEBEC R.

VT.

N.H.

SACO R.

CONNECTICUT R.

MERRIMAC R.

MASS.

CONN. | RI.

⊞ LOW COBALT AREAS

↓ GENERAL DIRECTION OF GLACIAL MOVEMENT

Figure 9-6   Broad areas of low-cobalt soils in New England. (From Kubota, 1964.)

is needed by ruminants (sheep, cattle, goats) for synthesis of vitamin B12. Cobalt deficiency in livestock was widespread in colonial days; the disease had many names associated with local reference, such as "albany ail" or "neck ail." The latter was used in southern Massachusetts and orginated from the fact that the disease was most prevalent on necks of land that extended into bays. The Spodosols of southern Massachusetts low in cobalt developed in parent materials derived from low cobalt grandiorite. In Maine and New Hampshire low-cobalt soils developed in glacial materials originating from low-cobalt rocks of the White Mountains. Intensive leaching in Spodosols also tended to remove the cobalt that was orginally present in parent materials.

Most of the Spodosol areas of the northeastern United States are in forest, and the major products are lumber, pulpwood, and Christmas trees. Mining is important in northern Wisconsin and Michigan. Cool summers and snowy winters attract urban people seeking recreation, and land values have greatly appreciated in recent years. The high soil permeability, up to 50 centimeters per hour, makes some Spodosols of limited use for septic tank filter fields because shallow water supplies might become contaminated with effluent.

Land Abandonment in the Spodosol Region

Farmers had abandoned considerable land in the east before 1800. The opening of the Erie Canal in 1825 was followed by a large influx of settlers moving west. Lumbering was a big industry on the frontier as loggers set about to harvest what seemed to be an inexhaustible stand of timber. The logging era was one of life at concert pitch. "It was a period of swank, of color, of absorption in the moment, the day and small thought of the morrow." Just as abruptly as the logging began in southern Michigan, it came to a sudden halt between 1855 and 1895. Timber that was orginally supposed to take 500 years to harvest was harvested in about 50 years.

In southern Michigan the soils were mainly Alfisols, and a prosperous agriculture was established after logging failed. North of a line running east and west from about Muskegon to Bay City, Spodosols were dominant soils. These soils were easy to clear and till and produced a good first crop of wheat. After 2 or 3 years yields dwindled, and things went from bad to worse for the pioneers. They turned to the agricultural college at East Lansing (now Michigan State University) for help.

The famous botanist W.J. Beal experimented with crop rotations and use of manure on the sands and concluded in 1889 that: "If the homesteader has to depend for his living from the start on what he can dig out of the soil and has no other business to help him, the plains are no place for him." The soils were infertile and droughty, and crops often froze in the late spring and early fall. Consequently, the great logging era in the north was followed by widespread abandonment of land, and ownership reverted back to the state (see Fig. 9-7). Tax delinquencies

Figure 9-7 Abandoned land that is gaining in value for pulp wood production (note planting in rear), Christmas trees, and recreation. Sandy nature of Spodosols is shown by soil exposed in "blowouts."

were common, and local governments had difficulty raising revenue. The Spodosol regions of New England and the lake states very appropriately have been called the "land of abandonment."

Alfic Haplorthods—Spodosols with Argillic Horizons

Land use on Spodosols, as we have seen, is mainly forestry and recreation. Some small acreages are farmed by part-time farmers mainly to support dairy cattle or for specialty crops. Within the large Spodosol region of the northern United States there are isolated important areas with good agriculture. These areas are dominated by two kinds of soils. Some of the soils developed in fine-textured parent materials and are Boralfs. The others developed in medium-textured parent materials, such as sandy loams, and are Alfic Haplorthods. Alfic Haplorthods are Spodosols with argillic horizons.

Alfic Haplorthods are bisequual soils with albic and spodic horizons (characteristic of Spodosols) in the upper part of the solum and A2 and argillic horizons (characteristic of Alfisols) in the lower part of the solum. When soils in the region form in loamy sand to sandy loam parent material, clay is translocated, and an argillic horizon is formed. The loss of clay in the A horizon results in a coarser "parent material" compared to the original material and enhances spodic horizon formation. As a consequence, albic and spodic horizons form in the upper part of

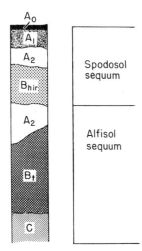

Figure 9-8 Horizon sequence of Alfic Haplorthod. (After Whiteside et al., 1968.)

the Alfisol A2, leaving the lower part of the A2 and argillic horizons below to form the lower sequum of horizons (see Fig. 9-8).

The greater productivity of Alfic Haplorthods compared to Typic Haplorthods for corn and wheat, with high-level management and the same climate, is shown in Table 9-2. Yields on Alfic Haplorthods are similar to yields on Typic Hapludalfs.

Alfic Haplorthods are the dominant soils of the well-known potato-growing area in northeastern Maine's Aroostook County (see Fig. 9-9). The potato area also extends into western New Brunswick, Canada. Caribou loam is one of the major soils; it developed from calcareous sandy loam till derived from limestone and shale. Selected properties of the Caribou loam are given in Table 9-3.

Caribou soils have a clay content several times greater than that of typical Haplorthods. The soil is quite acidic in the upper horizons, but base saturation is sufficient to provide for considerable bases. Considering the cation-exchange capacity and average exchangeable calcium, there are about 4 milliequivalents per 100 grams of soil. At a depth of only 36 centimeters, base saturation is 35 and the soil

Table 9-2 Relative Estimated Yields of Corn and Wheat With High-Level Management

| | Relative Yield | |
Soil	Corn	Wheat
Typic Haplorthod, Kalkaska sand	51	54
Alfic Haplorthod, Onaway loam	88	96
Typic Hapludalf, Miami loam	100	100

Figure 9-9 Potatoes growing on Alfic Haplorthods in Maine.

becomes calcareous in the upper C horizon. Even though the Caribou soil is a Spodosol, it has high potential for production of crops suited to the cool summers.

Temperature regimes of most Orthods in the northern United States are cryic or frigid, and crop choices are very limited. Hay and pasture to support dairying are common. White potatoes do well, since they require a cool growing season. The sandy nature of the soil favors the development of well-shaped tubers, and the acidity inhibits potato scab organisms. Maine, Wisconsin, and Minnesota are among the eight major potato-producing states.

Table 9-3   Selected Properties of Caribou Loam, Alfic Haplorthod

Depth, centimeters	Horizon	Percent Clay	Percent Organic Carbon	CEC	pH	Percent Base Saturation	Sequum
0–20	Ap	15	2.08	23	4.8	20	Spodosol
20–36	Bir	14	1.55	24	4.6	6	
36–84	A'2	13	0.16	7	4.8	35	
64–81	B'21t	25	0.14	11	5.1	57	Alfisol
81–102	B'22t	21	0.08	10	6.0	72	
102–125	B'23t	22	0.10	10	6.2	74	
125+	C1	22	0.08	—	7.7	Calcareous	

Adapted from Arno, 1964.

Figure 9-10 Prime fruitland, Alfic Haplorthods, on ridges near Great Lakes being converted into use as sites for houses.

Fruit is important on Alfic Haplorthods along the Great Lakes, where the water modifies air temperatures and reduces frost hazard in the spring and fall. Land forming is being used to improve air drainage so that cold, heavy air slides out onto the lakes instead of accumulating in pockets in the lower parts of the landscape. It is ironic that some of these choice fruit sites are unique and represent very limited locations where fruits can be produced, and yet these lands are rapidly disappearing as they are converted into subdivisions for summer and permanent homesites (see Fig. 9-10).

Aquods

Aquods are the Spodosols of wet places and have a shallow, fluctuating water table or a climate that is extremely humid. Many Aquods in warmer climates have a nearly white albic horizon that is thick enough to persist after cultivation. Some very wet Aquods have a black surface horizon that rests on a reddish-brown spodic horizon that has very little iron. Many of the Aquods have cemented spodic horizons.

Aquods have developed mainly in sandy materials of Pleistocene age and may have any temperature regime. Native vegetation ranges from spaghnum in cold places to palms in the tropics. About 5.1 percent of the soils in the United States are Spodosols, 4.4 percent are Orthods, and only 0.7 percent are Aquods. The soils were called Ground-Water Podzols in earlier classification systems in the United States.

Haplaquods of Florida and the Atlantic Coastal Plain

Haplaquods are the dominant soils in a few small areas of the Atlantic coastal plain in the Carolinas and Georgia. About 99 percent of the Aquods in the southeastern United States, however, occur in Florida, where soil associations dominated by Aquods make up about 33 percent of the state, as shown by areas S1a in Fig. 2-1. The major associated soils are Psammaquents and Humaquepts.

The Haplaquods have developed on the nearly level coastal plain in sandy sediments at low elevation. The natural water table is near the soil surface after heavy rains in the summer and occurs at a depth of about 125 centimeters during the dry spring months. Most of the soils occur in the flatwoods and are covered with pine forests that include false palmetto in the understory (see Fig. 9-11).

The thermic-hyperthermic temperature line cuts across the Florida peninsula

Figure 9-11   Typical pine and saw palmetto vegetation on the flatwoods where Aquods are dominant soils. Note wetness in foreground of roadside ditch indicative of shallow water table.

near the mainland. Leon sand is the dominant series with thermic soil temperature regime; the Myakka series has a hyperthermic soil temperature regime. The Leon and Myakka soils have spodic horizons considerably enriched with humus (B2h); they are slightly cemented and have less permeability than the horizons above and below them. The soils are devoid or nearly devoid of iron, and carbon and aluminum are the major active components of the spodic horizons. A seasonal high water table tends to cause reduction and removal of iron. Few weatherable minerals exist in the soil to supply plant nutrients. The low cobalt content of these soils is related to both low cobalt content of parent material and intensive leaching.

Much of the land is in open pine forest and has an understory of grasses and shrubs; most of it is used for cattle range. A considerable area with a hyperthermic temperature regime is intensively managed for winter vegetables and improved pasture. The choicest citrus lands are the well-drained deep sands (Quartzipsamments), but competition between agriculture and urban uses has caused more soils with aquic moisture regimes to be used for citrus and other crops. Citrus on Aquods require deep drainage to depths of 1 to 1½ meters and the planting of trees on beds to provide adequate rooting space above fluctuating water tables.

Drainage is essential for agricultural use of the land in order to remove excess water during wet seasons. The normally shallow water table and high soil permeability make Aquods well suited for subirrigation. Drainage ditches or subsoil drains can be used to remove excess water in wet seasons. Subirrigation in dry seasons is by control of water level in drainage ditches. Aquods of Florida are perhaps the best example of the use of subirrigation in the United States.

Lime and fertilizers are needed, depending on the crop. A recent census showed that St. Johns County, Florida, used the most fertilizer per hectare of cropland of any county in the United States. This high fertilizer usage reflects both the infertility of soils and intensive management. Sand soils and intensive rains mean frequent fertilizer application. Low phosphorus fixation results in leaching of soluble phosphorus fertilizers and the need for insoluble phosphorus fertilizer for infrequent forest application.

Geraldson (1962) developed a system using plastic mulch to reduce the frequency of fertilizer application and to create a more nearly constant root environment for vegetable crops during the winter growing season. Beds are formed, heavily fertilized, and covered with a sheet of plastic. The plants are planted in slits in the plastic. Maintenance of the water table at a constant level provides a nearly constant soil moisture tension in the root zone throughout the season. Fertilizer diffuses out from the fertilizer bands and keeps a nearly constant high fertility level in the root zone. Tomato and cucumber yields have been increased 20 to 30 percent. The plastic mulch also makes fumigants more effective for disease control and reduces the amount of virgin land that must be cleared and converted into "new" vegetable lands each year to achieve disease control.

Sideraquods of the California Coast

Along the California coast near Mendocino are about 20 irregular patches of cane-like cypresses that compose a pygmy forest growing on Spodosols. These areas are surrounded by luxurious giant hardwoods and Douglas firs. This great contrast in forest communities has been an enigma for many years. The larger forest trees are growing on younger soils with higher nutrient level that are thick and have no root-restricting layers. The pygmy forests are growing on Spodosols that are extremely weathered and have some of the lowest pH values recorded for any soil horizons. The pH of the surface soils is 2.8 to 3.9, and the soils are low in exchangeable cations, nitrogen, and phosphorus. The soil has a 35-centimeter-thick white A2 horizon that rests on a cemented iron pan (Birm). The spodic horizons are relatively rich in iron and have a ratio of 0.2 or more for free iron (dithionite-citrate) and carbon. The soils are Sideraquods (Gr., *sideros,* iron) and are rare in the United States. Normally, there must be some outside source of iron, since reduction and removal of iron tend to occur under conditions of high water table. A perched water table occurs above the iron pan in the wet season. Relationships of vegetation, topographic position, and soils are shown in Fig. 9-12.

The soils and pygmy forest represent an ecosystem that comes about as close to a terminal steady-state system with balanced inputs and outputs as can be expected in nature. For this reason the soils are of great scientific interest, even though the area of the soils is small. In fact, the area has been made into a national monument to preserve the ecosystem for posterity.

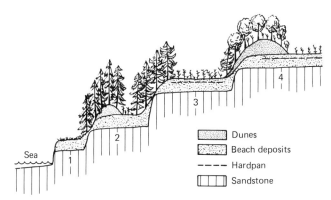

Figure 9-12 Schematic arrangement of soils and vegetation on four wave-cut terraces. Fresh dune sand blown up by wind continues to add nutrients to 1 and 2. Pygmy forest exists on 3 and 4, where Sideraquods occur. (Adapted from H. Jenny, R. J. Arkley, and A. M. Schultz, 1969, *Madrono, 20:*62.)

Tropaquods of Hawaii

Tropaquods are Aquods of intertropical areas. Tropaquods of Hawaii, like the Sideraquods of California, are of interest because of their uniqueness rather than their great extent. Only one series is recognized in the United States, Waialeale, which is a Histic Lithic Tropaquod. The soils are unique because they are Spodosols that have developed in the weathered residuum of basalt. They occur near Mount Waialeale in central Kauai, where the highest annual precipitation in the world was recorded. The associated Alakai soils are Oxisols.

The typical profile consists of a 5- to 10-centimeter-thick layer of mucky peat that overlies an albic (A2g) horizon up to about 10 centimeters thick that rests on a brown to reddish-brown Bir with a weak, fine subangular structure and a gravelly, silty, clay loam texture. Saprolite occurs at depth of 25 to 60 centimeters. The soils are used for water supply and wildlife.

Humods

Humods are not extensive soils in the United States. They are known to occur in the Adirondack Mountains and in Alaska. The soils are associated with coniferous forests at high latitude or high altitude. Spodosols with spodic horizons high in carbon and low in iron occur in the cool to cold perhumid climates in the Maritime Provinces of Canada. In Canada parent materials are low in iron, and vegetation is usually heath or includes heath in the understory. In southern Alaska Humods occur on volcanic ash, although not all Spodosols formed from ash are Humods. It seems that the weathering of the ash releases relatively more aluminum than iron and is a factor in their formation.

References

Alfred, S. D., and A. G. Hyde, *Soil Survey of Charlevoix County, Michigan,* USDA and Mich. Agr. Exp. Sta., Washington, D.C., 1974.

Allaway, W. H., "The Effect of Soils and Fertilizers on Human and Animal Health," *USDA Information Bulletin 378,* 1975.

Arno, J. R., *Soil Survey of Aroostook County, Maine Northeastern Part,* USDA and M. Agr. Exp. Sta., Washington, D.C., 1964.

Austin, M. E., *Land Resource Regions and Major Land Use Resource Regions of the United States,* USDA Handbook 296, Washington, D.C., 1965.

Ballard, R., and J. G. A. Fiskell, "Phosphorus Retention in Coastal Plain Forest Soils: I. Relationship of Soil Properties," *Soil Sci. Soc. Am. Proc., 38:*250–255, 1974.

Ballard, R., and W. L. Pritchett, "Phosphorus Retention in Coastal Plain Forest Soils: II. Significance to Forest Fertilization," *Soil Sci. Soc. Am. Proc.,* 38:363–366, 1974.

Beal, W. J., "Experiments and Observations on the Jack-Pine Plains," *Mich. Arg. Exp. Sta. Bull. 54,* 1889.

Bradford, W. *Of Plymouth Planatation, The Pilgrims in America,* Selected and edited with an introduction by H. Wish, Capricorn, New York, 1962.

Buol, S. W., Ed., "Soils of the Southern States and Puerto Rico," *Southern Coop. Series Bull. 174,* Agr. Exp. Sta. of Southern States and Puerto Rico Land Grant Universities and USDA, 1973.

Buol, S. W., F. D. Hole, and R. J. McCracken, *Soil Genesis and Classification,* Iowa State Press, Ames, 1973.

Ceci, Lynn, "Fish Fertilizer: A Native North American Practice?," *Science,* 188:26–29, 1975.

Clayton, J. S., W. A. Ehrlich, D. B. Cann, J. H. Day, and I. B. Marshall, *Soils of Canada,* Canada Department of Agriculture, Ottawa, 1977.

Danhof, C. H., *Change in Agriculture: The Northern States, 1820–1870,* Harvard University Press, Cambridge, Mass., 1969.

Day, C. A., *Farming in Maine 1860–1940,* University of Maine Press, Orono, 1963.

De Kimpe, C. R., and Y. A. Martel, "Effects of Vegetation on the Distribution of Carbon, Iron and Aluminum in the B Horizons of Northern Appalachian Spodosols," *Soil Sci. Soc. Am. Jour.,* 40:77–80, 1976.

Foote, D. E., E. Hill, S. Nakamura, and F. Stephens, *Soil Survey of Islands of Kauai, Oahu, Maui, Molokai, and Lanai, State of Hawaii,* USDA and Univ. Hawaii Agr. Exp. Sta., Washington, D.C., 1972.

Franzmeier, D. P., and E. P. Whiteside, "A Chronosequence of Podzols in Northern Michigan," *Mich. Agr. Exp. Sta. Quart. Bull.,* 46:2–36, 1963.

Furman, A. L., H. O. White, O. E. Cruz, W. E. Russell, and B. P. Thomas, *Soil Survey of Lake County Area, Florida,* USDA and Fla. Agr. Exp. Sta., Washington, D.C., 1975.

Geraldson, G. M., "Growing Tomatoes and Cucumbers with High Analysis Fertilizer Using Plastic Mulch," *Proc. Fla. State Hort. Soc.,* 75:253–260, 1962.

Geraldson, C. M., A. J. Overman, and J. P. Jones, "Combination of High Analysis Fertilizers, Plastic Mulch and Fumigation for Tomato Production on Old Agricultural Land," *Soil and Crop Sci. Soc. of Fla.,* 25:18–24, 1965.

Haystead, L., and G. C. Fite, *The Agricultural Regions of the United States,* Oklahoma University Press, Norman, 1955.

Hole, F. D., *Soils of Wisconsin,* University of Wisconsin Press, Madison, 1976.

Holzhey, C. S., R. B. Daniels, and E. E. Gamble, "Thick Bh Horizons in the North Carolina Coastal Plain: II. Physical and Chemical Properties and Rates of Organic Additions from Surface Sources," *Soil Sci. Soc. Am. Proc., 39:*1182–1187, 1975.

Jenny, H., R. J. Arkley, and A. M. Schultz, "The Pygmy Forest-Podzol Ecosystem and Its Dune Associates on the Mendocino Coast," *Madrono, 20:*60–74, 1969.

Kedzie, R. C., "Early Amber Cane as a Forage Crop," *Mich. Agr. Exp. Sta. Bull. 1,* 1883.

Kubota, J., "Cobalt Content of New England Soils in Relation to Cobalt Levels in Forages for Ruminant Animals," *Soil Sci. Soc. Am. Proc., 28:*246–251, 1964.

Kubota, J., "Distribution of Cobalt Deficiency in Grazing Animals in Relation to Soils and Forage Plants of the United States," *Soil Sci., 106:*122–130, 1968.

Messenger, A. S., E. P. Whiteside, and A. R. Wolcott, "Climate, Time, and Organisms in Relation to Podzol Development in Michigan Sands: I. Site Descriptions and Microbiological Observations," *Soil Sci. Soc. Am. Proc., 36:*633–638, 1972.

National Cooperative Soil Survey of United States, National Soil Survey, Committee of Canada and FAO, *Soil Map of the World, II. North America* UNESCO, Rome, 1975.

North Central Regional Committee on Soil Survey, "Soil Yield Potential," *Crops and Soils,* Madison, Wis., 1970.

North Central Regional Technical Committee on Soil Survey, "Soils of the North Central Region of the United States," *North Central Regional Pub. 76, Bulletin 544,* Univ. Wis. Agr. Exp. Sta., Madison, 1960.

North Central Regional Technical Committee 3, "Productivity of Soils in the North Central Region," *Univ. Ill. Agr. Exp. Sta. Bull., 710 North Central Regional Research Pub. 166,* 1965.

Raman, K. V., and M. M. Mortland, "Amorphous Materials in a Spodosol: Some Mineralogical and Chemical Properties," *Geoderma, 3:*37–43, 1969/1970.

Rieger, S., "Humods in Relation to Volcanic Ash in Southern Alaska," *Soil Sci. Soc. Am. Proc., 38:*347–351, 1974.

Rieger, S., and J. A. DeMent, "Cryorthods of the Cook Inlet-Susitna Lowland, Alaska," *Soil Sci. Soc. Am. Proc., 29:*448–453, 1965.

Russell, H. S., "Indian Corn Cultivation," *Science, 189:*944–946, 1975.

Schoephorster, D. B., *Soil Survey of Matanuska Valley Area, Alaska,* USDA, Washington, D.C., 1968.

Shetron, S. G. "Distribution of Free Iron and Organic Carbon as Related to Available Water in Some Forested Sandy Soils," *Soil Sci. Soc. Am. Proc., 38:*359–362, 1974.

Smith, F. B., R. G. Leighty, R. E. Caldwell, V. W. Carisle, L. G. Thompson, Jr., and T. C. Mathews, "Principal Soil Areas of Florida," *Fla. Agr. Exp. Sta. Bull. 717,* 1973.

Soil Conservation Service, *Classification of Soil Series of the United States,* USDA, Washington, D.C., 1977.

Soil Survey Division of Bureau of Chemistry and Soils, "Soils of the United States," in *Soils and Men,* USDA Yearbook, pp. 1019–1161, Washington, D.C., 1938.

Soil Survey Staff, "Field and Laboratory Data of Some Podzolic Soils of Northeastern United States," *Soil Survey Laboratory Memorandum 1,* USDA, Washington, D.C., 1952.

Soil Survey Staff, *Soil Taxonomy,* USDA Agriculture Handbook 436, Washington, D.C., 1975.

Stobbe, P. C., and J. R. Wright, "Modern Concepts of the Genesis of Podzols," *Soil Sci. Soc. Am. Proc., 23:*161–164, 1959.

Swanson, C. L. W., "The Road to Fertility," *Time,* January 18, 1954.

Titus, H., "The Land Nobody Wanted," *Spec. Bull. Mich. Agr. Exp. Sta., 332,* 1945.

Ugolini, F. C., H. Dawson, and J. Zachara, "Direct Evidence of Particle Migration in the Soil Solution of a Podzol," *Science, 198:*603–605, 1977.

Upham. C. W., *Soil Survey of Plymouth County, Massachusetts,* USDA and Mass. Agr. Exp. Sta., Washington, D.C., 1969.

U.S. Department of Agriculture, U.S. Department of Commerce, U.S. Department of the Interior, and the state of Alaska, *Alaska Conservation Needs Survey (mimeo),* Washington, D.C., 1968.

Veatch, J. O., *Soils and Land of Michigan,* Michigan State College Press, East Lansing, 1953.

Whiteside, E. P., I. F. Schneider, and R. L. Cook, "Soils of Michigan," *Extension Bulletin E-630,* Mich. Agr. Exp. Sta., 1968.

Wildermuth, R., and D. P. Powell, *Soil Survey of Sarasota County, Florida,* USDA and Fla. Agr. Exp. Sta., Washington, D.C., 1959.

Yuan, T. L., "Characteristics of Surface and Spodic Horizons of Some Spodosols," *Soil and Crop Sci. Soc. of Fla., 26:*163–174, 1966.

10

Oxisols of the United States

Central Concept

Oxisols are characterized by extreme weathering of minerals in tropical areas; they do not occur in the continental United States. The soils are mainly on stable land surfaces that are early Pleistocene or older and commonly occur on high places in landscapes where weathered materials could have been deposited. Some of the Oxisols of Puerto Rico began to form 20 million years ago on surfaces that have been essentially stable over this period. Oxisols develop in areas with a humid climate, and their presence in dry areas suggests climatic change since their formation. Present vegetation varies widely and is related to current climate. Although Oxisols tend to be red, many of the red soils of the tropics are not Oxisols. Most of these soils were formerly called Latosols.

Dominant Properties of Oxisols

For the most part Oxisols are nearly featureless and have no marked horizon boundaries. The soils are mixtures of quartz, free oxides, kaolin, and organic matter. Selected properties of a representative Oxisol are shown in Fig. 10-1. The soil

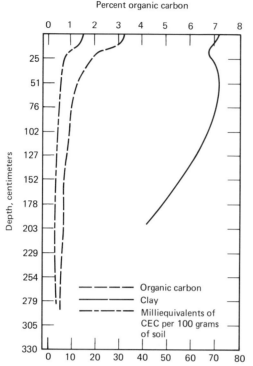

Percent organic carbon

Depth, centimeters

Organic carbon
Clay
Milliequivalents of
CEC per 100 grams
of soil

Percent clay and milliequivalents of CEC per 100 grams of soil

Figure 10-1   Selected properties of an Oxisol from
Puerto Rico. (Adapted from *Soil Taxonomy,* 1975.)

has a relatively high organic carbon or organic matter content; the clay content is
high throughout, but decreases with depth, although there is no distinct clay distribution pattern with depth in Oxisols. The low cation-exchange capacity decreases
with depth because of decreasing organic matter with depth. Since the differences
with depth are gradual, an arbitrary limit of 2 meters is considered the lower
boundary of most Oxisols (see Fig. 10-2). Soils are acid, and aluminum is commonly the most abundant exchangeable cation. Soil evolution resulting in Oxisols
could be represented by the sequence Entisols-Inceptisols-Alfisols-Ultisols-Oxi-
sols. Oxisols usually exist in landscapes where Ultisols occur on younger surfaces
that have been subjected to more erosion.

Properties of Oxic Horizons

The oxic horizons of most Oxisols are subsurface horizons that are intensely weathered and at least 30 centimeters thick. Oxic horizons consist of very insoluble
minerals, including kaolinite, quartz, and hydrated oxides of iron and aluminum.

The fine earth contains no or few minerals that weather to release plant nutrients. The clay fraction has a cation-exchange capacity of 16 milliequivalents or less. Even though the clay content may be high, clay skins are few or absent. Color ranges from red and brown to shades of gray. Where the soils have formed from underlying bedrock, the long period of formation has resulted in obliteration of any rock structure of the saprolite due to root and animal activity. In the field the oxic horizon resembles many cambic horizons.

Relation of Oxisol Properties to Plant Growth

The high content of hydrated oxides, such as iron oxide, results in the formation of a very stable soil structure. Soils tend to consist of fine, stable aggregates that are about the size of sand grains. The soils retain water that is available to plants

Figure 10-2 Oxisol in Hawaii (Oahu) without marked horizon boundaries. Underlying rock is basalt. (Photograph courtesy Dr. W. W. McCall.)

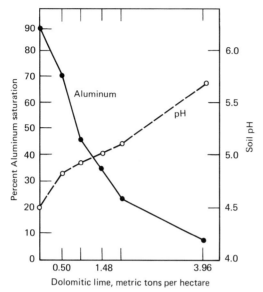

Figure 10-3   Effect of liming on soil pH and exchangeable aluminum saturation in a Brazilian Oxisol. (From "Annual Report of the Tropical Soils Research Program," Soil Science Department, North Carolina State University, 1976.)

mostly at very low tension and, unless there is frequent rain, the soils are droughty. There is rapid permeability, and soils are resistant to erosion on steep slopes. Moist or wet soils can be tilled without puddling.

Many Oxisols have low pH and high exchangeable aluminum that exceeds 90 percent saturation in some cases. Aluminum toxicity is the main problem in using the soils for agriculture. There is, however, great variation in plant response to aluminum in regard to species and variety. The major role of liming is a modest increase in soil pH to about 5 to 5.5, or about 20 to 40 percent exchangeable aluminum (see Fig. 10-3). Liming of some Oxisols also reduces manganese toxicity. Liming above a pH of about 5.5, or overliming, can greatly reduce yields by creating micronutrient deficiencies. Lime is also a source of calcium and magnesium, which crops need but which are deficient in some Oxisols.

Aluminum toxicity in subsoils or oxic horizons restricts rooting depth and reduces drought tolerance of plants. Some crops that can tolerate high aluminum include pineapple, cassava, rice, and cowpeas. For aluminum-sensitive crops the deep placement of lime is beneficial but may not be economically feasible. More soluble forms of lime [$Ca(OH)_2$] are more mobile. Overliming the surface horizon to promote the downward movement of lime is not recommended because the high pH depresses yields. However, soils limed sufficiently to rid the soil of exchangeable aluminum are likely to have significant downward movement of calcium from weakly bonded, pH-dependent sites.

As Oxisols weather, the total phosphorus content of the soil decreases and phosphorus is converted to insoluble iron and aluminum phosphates. Crop response to phosphorus is common and, in many cases, dramatic. Liming very acid Oxisols increases phosphorus availability by increasing organic amatter decomposition and reducing exchangeable aluminum. On high phosphorus fixing soils insoluble fertilizers such as rock phosphate are recommended. As with soil pH and exchangeable aluminum levels, there is considerable plant tolerance to low phosphorus levels related to species and variety. Selecting plants that can tolerate low soil phosphorus can reduce phosphorus fertilizer needs. The low soil phosphorus and hence plant phosphorus content is low, and many native pasture grasses are inadequate for proper growth and reproduction of cattle.

Potassium exists in Oxisols in solution and as an exchangeable cation. Potassium needs of plants are related to the amount of exchangeable potassium. A minimum level of 0.10 milliequivalent per 100 grams of soil has been established for many crops. In addition, the amount of exchangeable potassium should represent 1.5 to 2 percent of the cation-exchange capacity, or 2 to 3 percent of the sum of exchangeable bases. Attention must also be given to the ratios of magnesium and calcium to potassium.

Oxisols of the humid tropics have an organic matter content comparable to soils of the temperate region, and the guidelines for the use of nitrogen fertilizers parallel those of the temperate region. Nitrogen fertilizer use is greatly dependent on type of crop and yield expectation.

Most of the plant nutrients in Oxisols are associated with the organic fraction. In the absence of fertilizers, agriculture is shifting cultivation. On the other hand, the excellent physical properties of Oxisols, combined with irrigation, fertilization, and other production inputs, have converted Oxisols into some of the world's most productive soils. A level of management between shifting cultivation, and intensive management consists of a modest use of fertilizer, minimum tillage, mulching with organic matter, and multiple cropping. Such a program is within the means of some shifting cultivators; it enables the land to be used for continuous cropping and eliminates the need for the long fallows of the shifting cultivation systems.

Plinthite and Laterite in Oxisols

Some Oxisols contained plinthite that has hardened into laterite. The uniqueness of laterite and the dramatic manner in which it has been used for building purposes has likely resulted in an overemphasis of laterite in Oxisols. Where laterite is exposed at the soil surface and limits crop production, it is important locally. Plinthite formation is associated with shallow water tables at some time during the year, and soils with continuous plinthite are in the suborder Aquox.

Oxisols Suborders

Oxisol suborders are determined by moisture and temperature regimes. The five suborders are Aquox, Humox, Orthox, Ustox, and Torrox.

Aquox

The wet Oxisols are in the suborder Aquox. They lie in shallow depressions that are flooded during the rainy season or occur at the base of slopes and receive seepage water. Many Aquox soils at the base of slopes have plinthite at the surface that is recementing transported ironstone fragments, which are one form of laterite. Aquox soils must meet one of two requirements. Some Aquox soils have plinthite, which forms a continuous phase within a depth of 30 centimeters, and are saturated at some time during the year. These Oxisols are not required to have an oxic horizon. The other Aquox soils are also saturated at some time during the year unless they are drained and have either a histic epipedon or an oxic horizon, which have characteristics associated with wetness.

Humox

Humox soils are the Oxisols of relatively cool, humid climates at high altitudes or at relatively high latitudes for Oxisols. Most of these soils have a reddish hue and a high content of organic matter. Humox have 16 kilograms or more of organic carbon per square meter to a depth of 1 meter, excluding surface organic litter. The supply of bases is low, and base saturation is less than 35 percent. As a rule, they are dark colored and have low chroma in the surface horizons. They are rare in the United States but are extensive in South America and Africa.

Orthox

Orthox soils are the Oxisols that have traditionally been associated most with the humid tropical rain forests. These Oxisols have a short dry season or none at all, and they are most common near the equator. They are yellowish to red in color, and high chroma is common. The natural vegetation is a rain forest, although anthropic savanna commonly is the present cover. The soils are very extensive in the Amazon and Congo basins. Orthox may have as much organic matter as Humox soils, but they have higher temperatures.

Ustox

Ustox soils occur in areas with dryer climates than Orthox and have ustic soil moisture regimes. The soils are mostly red and dry in all parts of the moisture control section for extended periods. The soils, however, are moist during at least 90 days of a rainy season and tend to occur near the Tropics of Cancer and Capricorn. Their vegetation is savanna, deciduous forest, or semideciduous forest. They can produce some annual crops without irrigation.

Torrox

Torrox soils have a drier climate than Ustox and have a torric (aridic) soil moisture regime. The soils are too dry for cropping unless they are irrigated. They are dominantly red and apparently are relics preserved from development during an earlier more pluvial period. Their soil temperature regime is believed to be isohyperthermic. The soils of this suborder are very similar, and there are no subgroups.

Oxisols of Hawaii

Hawaii consists of mountainous islands on which about one-half the area is nonsoil and covered with rockland, steep, eroded slopes, or recent sediments. The most abundant soil order is Inceptisol, which reflects the dominance of steep slopes and youthful soils. On the island of Hawaii there are also large areas of ash soils, which are called Andepts. Oxisol is the third most abundant order and covers about 5 percent of the land (see Table 10-1).

Factors Affecting Oxisol Development

The two main factors affecting the occurrence of Oxisols in Hawaii are age and rainfall. Rainfall offshore, over the ocean, averages 60 to 75 centimeters annually. The winds are mainly northeast trades, and rainfall is mainly orographic. The low-

Table 10-1   Abundance of Soil Orders
in Hawaii

Order	Percent Abundance
Inceptisols	15.8
Histosols	14.1
Oxisols	5.14
Mollisols	4.28
Ultisols	2.54
Aridisols	1.19
Entisols	1.1
Vertisols	0.84
Spodosols	0.64
Alfisols	0.2
Nonsoil areas	45.6

Adapted from Mcall, 1973.

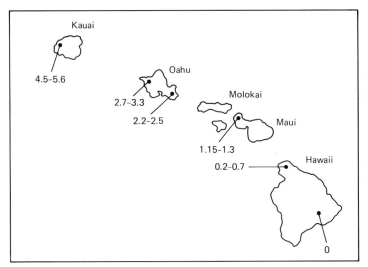

Figure 10-4   Ages of rock as determined by potassium-argon dating in millions of years.

est rainfall is near sea level and increases on windward slopes with increasing elevation to a maximum at 600 to 1200 meters. Rainfall is low in the rain shadow or leeward sides of mountains and changes rapidly within short distances. On Kauai, Mount Waialeale (elevation 1778 meters) is the wettest place on earth, with 835 centimeters of annual rainfall; Mount Waialeale is only 24 kilometers from Barking Sands, whose annual rainfall is less than 50 centimeters.

The age or time available for soil evolution ranges from 0 years on Hawaii, where volcanoes are still active, to about 5 million years on Kauai, as shown in Fig. 10-4. Increasing age from southeast to northwest is associated with the direction of the movement of the earth's plates relative to a "hot spot," or melting spot, in the mantle. The youngest island, Hawaii, has no Oxisols and the next youngest, Maui, has less than 5 percent Oxisols. Oxisols occur on about 18 percent of the land on Oahu and 22 percent on Kauai, the oldest of the four largest islands.

Parent material for Oxisols is quite uniform from island to island and is composed of thin-bedded *pahoehoe* and aa basaltic lavas. The lavas have weathered completely, producing soils that uniformly have a clay texture. Some quartz exists in the surface horizons, which is believed to be of aeolian origin.

Oxisols on Oahu

Oahu was formed by the merging of two mountain ranges. The western Waianae range is older than the eastern Koolau range. The Schofield Plateau lies between the two ranges and is dissected by steep gulches. Dominant soils on the plateau are Oxisols (see Figs. 10-5 and 10-6). The elevation rises from sea level on the north

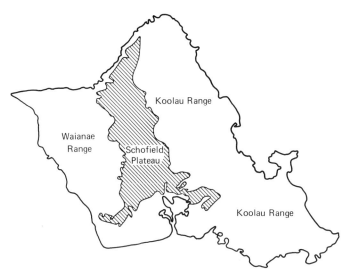

Figure 10-5 The Oxisols of Oahu are on the Schofield Plateau between the Waianae and Koolau ranges.

Figure 10-6 View on Schofield Plateau looking north with Waianae Range along left margin. Wahiawa silty clay, Tropeptic Eutrustox, is dominant soil in the landscape used for pineapples.

and south coasts and attains an elevation of about 366 meters in the saddle. Wahiawa are the dominant soils; they developed from residum and old alluvium derived
from basic igneous rocks (see Fig. 10-2). The elevation ranges from 150 to 366
meters, and rainfall ranges from 60 to 150 centimeters annually. The soils are in
the rain shadow of the Koolau range, have an ustic moisture regime, and are Tropeptic Eutrustox; pineapple has been the major land use (see Fig. 10-6). The clay
fraction of the oxic horizon has a cation-exchange capacity exceeding 1.5 milliequivalents per 100 grams, and base saturation is 50 percent or more. Pineapple
needs about 45 centimeters of water a year and can produce well on the soil without irrigation; sugarcane is grown with irrigation.

At lower elevations on the Schofield Plateau and at lower rainfall the soil moisture regime is torric, and Torrox soils such as the Molokai occur. At lower elevations there is more sunlight (less cloudiness) and higher temperatures but less rainfall; irrigated sugarcane is the major agricultural crop. During the early years of
agricultural development, pineapple was grown on land where irrigation water was
not available. Sugarcane requires over 150 centimeters of water per year and is
generally grown at lower elevations with gravity flow irrigation. The water use is
about 2000 grams of water for each gram of sugar produced.

Over 80 percent of the population of Hawaii live on Oahu, which has a population density of over 3.1 per hectare; migration from the mainland is causing rapid

Figure 10-7 Space is a premium on Oahu, where there is high population density and a
finite and small amount of land well suited for building sites.

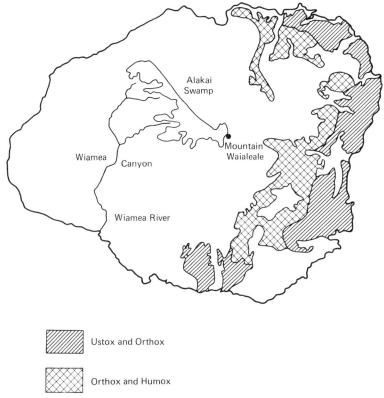

Alakai
Swamp

Wiamea
Canyon

Mountain
Waialeale

Wiamea River

Ustox and Orthox

Orthox and Humox

Figure 10-8 Distribution of Oxisol suborders on Kauai. (After Foote et al., 1972.)

population growth. Pressure for good home building sites has resulted in subdivision development up the undissected portions of the Koolau dome near Honolulu (see Fig. 10-7). The topography and presence of Oxisols on the Schofield plateau is a great attraction for homesites; in a short time the best and most extensive agricultural area of Oahu will be in urban use, and sugarcane and pineapple will be unimportant to the economy of the island.

Oxisols on Kauai

Kauai is the oldest and smallest of the four main islands. Rough, mountainous land dominates the island where the Wiamea Canyon is located. The Alakai Swamp in the center of the island is the remnant of an ancient volcanic dome of the Pliocene. The Oxisols on Kauai are located on the eastern side on windward slopes of the northeast trades (see Fig. 10-8).

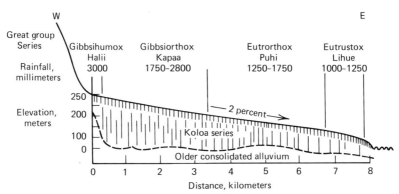

Figure 10-9   Climosequence of soils developed on Koloa volcanic series on Kauai. (Reproduced from *Soil Sci. Soc. Am. Proc.,* 38:129, 1974 by permission of the Soil Science Society of America.)

The area of Oxisols is underlain by the most recent volcanic rocks of Kauai of the Koloa series, which are dated 0.5 to 1 million years old. The volcanic surface was all formed at the same time and slopes gradually down to the sea. Oxisols have formed in this relatively short period of time because of the readily weatherable basic rocks such as olivine and melilite basalts. Rainfall increases with elevation, and the following sequence of Oxisols has been produced: Eutrustox-Eutrorthox-Gibbsiorthox-Gibbsihumox (Fig. 10-9). The formation of Eutrustox is likely related to a paleoclimate with higher rainfall than that of today, and the high base saturation is partially due to splash from the ocean carried by winds to the land.

Ustox (Lihue) and Orthox (Puhi) soils are the dominant Oxisols at the lowest elevation along the coast (see Fig. 10-9). The Lihue (Ustox) has 1000 to 1250 millimeters annual rainfall, and the Puhi (Orthox) has 1250 to 1750 millimeters. Solar insolation is high on both soils, which are well drained and have the highest rating for sugarcane production. Irrigated sugarcane is the major crop, along with pineapple, pasture, truck crops, orchards, wildlife habitat, woodland, and homesites.

At higher elevations there is greater rainfall, less solar insolation, and lower temperature. Leaching and weathering are more intense, and organic matter accumulation is enhanced by increasing rainfall. Kapaa soils (Gibbsiorthox) have 1750 to 2800 millimeters of annual rainfall, and the Halii soils (Gibbsihumox) have over 2800 millimeters. Both soils have cemented sheets or gravel-sized cemented aggregates that are rich in gibbsite. Gibbsite formation in these soils is favored by silica removal under conditions of high rainfall and continuous wet soils that are well-drained and allow free flow of water downslope toward the coast through clinker lava strata. Some of the soils have been assayed for potential use as aluminum ore. The soils are used for sugarcane, pineapple, woodland, wildlife habitat, and water supply. Game birds and wild pigs are the principal wildlife.

In association with the Halii and Kapaa soils are soils of the Pooku series, which are Acrohumox—most weathered Humox. These soils have lost nearly all their capacity to retain cations, and the clays in some part of the oxic horizon have a cation-exchange capacity of 1.5 or less. Nearly all the exchange capacity is associated with the organic fraction. In the lower part of the oxic horizon where the organic matter content is lowest, the oxic horizon typically is positively charged.

Oxisols on Molokai and Lanai

Molokai and Lanai are fifth and sixth in size, respectively, of the Hawaiian Islands. The Molokai-Lahaina soil association makes up 25 percent of the two islands, and the soils are Torrox. There is generally less elevation on these two islands and, consequently, less rainfall. Irrigation water is lacking for sugarcane. The Oxisols are important for pineapple production. The largest pineapple plantation in the world exists on the central plateau of Lanai.

Oxisols of Puerto Rico

Puerto Rico is a volcanic mountainous island in the trade winds belt and is somewhat similar to the older islands of Hawaii. The primary volcanic rocks, however, are much older and more andesitic in composition. Peneplanation of the interior occured and was followed by dissection. Limestone was deposited during Tertiary times on the north and south flanks of the central east-west mountains. Over 70 percent of the island is hills and mountains, and Inceptisols are the most extensive soils. Soil associations in which Oxisols occur exist on 6.7 percent of the land, resulting in about 5 percent or less of the soils in Puerto Rico being Oxisols.

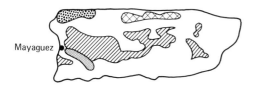

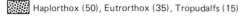

Haplorthox (50), Eutrorthox (35), Tropudalfs (15)

Eutrorthox (50), Paleudults (45), Limestone Rockland (5)

Serpentine Rockland (70), Haplorthox, (25), Acrorthox (5)

Tropohumults (70), Eutropepts (25), Haplorthox (5)

Figure 10-10 Distribution of soil associations with Oxisols on Puerto Rico including percentage of soils in the associations. (Adapted from Soils of the Southern States and Puerto Rico (map) October 1972.)

Oxisol Development and Distribution

Oxisols occur in several distinct situations in Puerto Rico, as shown in Fig. 10-10. There is a discontinuous band along the northern coastal plain. The most extensive Oxisols in the northwestern corner are Haplorthox and further east, extending almost to San Juan, Eutrorthox are most common. These Oxisols have developed in clayey to sandy Quarternary deposits. Oxisols have developed on these relatively recent sediments because the sediments were derived from old and presumably laterized peneplains. Oxisols could develop rapidly in these preweathered, well-drained sediments.

Southeast of Mayaguez in western Puerto Rico is an area of largely serpentine

Figure 10-11   Laterite, hardened plinthite, in the Nipe clay of Puerto Rico. The soil is one of the most weathered soils in the world and is an Acrorthox. Note strong granular soil structure.

rockland where Oxisols exist on the few remaining old undissected surfaces. The Nipe clay, Acrorthox, is an extremely weathered soil that has developed from ultrabasic serpentinite rock of early Cretaceous age (see Fig. 10-11).

Finally, Oxisols occupy about 5 percent of the steeply dissected interior in soil associations where Tropohumults are the dominant soils. The soils occur on moderately sloping, benchlike positions. These soils are Haplorthox.

Land Use on Oxisols in Puerto Rico

The Oxisols of Puerto Rico are mainly used for agriculture. Sugarcane, truck crops, and pasture are the major uses on the more humid soils of the northwest corner. Along the central-northern coast, commercial pineapple production is important. In the interior mountains Oxisols are used mainly for subsistance crops such as plantains, bananas, yams, and taniers. Minor crops are coffee, sugarcane, and tobacco. The Oxisols southeast of Mayaguez are in forest, native pasture, or idle land. The mesas or ridge tops where the Oxisols occur are ideal homesites, and urbanization is rapidly infringing on the Oxisol areas.

References

Alexander, L. T., and J. G. Cady, "Genesis and Hardening of Laterite in Soils," *USDA Tech. Bull. 1282,* Washington, D.C., 1962.

Allen, V. T., and G. D. Sherman. "Genesis of Hawaiian Bauxite," *Econ. Geol. 60:*89–99, 1965.

Bartholomew, W. V., "Soil Nitrogen and Organic Matter," in *Soils of the Humid Tropics,* National Academy of Sciences, Washington, D.C., 1972.

Baver, L. D., "Physical Properties of Soils," in *Soils of the Humid Tropics,* National Academy of Sciences, Washington, D.C., 1972.

Beinroth, F. H., "An Outline of the Geology of Puerto Rico," *Agr. Exp. Sta. Bull. 213,* University of Puerto Rico, 1969.

Beinroth, F. H., "Oxisols-Highly Weathered, Red Soils of the Tropics," *Southern Coop. Series Bull. 174,* USDA and Agricultural Experiment Stations of Southern States and Puerto Rico Land Grant Universities and USDA, 1973.

Beinroth, F. H., G. Uehara, and H. Ikawa, "Geomorphic Relations of Oxisols and Ultisols of Kauai, Hawaii," *Soil Sci. Soc. Am. Proc. 38:*128–131, 1974.

Bornemisma, E., and A. Alvarado, Eds., *Soil Management in Tropical America,* Proceedings of seminar at Cali, Colombia, 1974, North Carolina State University, 1975.

Boyer, J., "Soil Potassium," in *Soils of the Humid Tropics,* National Academy of Sciences, Washington, D.C., 1972.

Buol, S. W., Ed., "Soils of the Southern States and Puerto Rico," *Southern Coop.*

Series Bull. 174, Agricultural Experiment Stations of Southern States and Puerto Rico Land Grant Universities and USDA, 1973.

Buol, S. W., F. D. Hole, and R. J. McCracken, *Soil Genesis and Classification,* Iowa State University Press, Ames, 1973.

Buringh, P., *Introduction to the Study of Soils in Tropical and Subtropical Regions,* Centre for Agricultural Publishing and Documentation, Wageningen, 1968.

Committee on Tropical Soils, National Research Council, *Soils of the Humid Tropics,* National Academy of Sciences, Washington, D.C., 1972.

Coulter, J. K., "Soil Management Systems," in *Soils of the Humid Tropics,* National Academy of Sciences, Washington, D.C., 1972.

Dalrymple, G. B., E. A. Silver, and E. D. Jackson, "Origin of the Hawaiian Islands," *Sci. Am., 61:*294–308, 1973.

Foote, D. E., E. Hill, S. Nakamura, and F. Stephens, *Soil Survey of the Islands of Kauai, Oahu, Maui, Molokai, and Lanai, State of Hawaii,* USDA and Univ. Hawaii Agr. Exp. Sta., Washington, D.C., 1972.

Gierbolini, R. E., *Soil Survey of Mayaguez Area of Western Puerto Rico,* Washington, D.C., 1975.

Kamprath, E. J., "Soil Acidity and Liming," in *Soils of the Humid Tropics,* National Academy of Sciences, Washington, D.C., 1972.

Martini, J. A., R. A. Kochmann, O. J. Siqueira, and C. M. Borkert, "Response of Soybeans to Liming as Related to Soil Acidity, Al and Mn Toxicities, and P in Some Oxisols of Brazil," *Soil Sci. Soc. Am. Proc., 38:*616–620, 1974.

McCall, W. W., "Agriculture in Hawaii," *Ext. Leaflet 160,* University of Hawaii, 1972.

McCall, W. W., "Soil Classification in Hawaii," *Univ. Hawaii Coop. Ext. Circular 476,* 1973.

McNeil, M., "Lateritic Soils," *Sci. Am.,* November 1964, pp. 97–102.

Olson, R. A., and O. P. Engelstad, "Soil Phosphorus and Sulfur," in *Soils of the Humid Tropics,* National Academy of Sciences, Washington, D.C., 1972.

Ruhe, R. V., "Relation of Fluctuations of Sea Level to Soil Genesis in the Quarternary," *Soil Sci., 99:*23–29, 1965.

Sanchez, P. A., *Properties and Management of Soils in the Tropics,* Wiley, New York, 1976.

Sanchez, P. A., ed., "A Review of Soils Research in Tropical Latin America," *Tech. Bull. 219,* North Carolina Agricultural Experiment Station, 1973.

Sato, H. H., et al., *Soil Survey of the Island of Hawaii, State of Hawaii,* USDA and Hawaii Agricultural Experiment Station, Washington, D.C., 1973.

Sherman, G. D., and H. Ikawa, "Soil Sequences in the Hawaiian Islands," *Pacific Sci., 22:*458–464, 1968.

Soares, W., V. E. Lobato, E. Gonzalez, and G. C. Naderman, Jr., "Liming Soils of the Brazilian Cerrado," in *Soil Management in Tropical America,* Proceedings of seminar at Cali, Colombia, 1974, North Carolina State University, 1975.

Soil Science Department, "Agronomic-Economic Research on Tropical Soils," *Annual Report,* North Carolina State University, 1975.

Soil Survey Staff, *Soil Taxonomy,* USDA Agriculture Handbook 436, Washington, D.C., 1975.

Uehara, G., H. Ikawa, and H. H. Sato, "Guide to Hawaii Soils," *Hawaii Ag. Exp. Sta. Misc. Pub. 83,* 1971.

Vageler, P., *An Introduction to Tropical Soils* (translated from the 1930 German edition by H. Greene), Macmillan, London, 1933.

11

Histosols of the United States

Central Concept

Histosols are the "tissue" or organic soils. Histosols form where the production of organic matter exceeds mineralization over long periods of time. Usually the limited mineralization of organic matter is due to water-saturated conditions but, in some cases, blanket peats occur on the uplands, as in Ireland, where the climate is perhumid and cool along the coast.

Genesis and Properties

Histosols are excellent preservators of animal as well as plant materials. Human remains in mineral soils generally consist only of scattered bones or skeletons but, in the case of organic soils, nearly perfect preservation may occur. The bristle of the beard and skin pores were clearly discernible on a 2000-year-old human body exhumed from a peat bog in Denmark in 1950. The last meal that the person ate was analyzed and was found to contain a porridge made from grains and seeds and parts of both cultivated and native wild plants. About 2 meters of organic soil had formed over the body since burial; the soil had accumulated at the rate of about 1 centimeter every 10 years.

Genesis of Histosols

Most Histosols develop where the soil is saturated from at least 1 month each year to continuous saturation. The characteristics of Histosols depend primarily on the nature of the vegetation that was deposited in the water and the degree of decomposition. In relatively deep water the remains of algae and other aquatic plants give rise to highly colloidal material that shrinks greatly on drying. As the lake fills, rushes, wild rice, water lilies, and similar plants flourish. Gradually, sedges, reeds and eventually grasses are able to grow. Peat from such plants is much more fibrous than that produced from plants that grow in deeper water. Shrubs and trees follow in time and produce a woody peat. Changes in water depth may cause a recurrence of deeper-water plants; therefore layers of more pulpy material may occur over fibrous peat and the like.

Changes in the water level are correlated with climatic changes that cause vegetation changes. Pollen from surrounding vegetation drift into the bog areas and are preserved. A study of the changes in pollen provides the basis for interpreting the past climate and vegetation during genesis of Histosols. The peat profile shown in Fig. 11-1 developed in Sweden and shows three major periods of climate change that occurred since glaciation. The upper layer, composed of sphagnum peat, formed during the past 2400 years in a wet and cool period when spruce was abundant in the area. Pine stumps in the next lower layer indicate that the water level in the bog area was low and that pine trees were able to grow. The period was dry and warm and lasted about 2000 years. This period was preceded by another cool and wet period of about 3500 years, and spaghnum peat accumulated. The lake mud was deposited during a time when willow and birch were abundant and conditions were quite cool just after or near the end of glaciation. During this period, pine increased and birch and willow declined as the temperature increased. The soil at the bottom of the Histosol is 7000 years old and accumulated at a rate of 1 centimeter every 15 years. The range of accumulation for many Histosols is 1 centimeter every 7 to 23 years.

Histosols can form in virtually any climate, even in arid regions, as long as water is available. Some are underlain by permafrost and some are on the equator. They may be closed depressions, in coastal marshes, or on steep slopes where there is seepage, or they may blanket a dissected landscape. The common feature is water, which may come from any source. The blanket bogs are present only in cool and very humid climates, but no other climatic feature is necessary for the presence or

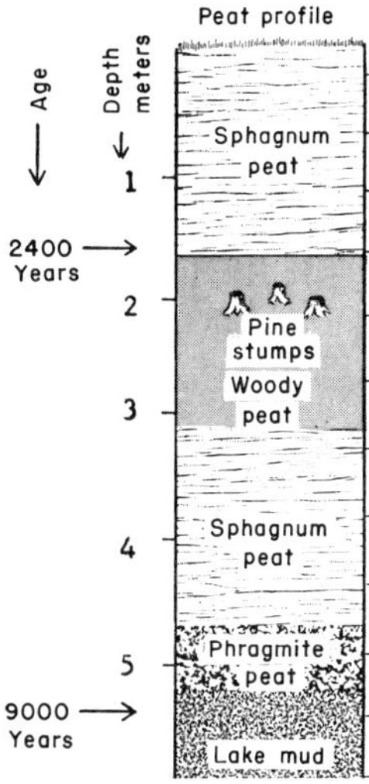

Percent pollen count

Peat profile	Willow	Birch	Pine	Alder	Linden oak elm	Spruce
	0	10	56	3	0	31
Sphagnum peat	0	20	42	8	2	28
	0	33	43	8	5	11
	0	25	53	12	8	2
Pine stumps / Woody peat	0	28	43	15	11	1
	0	32	40	18	9	0
	0	30	41	19	10	0
Sphagnum peat	0	34	30	16	20	1
	0	31	40	13	16	0
Phragmite peat	1	49	37	10	3	0
Lake mud	6	62	32	0	0	0

Figure 11-1 Stratification of peat soil in Sweden caused by climatic change accompanied by vegetation changes. Pollen analysis indicates that sphagnum was deposited during cool, wet periods; the pine stumps indicate a warm, dry period. (Adapted from Davis and Lucas, 1959.)

absence of Histosols. Vegetation consists of a very wide variety of water-loving or water-tolerant plants. The extreme is the soil that may be only an organic mat that floats on water.

Dominant Properties of Histosols

Organic soil materials are defined on the basis of degree of water saturation. Organic soil materials saturated with water for long periods of time, unless they are drained, have:

1. Eighteen percent or more organic carbon if the mineral fraction is 60 percent or more clay.
2. Twelve percent or more organic carbon if the mineral fraction has no clay.
3. A proportional content of organic carbon dependent on clay content of mineral fraction between 0 and 60 percent.

Some organic soils or Histosols include soils where the soil is rarely if ever water saturated, and these soils must have 20 percent or more organic carbon. Some of these soils are quite shallow to rock.

Figure 11-2 Lightness of Histosols makes the soils ideal for sod production where large quantities of soil are moved great distances. Windbreaks of trees to control wind erosion are common in Histosol landscapes used for agriculture.

One of the most outstanding characteristics of Histosols is their generally low bulk density. Most have a bulk density less than 0.5 grams per cubic centimeter, or less than 0.1. The lightness of the soil makes Histosols ideal for grass sod production (see Fig. 11-2). Water-holding capacity on a weight basis is very high, ranging from 300 to 3000 percent. Large changes in volume are associated with wetting and drying.

The pH is a function of the nature of the water that flowed into the bog area and the vegetation. Histosols in landscapes dominanted by acid, sandy Spodosols are typically very acid and have a pH of 4.5 or less; they are classified as low-lime peats. Dominant vegetation includes sphagnum moss, leather leaf, cranberry, Laborador tea, black spruce, tamarack, and swamp blueberry. High-lime peats tend to form where drainage waters carry in large amounts of bases and have a pH of 4.6 to 7. Characteristic vegetation includes black ash, elm, maple, white cedar, alder, reeds, and rushes. Alkaline peats have a pH above 7 and are dominated by grass vegetation.

The cation-exchange capacity is due to carboxyl, phenolic, and other functional groups and is high compared to that of mineral soils. The cation-exchange capacity is pH dependent and ranges from 10 to 20 milliequivalents at a pH of 3.5 to 4 to 100 to 200 at a pH of 7. The soils are high in nitrogen content, and nitrogen availability for plants is related to the rate of organic matter mineralization. The low content of minerals in Histosols is related to low total potassium and generally low availability of potassium for crops. Micronutrient availability is importantly related to pH, and acid Histosols are not limed above a pH of about 5.5 in order to reduce the likelihood of creating micronutrient deficiencies in crops.

Suborders of Histosols

The Histosol suborders are defined by the moisture regime and degree of decomposition of the organic materials. The degree of decomposition is closely related to bulk density and to the initial subsidence that takes place within 2 or 3 years after drainage. The Histosols were called Bog soils and Lithosols in earlier classification systems.

Fibrists

The Fibrists consist largely of plant remains that are so little decomposed that they are not destroyed by rubbing and their botanic origin can be readily determined. They may consist either of partly decomposed wood, the remains of mosses and herbaceous plants such as grasses, sedges, and papyrus, or a mixture of these materials. The reasons why the plant remains have been preserved probably vary, but the absence of oxygen is perhaps the most important. If the water level fluctuates appreciably within the soil, decomposition rapidly destroys the fibers. Some plant remains seem more resistant to decomposition than others. The wood of cypress (Taxodium distichum), for example, is more resistant than many woods and most grasses and sedges.

The Fibrists tend to have the lowest bulk density and the lowest ash content of the Histosols, although there are some exceptions. Fibrists may have a high ash content, particularly if they receive repeated ashfalls from nearby volcanoes. If there is no source of eolian sediments, however, the ash content normally is low. The bulk density usually is less than 0.1 gram per cubic centimeter.

Hemists

These are primarily Histosols in which the organic materials have been decomposed enough that the botanic origin of as much as two-thirds of the materials cannot be readily determined or the fibers can be largely destroyed by rubbing between the fingers. The bulk density usually is between 0.1 and 0.2 gram per cubic centimeter. Hemists have an aquic or peraquic moisture regime; that is, groundwater is at or very close to the surface nearly all the time unless artifical drainage has been provided. The level of groundwater may fluctuate but seldom drops more than a few centimeters below the surface tier (upper 60 centimeters).

Saprists

The Saprists consist of almost completely decomposed plant remains. The botanic origin of the materials cannot be directly observed for the most part. The soils usually are black, and they tend to have a bulk density of more than 0.2 gram per cubic centimeter.

The Saprists occur in areas where the groundwater table tends to fluctuate within the soil. They consist of the residue that remains after aerobic decomposition of organic matter. Fibric and hemic materials, when drained, decompose to form sapric materials; if they are deeply drained either artifically or naturally, the Fibrists and Hemists are converted to Saprists after some decades.

Folists

These are the more or less freely drained Histosols and are comprised primarily of O horizons derived from leaf litter, twigs, and branches resting on rock or on fragmental materials that consist of gravel, stones, and boulders in which the interstices are partly filled or are filled with organic materials. Plant roots grow only in the organic materials. These soils occur mostly in very humid climates from the tropics to the high latitudes. In the United States they occur mostly in Hawaii and Alaska. There were some Folists in the northern lake states, but they were destroyed almost completely by fires, especially those that followed the first commercial timber cutting in the area.

Horizon Designations for Histosols

The suborders Fibrist, Hemist, and Saprist represent organic materials of increasing degree of decomposition, increasing bulk density, and increased aeration associated with fluctuation of water table. Horizon symbols used to designate the various kinds of organic materials are:

    Oi,   organic layer fibrist in character.
    Oe,   organic layer hemic in character.
    Oa,   organic layer saprist in character.

Limnic layers occur in Histosols; they consist of both mineral and organic materials that are produced by precipitation or through the action of aquatic organisms as diatoms, or they are derived from underwater and floating aquatic plants subsequently modified by aquatic animals. Limnic materials include marl, coprogenous earth, and diatomaceous earth. Marl is light colored and reacts with hydrochloric acid to evolve carbon dioxide and leave disintegrated plant remains. Coprogenous earth is a sedimentary peat layer that contains many fecal pellets of a few hundreths to a few tenths of a millimeter in diameter. Colors are characteristically olive or olive-brown. When moist, the material feels slightly plastic but not sticky; when dried, it forms clods that resist rehydration. Diatomaceous earth is more mineral than organic and is formed by the accumulation of diatom skeletons that can be identified with a 440-power microscope. Limnic layers are designated as follows: marl, Lca; coprogenous earth, Lco; and diatomaceous earth, Ldi.

Histosols of the United States

Histosols make up only 0.50 percent of the land area and are less extensive than any other soil except Oxisols in the United States. Nearly 10 percent of the acreage is in Alaska, where soil wetness is associated with permafrost, which contributes to retention of water within the soils instead of loss by leaching. There are no large areas of Histosols in Alaska shown on the general soil map because the soils occur mainly in landscapes dominated by Cryaquepts. The extent of Histosols in the 10 states with the largest area is shown in Table 11-1. The extent of the Histosol suborders in the United States is as follows: Fibrists 0.2, Hemists 0.2, and Saprists 0.1 percent. The Folists are less than 0.1 percent (see appendix Table 2).

Histosols of Northern Minnesota

Minnesota ranks second after Alaska in area of Histosols in the United States. The map in Fig. 2-1 shows two large areas (H1a) in the northern part of the state. The Histosols are on broad, level areas of wet uplands, on lake plains, and in depressions. The soils are Hemists and, with mean annual soil temperature lower than 8°C, they are Borohemists. Although the upper soil layers are frozen in the winter, the ice melts completely in the summer. The principal associated soils are Udipsamments on the better-drained sites and Psammaquents in the low, wet areas. Most of the vegetation consists of sedges or water-tolerant grasses.

Nearly all of the area is in national forest and is used for recreation and lumbering. Timber production is low because of unfavorable soils. Frost is a serious problem when cultivated crops are grown. Water from lakes and streams is ample, and the many lakes are suited for recreational use, but there has been little development.

Table 11-1   Area of Histosols in the 10 Leading States

State	Acres	Hectares
Alaska	27,000,000	10,935,000
Minnesota	6,377,000	2,582,685
Michigan	4,530,000	1,834,650
Florida	3,000,000	1,215,000
Wisconsin	2,831,000	1,146,555
Louisiana	1,800,000	729,000
North Carolina	1,200,000	486,000
Maine	771,800	312,579
New York	648,000	262,480
Hawaii	486,500	197,033

Adapted from data of William E. McKinzie, Regional Technical Center, USDA, 1972.

Histosols of Louisiana

The Histosols of Louisiana are located in the southern part of the state in marshes on the flat, deltaic plain of the Mississippi River (see area H2a Fig. 2-1). The plain is 320 kilometers wide and developed over the past few thousand years in response to shifts in the site of the Mississippi River. Natural vegetation is either freshwater marsh grasses or salt grasses. The soils are mainly Medihemists and Medisaprists. Principal associated soils are Fluvaquents and Haplaquents at higher elevations on levees. The general elevation of the area varies from sea level to about 2 meters above sea level. The soils are underlain by soft, gray Mississippi River clay. The areas are naturally wet and subject to frequent overflow. Most of the land is unsuited for cultivated crops or pasture. Some areas with mineral soil materials in the profile are in the terric subgroups and will support cattle; they are used for grazing, as shown in Fig. 11-3. These coastal marshlands provide a habitat for many animals and birds. Nearly one-half of the ducks and geese of North America spend winters in the marshes along the Louisiana coast. In areas of open water there are some floating organic soils.

New Orleans was established on mineral soils on the highest elevations in the area of the levees. The population of the city expanded so that in 1846 swamp and marsh soils, Histosols, were diked and the water pumped out. Now many residential and commercial developments are on Histosols and below sea level. Because

Figure 11-3 Cattle grazing on Saprists near New Orleans, Louisiana. The soil is in the terric subgroup and contains sufficient mineral material to support cattle hooves.

Figure 11-4 Subsidence around a house built on Histosol. Note planks to enable car to get into garage and props under the stairs alongside the house.

of the low bearing strength of the organic soils, pilings must be used to support foundations. Many homes have driveways that have subsided due to shrinkage and oxidation of organic matter following drainage. It is common for carports to become unusable because of subsidence of the driveway. Many of these carports are then converted into another room, since both carport and house are built on pilings and resist subsidence. In some residential areas one can approximate the time the houses were built by noting the extent of subsidence around the foundations. Settling of driveways and sidewalks is a common phenomenon in areas with Histosols, as shown in Fig. 11-4.

Histosols of the Florida Everglades

The Everglades consists of an area 64 kilometers wide that surrounds the southern end of Lake Okeechobee and extends southward 160 kilometers. Histosols have developed in a depression underlain by limestone; sandy flatlands and the Big Cypress Swamp are on both the east and west as shown in Fig. 11-5. The Everglades receive water from several rivers, including the Kissimmee, which in essence is a shallow, freshwater lake. The area also receives about 5 centimeters more water as precipitation per year than is evaporated. The natural vegetation is mainly sawgrass; occasional higher elevations have trees, shrubs, and open water areas (see

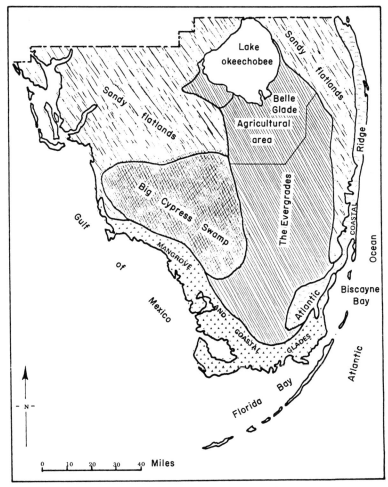

Figure 11-5   Location of the Everglades in southern Florida and surrounding features. (From Beardsley, Casselman, and Volk, 1972.)

Fig. 11-6). The excess water from the inflowing rivers and natural precipitation slowly flows southward through water-saturated soils at about ½ kilometer per day. The excess water eventually reaches the mangrove swamps of the Everglades National Park at the southern tip of Florida.

Histosols of the Everglades   The dominant soils of the Florida Everglades are the Medisaprists or Saprists of the midlatitudes (medi). The soils have formed in a limestone depression that slopes southward from Lake Okeechobee. Soils are deepest, 5 to 6 meters, just south of the lake, and decrease in thickness to about 1 meter in the southern Everglades. Some properties of an Everglades Medisaprist are given in Table 11-2.

The soil was located in the experimental area of the Belle Glade Agriculture and Education Center, which accounts for the higher bulk density in the surface layer as compared to the underlying organic horizons. The soil is highly organic and has high water retention, as measured by percent of oven-dry weight. The pH shows that the soil is only mildly acid, and this is confirmed by the high extractable calcium. Generally, the soil is rich in extractable bases. It is underlain by limestone.

Studies of Histosols near Belle Glade show that 300 centimeters of soil formed in the period 4400 B.P. (before present) to 1914. This results in 6.8 centimeters per century. After drainage in 1914, the subsidence in 50 years was equal to the amount of soil formed in 1200 years, or about 2.5 centimeters every 10 years. In 1970 66 percent of the soil had subsided, and in the year 2000 only 35 centimeters of soil are estimated to be left. Subsidence follows the drainage of Histosols and is the result of oxidation, shrinkage, compaction, erosion, and dehydration. The rate of subsidence in the Everglades is related to the depth of water table, as follows.

$$\text{centimeters subsidence per year} = \frac{\text{depth of water table (centimeters)} - 2.45}{14.77}$$

If the water table is 17.22 centimeters below the soil surface, the subsidence is 1 centimeter per year (see Fig. 11-7).

Land Use in the Everglades   The thickest Histosols occur just south of Lake Okeechobee, where 283,000 hectares have been set aside for agriculture (see Fig. 11-5). The remainder, or two-thirds of the area, is designated for water conservation. The water supply of southern Florida is importantly related to the southerly flow of water through the Everglades.

Figure 11-6 Everglades landscape dominated by sawgrass (Mariscus jamaicensis). Slightly elevated hammocks support trees and shrubs.

Table 11-2 Selected Properties of a Typic Medisaprist, Terra Ceia Muck

Depth, centimeters	Horizons	Color	Bulk Density	Percent Organic Matter	Percent 1/3-Bar Water	pH	Extractable Bases, Milliequivalents per 100 Grams				
							Calcium	Magnesium	Sodium	Potassium	Sum
0–30	Oap	Black	0.36	85	179	5.6	151	13	0.3	0.3	165
30–60	Oa2	Black	0.11	88	627	5.7	170	22	0.5	0.2	193
60–90	Oa3	Black	0.12	84	547	5.4	145	29	0.5	0.1	175
90–120	Oa4	Black	0.16	76	445	5.7	149	35	0.9	0.1	185
120–150	IIC	Black (clay)	—	17	—	7.2	76	12	0.6	0.3	89
150+	IIR	Limestone									

Adapted from data for soil number S71Fla-50-18, USDA.

Figure 11-7 In 1924 the bottom of this 9-foot concrete post was placed on the underlying limestone at Belle Glade and the top of the post was at ground level. This photograph, taken in 1972, shows that subsidence has been about 1 inch per year.

Rainfall occurs mainly in the summer, and winter is the driest season. The agricultural area is intensively used for a wide variety of crops, since the soil temperature regime is hyperthermic (see Fig. 11-8). Palm Beach County is near the center of the Everglades agriculture; descending rank order for dollar income is: sugarcane, winter vegetables, sod, dairy, and beef. The production of sugarcane in the Everglades results in Florida ranking third after Hawaii and Louisiana. It is interesting that most of the sugarcane in Hawaii is grown on Andepts, in Louisiana on Aquepts, and in Florida on Histosols. This highlights how greatly the environment in which a soil is located affects the use of the soil.

The Everglades soils have high nitrogen-supplying power, since the organic matter decomposes and supplies enough nitrogen so that nitrogen fertilizers are not needed. Drainage is obviously needed and results in subsidence. The soils represent a nonrenewable or fund resource that is disappearing at the rate of about 2.5 centimeters per year, resulting in $250 million of agricultural production. On this basis each year's loss of soil is equal to $250 million. The gradual disappearance of agriculture as the hard limestone emerges close to the surface is an important concern for landowners. Perhaps conversion to sod crops that include pasture will enable these lands to remain productive long after they have become too shallow for normal tillage of crops such as sugarcane.

Figure 11-8 Histosols near Lake Okeechobee are intensively used for agriculture with sugarcane as the dominant crop in a plantation agriculture. (USDA photograph)

Histosols of Hawaii

Histosols of Hawaii are of two major kinds, Troposaprists and Tropofolists. Troposaprists occur in depressions in the centers of craters of old volcanoes where water is held and plant materials do not decompose. Still others form at high elevations where cloud cover keeps soils moist and prevents or greatly retards decomposition. The largest area of Troposaprists is in the Alakai Swamp near Mount Waialeale on Kauai. Another area is on top of Mount Kaala on Oahu. These soils are used mainly for water supply and wildlife.

Tropofolists are abundant on the large island of Hawaii on mountain slopes where high rainfall and rain forest vegetation cause large amounts of plant material to accumulate on top of lava rock. The major locations are shown in Fig. 11-9. Histosols are the second most dominant soil in Hawaii because 60 percent of the land in the state is on Hawaii and 16 percent of the land has Histosols. Tropofolists are used for pasture, wildlife, recreation (especially hog hunting), and macadamia, papaya, and coffee orchards. Tree crops are established in the clinkery lava, and the roots find water and nutrients from the fines, which are made up mainly of organic matter and some volcanic ash. In time the organic layer at the soil surface decomposes and the bare lava rock is exposed. Some Folists have also been reported on Whiteface Mountain in New York.

Engineering Problems Associated with the Use of Histosols

The major engineering problem associated with the use of Histosols is compressibility. The compressibility is related to the nature of the organic material, water content, structure, and density.

The kind of engineering problem described next exists for road and highway construction, building foundations, transmission power foundations, airstrips, and

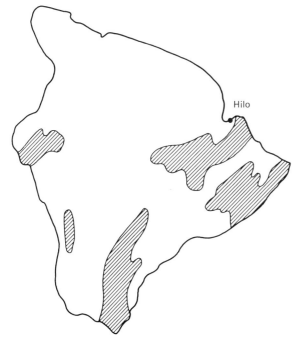

Figure 11-9 Major areas of Tropofolists on the island of Hawaii.

pipelines. Piling is commonly used to give buildings support. Road construction utilizing floating designs has been used in Canada for four-lane expressways. If the area is not large, the organic material can be removed and replaced with sand and gravel.

References

Aandahl, A. R., S. W. Buol, D. E. Hill, and H. H. Bailey, Eds., "Histosols, Their Characteristics, Classification and Use," *Soil Sci. Soc. Am. Spec. Pub. Series, 6,* Madison, Wis., 1974.

Austin, M. E., *Land Resource Regions and Major Land Resource Areas of the United States,* USDA Handbook 296, 1965.

Beardsley, D. W., T. W. Casselman, and B. G. Volk, "Organic Soil Subsidence in the Upper Everglades," *Sunshine State Agr. Res. Report, 17:2,* March–April 1972.

Buol, S. W., F. D. Hole, and R. J. McCracken, *Soil Genesis and Classification,* Iowa State University Press, Ames, 1973.

Carter, L. J., "Agriculture: A New Frontier in Coastal North Carolina," *Science, 189:*271–275, 1975.

Davis, J. F., and R. E. Lucas, "Organic Soils," *Mich. Agr. Exp. Sta. Spec. Bull. 425,* 1959.

Dawson, J. E., "Organic Soils," in *Advances in Agronomy,* Vol. 8, pp. 378–401, Academic, New York, 1956.

Dolman, J. D., and S. W. Buol, "A Study of Organic Soils (Histosols) in the Tidewater Region of North Carolina," *N. C. Agr. Exp. Sta. Tech. Bull. 181,* 1967.

Glob, P. V., "Lifelike Man Preserved 2,000 Years in Peat," *Natl. Geog. 105:*419–430, 1954.

Jones, L. A., "Soils, Geology and Water Control in the Everglades Region," *Fla. Agr. Exp. Sta. Bull. 442,* 1948.

Lucas, R. E., "Peat, Nature's Own Diary," *Crops and Soil, 11,* November 1958.

Lytle, S. A., "The Morphological Characteristics and Relief Relationships of Representative Soils in Louisiana," *La. Agr. Exp. Sta. Bull. 631,* 1968.

Macfarlane, I. C., and G. P. Williams, "Some Engineering Aspects of Peat Soils," *Soil Sci. Soc. Am. Spec. Pub. 6,* pp. 79–93, Madison, Wis., 1974.

McCall, W. W., "Soil Classification in Hawaii," *Univ. Hawaii Coop. Ext. Circular 476,* 1973.

McDowell, L. L., J. C. Stephens, and E. H. Stewart, "Radiocarbon Chronology of the Florida Everglades Peat," *Soil Sci. Soc. Am. Proc., 33:*743–745, 1969.

McKinzie, W. E., "Acreage of Organic Soils in the United States" (mimeo), *Regional Tech. Service Center, Midwest Region, USDA,* 1972.

National Cooperative Soil Survey of United States, National Soil Survey Committee of Canada, and FAO, *Soil Map of the World,* II: North America, Rome, 1975.

Pons, L. J., and W. J. Van Der Molen, "Soil Genesis under Dewatering Regimes During 1000 Years of Polder Development," *Soil Sci., 116:*228–235, 1973.

Slusher, D. F., Chairman Tour Committee, Soil Science Society of America Tour Guide, New Orleans, Louisiana, November 13, 1968.

Slusher, D. F., W. L. Cockerham, and S. D. Mathews, "Mapping and Interpretation of Histosols and Hydraquents for Urban Development," *Soil Sci. Soc. Am. Spec. Pub. 6,* 1974.

Soil Survey Staff, *Soil Taxonomy,* USDA Agriculture Handbook 436, Washington, D.C., 1975.

Soper, E. K., "The Peat Deposits of Minnesota," *Minn. Geol. Sur. Bull. 16,* 1919.

Stephens, J. C., "Drainage of Peat and Muck Lands," in *Water,* USDA Yearbook, pp. 539–557, Washington, D.C., 1955.

Truslow, F. K., and F. G. Vosbourgh, "Threatened Glories of Everglades National Park," *Natl. Geog. 132:*509–533, 1967.

Uehara, G., H. Ikawa, and H. H. Sato, "Guide to Hawaii Soils," *Hawaii Agr. Exp. Sta. Mesc. Pub., 83,* 1971.

Witty, J. E., and R. W. Arnold, "Some Folists on Whiteface Mountain, New York," *Soil Sci. Soc. Am. Proc., 34:*653–657, 1970.

Zelazny, L. W., and V. W. Carlisle, "Physical, Chemical, Elemental and Oxygen-Containing Functional Group Analysis of Selected Florida Histosols," *Soil Sci. Soc. Am. Spec. Pub. 6,* 1974.

12

Canada

Environment and Kinds of Soils

The nature and distribution of soils is affected by the physiography. Canada is dominated by the Canadian Shield, which is a large, massive, old surface of Precambrian crystalline rocks that covers almost one-half of the country (Fig. 12-1). The shield resembles a huge saucer that has a monotonously even erosion surface characteristic of an ancient peneplain. Glacial sediments on the shield tend to be sandy and thin, with extensive rockland areas. The shield is surrounded on all sides except the east by borderlands that consist of plains and lowlands composed of younger rocks. The westward-moving Wisconsin ice moved off the shield and onto the upward-sloping borderland and created extensive pounding of meltwater in front of the ice. Consequently, extensive lacustrine deposits exist on the Interior Plains. Since most of Canada has been glaciated, most soils have developed in glacial related parent materials for about 10,000 years or less. Canadian soils are generally characterized by soil temperature regimes colder than mesic, while U.S. regimes are mainly mesic or warmer and soil moisture regimes vary from subarid to perhumid. Vegetation ranges from short grass prairie to forests to tundra. The northern high Arctic is a polar desert with little biotic activity and large expanses

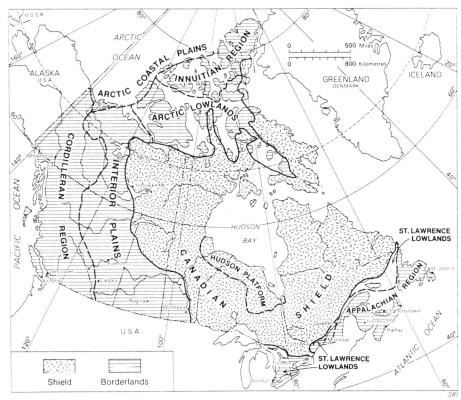

Figure 12-1 Physiographic regions of Canada. (From *Soils of Canada*, 1977, and reproduced by permission of the Minister of Supply and Services, Canada.)

of bare land. These conditions have resulted in a wide range of soils; however, soils comparable to Ultisols and Oxisols are absent. The major soil regions are shown in Fig. 12-2.

Cryosols (soils with permafrost) are the most extensive soils and occur on 40 percent of the land. Podzols are the second most extensive soil and produce most of the commercial forest products. Chernozemic soils make up only 5.1 percent but account for about one-half of the agricultural land and much of the agricultural production. The extent of the soil orders and Rockland is given in Table 12-1. (Their approximate equivalents to the U.S. orders are given in Table 1-11.)

The discussion of Canadian soils is in accordance with terminology used in Canada. Occasionally refer to appendix Table 1 for meanings of horizon symbols and to Chapter 1 for definitions of diagnostic horizons and a simplified key for classification of the soil orders.

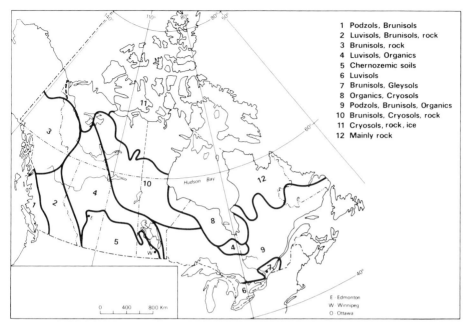

1 Podzols, Brunisols
2 Luvisols, Brunisols, rock
3 Brunisols, rock
4 Luvisols, Organics
5 Chernozemic soils
6 Luvisols
7 Brunisols, Gleysols
8 Organics, Cryosols
9 Podzols, Brunisols, Organics
10 Brunisols, Cryosols, rock
11 Cryosols, rock, ice
12 Mainly rock

Figure 12-2 Major soil regions of Canada. (Courtesy Land Resource Research Institute, Agriculture Canada.)

Table 12-1   Extent of Soils and Rockland in Canada

| | Area | | |
Order	Square Kilometers	Square Miles	Percent of Total Land Area
Brunisolic	789,780	304,855	8.6
Chernozemic	468,190	180,721	5.1
Cryosolic	3,672,080	1,417,431	40.1
Gleysolic	117,143	45,217	1.3
Luvisolic	809,046	312,293	8.8
Organic	373,804	144,286	4.1
Podzolic	1,429,111	551,633	15.6
Regosolic	73,442	28,349	0.8
Solonetzic	72,575	28,014	0.7
(Rockland)	1,375,031	3,543,548	15.0
Total	9,180,202	6,556,347	100.0

From Land Resource Research Institute, Agriculture Canada, Ottawa.

Regosolic Soils

Until 1978 over 90 percent of the soils in the Regosolic order had permafrost, however, these soils are now included in the Cryosolic order. Therefore Regosols are now minor soils and occur on less than 1 percent of the land in Canada (see Table 12-1). They do not occur as a major soil region on the map in Fig. 12-2.

General Features of Regosols

Regosols in Canada are weakly developed as a result of recent formation of parent material, quartzitic nature of sand parent material, instability of soil material, active erosion, or dry-cold soil climate. Regosols are generally well-drained to imperfectly drained soils without strong gleying. They occur under a wide range of climate and vegetation.

Regosols cannot have illuvial horizons and, if a Bm (cambic) horizon is present, it must be less than 5 centimeters thick. Organic horizons at the surface are permitted. If an Ah is present it does not satisfy the requirements of a chernozemic A. In general, the concept of Regosols in Canada relates to the concept of Entisols in the United States, excluding Aquents.

Regosolic Great Groups and Subgroups

There are two great groups in the Regosolic order based on degree of development of the Ah horizon. Humic Regosols have an Ah at least 10 centimeters thick. All other Regosols are in the Regosol great group.

Subgroups are based on evidence such as degree of stability of soil material, periodic deposition, and gleying. Orthic subgroups are the so-called typical or stable members that lack evidence of periodic deposition or gleying. Orthic Regosols may have a C horizon exposed at the surface or may have an Ah less than 10 centimeters thick and organic layers. The C horizon can be acid or calcareous. Orthic Humic Regosols have Ah horizons at least 10 centimeters thick that do not qualify as chernozemic A (see Fig. 12-3). Orthic subgroups may occur on any unconsolidated parent material, but most are common on coarse-textured materials or on actively eroding slopes. Orthic Regosols relate to Orthents and Psamments in the United States.

Cumulic subgroups have one or more buried Ah horizons (see Fig. 12-3). These horizons generally represent a former soil surface buried by fresh material deposited by wind or water. One type of Cumulic Regosol occurs in sand dune areas where wind erosion buries the surface soil and is comparable to Psamments. The other type occurs in alluvial floodplains where soils are subjected to periodic inundation and burial of surface soils by fresh alluvium. These soils are comparable to

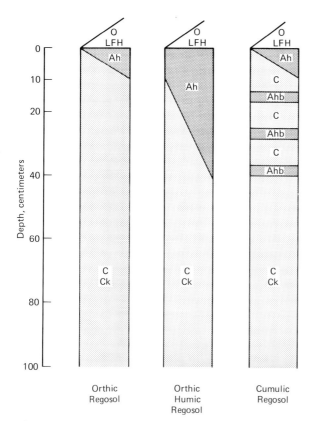

Figure 12-3   Diagrammatic horizon pattern for three subgroups of the Regosolic order. (Adapted from the *Canadian System of Soil Classification,* 1978.)

Fluvents. Gleyed subgroups of Regosols have faint to distinct mottling within 50 centimeters of the soil surface.

Distribution and Use of Regosols

About 1 percent of Canada soils are Orthic Regosols. They occur as dominant map unit areas mainly within the Interior Plains of Manitoba, Saskatchewan, and Alberta. Orthic Regosols are widely distributed as subdominant soils or minor inclusions in many areas of Canada. Most of these areas are in the boreal and cryoboreal climatic regions on coarse-textured glaciofluvial and alluvial deposits and dune deposits. The remainder of the Orthic Regosols occur on eroded slopes and colluvial positions of rolling, morainic landscapes. Over 90 percent of the

Orthic Regosols are eutric in character; the remainder are mainly dystric. A few of the Orthic Regosols are saline.

Productivity of Orthic Regosols for agriculture and forestry depends greatly on texture and topography as well as climate. Some of the more favorable soils are used for rangeland, improved pastures, supplemental crops, and commercial forestry.

Cumulic Regosols occur within the alluvial floodplains of many rivers; it is difficult to estimate their extent because many of the areas are narrow belts along river channels. Cumulic Regosols of high base status, however, occur as three large areas (36,716 square kilometers) on the floodplains of the Mackenzie and Slave river systems. A much smaller area (2975 square kilometers) of low base status and acidic Cumulic Regosols occurs in the tidal marshlands adjacent to the Bay of Fundy in the Maritime Provinces.

Cumulic Regosols occur in a wide range of climates. Parent materials are typical of alluvial deposits ranging in texture from fine sandy loam to clay and noncalcareous or calcareous material. Topography is generally level to very gently undulating. Land use varies with climate and with occurrence of flooding. In the Mackenzie delta the combination of arctic climate and frequent spring flooding limits land use to wildlife habitat. In the Saskatchewan and Slave deltas there is a potential for agriculture based on improved pasture and grazing with supplemental cropping. In the climates warmer than arctic there is potential for commercial forestry, which now occurs on the favored sites. Within the Saskatchewan and Qu'Appelle valleys in Saskatchewan, Cumulic Regosols are used for cropping, improved pasture, and grazing under semiarid to subhumid climate; water from rivers is used in some cases for irrigation. Extensive drainage of the tidal marshlands of New Brunswick has been undertaken for many years; however, the area has been only partially developed for agricultural use. Where Cumulic Regosols are used for agriculture the major problems are flooding and maintenance of adequate drainage. Most of the soils have developed in high base status materials, and fertility is rarely a significant limitation.

Brunisolic Soils

Brunisols are mineral soils that are genetically older than Regosols, but development of diagnostic horizons is minimal. Brunisols are sufficiently developed to exclude them from the Regosolic order but lack the degree of development to be included in some other soil order. The Brunisolic order in Canada approximates the Inceptisol order in the United States.

In many landscapes of the discontinuous permafrost zone Cryosols exist on areas with permafrost and Brunisols occur on areas without permafrost. Before 1978 some soils were classified as Cyric Brunisols and had permafrost. These soils are now classified as Cryosols. As a result, the percentage of Brunisolic soils in Canada was reduced from 10.5 to 8.6.

Central Concept

The central concept of Brunisols is soils under forest vegetation that have brownish-colored Bm horizons. Other colored soils are also included that have Ae horizons and weakly developed Bt and Bf (podzolic) horizons; they are, respectively, Btj and Bfj (j means weak or juvenile). Brunisols may also have Bf horizons if they are less than 10 centimeters thick.

The presence of the brownish-colored horizon indicates some alteration of minerals by weathering but some weatherable minerals still remain. Most Brunisols are well-drained to imperfectly drained and occur in a wide range of climate and vegetation environments. They do not have permafrost and lack evidence of strong gleying.

Brunisol Great Groups

Brunisols are placed into great groups on the basis of pH and thickness of Ah horizon. Melanic and Sombric Brunisols have Ah horizons over 10 centimeters thick that do not qualify as chernozemic. Melanic Brunisols develop in high base parent materials and have a pH of 5.5 or more in all or some part of the B horizon and the underlying material to a depth of either 25 centimeters from the soil surface or a lithic contact shallower than 25 centimeters. The pH requirement for Sombric Brunisols is less than 5.5. Eutric and Dystric Brunisols have Ah horizons less than 10 centimeters thick with pH requirements of 5.5 or more for Eutric Brunisols and less than 5.5 for Dystric Brunisols. A Dystric Brunisol is shown in Fig. 12-4.

Land Use on Brunisols

Brunisols occur on 8.6 percent of the land in Canada and comprise the fourth most abundant order. The soils are associated with almost all major soil groups; however, they have little association with Chernozemic and Solonetzic soils of the Interior Plains. Brunisols are found most frequently in association with Podzolic, Luvisolic soils and in Rockland areas (see areas 1, 2, and 9 of Fig. 12-2). Brunisols are dominant soils in northern British Columbia, the southern Yukon, and the western Northwest Territories. A smaller area dominated by Brunisols occurs surrounding Ottawa. The areas dominated by Brunisols are 3, 7, and 10 of Fig. 12-2.

Less than 1 percent of the Brunisols are cultivated. Agricultural development

Figure 12-4 Dystric Brunisol (thin Ah, acid) developed in thin loess cap over colluvium from metamorphoric rocks in the discontinuous permafrost zone in the Yukon Territory. On opposite north-facing slopes, Cryosols are common.

has been mainly in areas of relatively favorable climate, topography, and texture. Most of the agriculture is on Melanic Brunisols in the boreal and mesic areas of Ontario and Quebec. Cattle production is popular on the shallow Melanic Brunisols developed from limestone materials on Manitoulin Island and the Bruce Peninsula (see Fig. 12-5). Some Sombric Brunisols are cultivated in the lower Fraser Valley and on Vancouver Island, where the climate is mesic.

About 93,302 square kilometers, or 12 percent of the Brunisols, are supporting productive forests. Limiting factors in their use for forestry include severe climate, steep topography, stoniness, and shallowness to bedrock.

Figure 12-5 Melanic Brunisol underlain by limestone on Manitoulin Island.

Cryosolic Soils

The Cryosolic order was created in 1978 to include all soils with permafrost. These soils were previously in the cryic subgroups of other orders. Cryosols are the most extensive soils in Canada because of the large extent of permafrost. Many Cryosols would be Pergelic Cryaquepts in the United States and were formerly called Tundra and Arctic Brown.

Nature and Distribution of Cryosols

Cryosols have permafrost within 1 meter of the soil surface or, if strongly cryoturbated, the permafrost is within 2 meters of the soil surface. Mean annual soil temperature is below 0°C, and both mineral and organic soils are included. Cryosols are the dominant soils in the continuous permafrost zone. South of the continuous permafrost boundary the presence of Cryosols is favored by north-facing slopes, fine-textured and wet parent materials, and in peatlands. The distribution of permafrost is shown in Fig. 12-6. Cryosols occur on 40 percent of the land; the major area is in the north (area 11 of Fig. 12-2). Cryosols are subdominant with Brunisols in area 10 and with Organic soils in area 8.

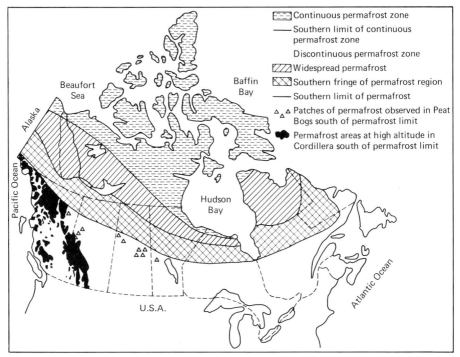

Figure 12-6 Permafrost distribution in Canada. (Used by permission of R. J. E. Brown, National Research Council, Canada.)

Vegetation of the Arctic and Subarctic

Northern Canada is dominated by three vegetation regions: arctic, subarctic, and boreal, as shown in Fig. 12-7. These regions occur as belts that roughly parallel the permafrost zones. Vegetation similar to those of the arctic and subarctic occur as alpine tundra and subalpine in the mountains of western Canada.

The arctic, where permafrost is continuous, is generally characterized by low precipitation and sparse plant growth. There is, however, great variation from the low to high arctic. In the low arctic water is generally sufficient, and the vegetative cover is continuous. Lichens, mosses, and ericaceous species are dominant. Shrubs, mainly willows, occur in sheltered positions. The arctic is considered north of the tree line but, on favored sites in the low arctic, small and scattered very slow-growing trees exist.

The biological potential decreases northward toward the high arctic. The mid-arctic has sparse vegetation, except on poorly drained sites. Precipitation decreases northward, and the most productive sites occur where snowbanks provide protective cover in the winter and a continuous supply of water in the summer. Dry,

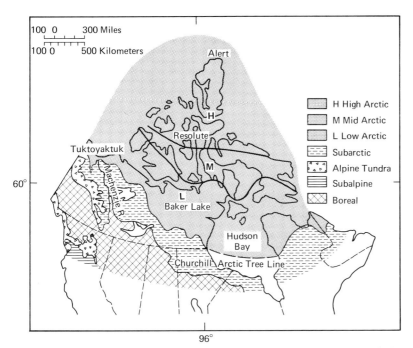

Figure 12-7 Vegetation zones of northern Canada. Cryosols are mainly in the arctic and subarctic. (Reproduced by permission of Charles Tarnocai, from "Distribution of Soils in Northern Canada and Parameters Affecting their Utilization," 11th International Congress Soil Science Symposia Papers, Volume 3, 1978.)

windswept ridges have little vegetation. In the high arctic the vegetation is minimal, and large unvegetated areas occur. Plant distribution is closely related to slope aspect, water supply, and soil texture. Much of the land is a polar desert with considerable Rockland.

South of the tree line in the subarctic vegetation is mainly open lichen-woodland. Dominant trees are black and white spruce, with lesser amounts of jack pine on the better-drained sites and tamarack in the wet areas. Willows and alder are common shrubs. Lichens dominate the ground cover, forming a nearly complete, continuous, thick carpet. The trees are slow growing and unproductive for commercial forests where Cryosols exist.

Cryosolic Great Groups

The Cryosolic order is subdivided into three great groups on the basis of cryoturbation and nature of soil material; they are Turbic Cryosols, Static Cryosols, and Organic Cryosols. Turbic Cryosols are the dominant soils north of the tree line.

Frost action is a major process in soil genesis; it creates strongly cryoturbated soils that contain inclusions of involuted organic matter. Permafrost is within 2 meters of the soil surface. Pattern ground, such as hummocky terrain, is commonly associated with these soils, which have disrupted horizons, as shown in Fig. 12-8. Soils are dominantly mineral, but they contain considerable organic matter that has been mixed into the soil. Frequently an organic-rich horizon exists near the permafrost table. The pedon of Turbic Cryosols includes all elements of the microtopography in cycles less than 7 meters (see Fig. 12-9).

The soil in Fig. 12-9 is a Brunisolic Turbic Cryosol with a Bm horizon over 10 centimeters thick that is continuous over the well-drained portion of the hummock, which is unaffected by frost action. The orthic subgroup includes soils with Bm horizons less than 10 centimeters thick; the regosolic subgroup members do not have Bm horizons but have C horizons on top of the hummock. Gleysolic Turbic Cryosols have evidence of gleying in the form of low chromas or mottling in the mineral surface. For all subgroups a continuous organic layer over the hummocks is common, expecially where the biological potential is relatively high.

Frost action in Turbic Cryosols causes soil heaving and produces trees that lean or tilt in all directions. The trees are predisposed to tilting because of the thin

Figure 12-8 Turbic Cryosol with hummocky terrain, showing downward movement of organic matter in troughs between hummocks. Note continuous plant cover and some small, black spruce trees in the distance at this site in the low arctic.

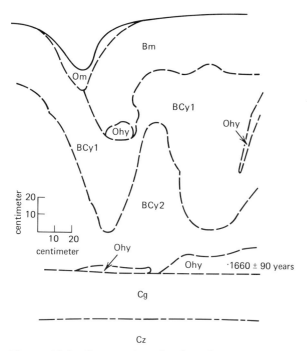

Figure 12-9 Cross section of pedon of (Brunisolic) Turbic Cryosol developed in clayey colluvium. Cryoturbation
is indicated with symbol y. (From *International Soil Science
Society Tour Guidebook for Tour 18,* 1978.)

active layer and shallow rooting depth. The tops of the hummocks are frequently
well drained and too dry for establishment of black spruce seedlings, so the trees
tend to grow on the sides of the hummocks or in troughs and lean away from the
hummocks. Compression (dense) wood is produced by conifers in the direction
that the trees lean. If frost action gradually moves trees to an upright position,
compression wood is no longer produced. Subsequent tilting may again result in
the formation of more compression wood in the direction of tilt. The basal cross
section of a 119-year-old black spruce showed eight distinct leaning periods. The
correlation of compression wood with climate records suggest above average frost
heaving between 1847 and 1943. Compression wood has also been used to study
a variety of events including slope and stream bank erosion.

Static Cryosols are mainly mineral Cryosols that are not strongly cryoturbated
and have permafrost within 1 meter of the soil surface. Static Cryosols develop
mostly from coarse-textured parent materials and are the second most abundant
soils of the arctic. Patterned ground features such as polygons may occur. The
subgroups are similar to those for Turbic Cryosols.

Figure 12-10   Ice wedge polygons, a type of patterned ground, in the depression in center of foreground where Organic Cryosols exist. Note lichen covered (yellowish) hummocks on Turbic Cryosols that surround the Organic Cryosols.

Organic Cryosols are Cryosols that have developed from organic material. The surface layer has more than 17 percent organic carbon by weight (about 30 percent organic matter) and are more than 40 centimeters thick (unless there is a lithic contact). Permafrost occurs within 1 meter of the soil surface, and patterned ground may occur (see Fig. 12-10). The location is typically low arctic, and vegetation is mainly dwarf tundra shrubs. Organic Cryosols are the least extensive Cryosols in the arctic and occur mainly in the low arctic, where water is most abundant. Organic Cryosols are an important component of area 8 in Fig. 12-2; they are located along the southwestern side of Hudson Bay in northern Manitoba and Ontario.

Land Use on Cryosols

Cryosols have shallow root zones determined by the thickness of the active layer and cold climate. The biological potential is low. Some gardens, however, have been grown as far north as 68 degrees north latitude at Inuvik in the Northwest Territories. Good gardens are seen in Dawson, which is located near the subarctic-boreal boundary. The soil is mounded up to hasten soil heating in the spring, and greenhouses are used to start plants before the danger of frost has passed. Some trees growing on favored sites of river valley terraces supplied fuel for steamships and timber for mining during the gold rush of 1898. Typically, few trees grow on

Figure 12-11 Utilidor system for servicing buildings, including the disposal of sewage, in Inuvik, Northwest Territories.

Cryosols; those that do grow slowly and have no commerical value. Some hay was also produced for miners' horses near Dawson during the gold rush. Increased transportation in recent years has reduced the need to grow food locally, and farms in the Yukon decreased from 41 in 1931 to 6 in 1966. There are no major farm operations in the Northwest Territories.

Interest in the development of the mineral and oil resources of northern Canada has produced increased traffic and activity in recent years. Disturbance of the insulative peaty layer on Cryosols results in the melting of the upper permafrost and mobilizes more water than is normal in summer. On hummocky terrain in lichen-black spruce landscapes a subsidence of 50 to 100 centimeters can be expected if the organic surface layer is removed by disturbance or fire. Wet depressions form if the water does not drain away, and the water transports heat to surrounding soils, increasing thermokarst subsidence. Regrowth of the vegetation and development of another organic surface organic layer restore the original insulative properties of the soil and development of the original permafrost table. Restoration is slow and may require 100 years or more.

Roads and buildings are constructed with techniques that prevent melting of permafrost to prevent thermokarst subsidence. For roads a gravel base about 1 meter thick is required, which makes road construction expensive, especially in areas without gravel deposits. Experiments are being conducted with insulative materials to replace the use of some of the gravel.

During 1955 to 1961 a modern arctic town was built at Inuvik along the eastern side of the Mackenzie delta to enlarge school, hospital, and other services for the people in northwestern Canada. A year before the construction of buildings, wooden pilings were placed deep in the permafrost and allowed to freeze solid over one winter. Buildings are constructed on these pilings and elevated above the ground so that heat from the buildings does not melt the permafrost. Elevated construction also permits maximum opportunity for the soil to refreeze each winter. The utilities are distributed with the "utilidor" system, which was also elevated and built on pilings (see Fig. 12-11). In the high arctic there is very little plant cover and lower temperature, and thermokarst formation is less likely to occur.

By 1980 a road will be finished to Inuvik (population 4000) and one can then essentially drive to the Arctic Ocean. There is an excellent airport and jet service. As a result, a tourist industry is developing. The Yukon and the Northwest Territories equal an area about one-half the size of the United States and have an estimated population of 76,600 (1980). The largest city, at Whitehorse, has a population of 13,400. Some of the inhabitants make a living by fishing and trapping.

Podzolic Soils

Podzolic soils occur in all the provinces of Canada and are the second most extensive soils. Podzols have a podzolic B horizon (podzol B), and their development is favored by an intense leaching regime and sandy parent materials. These conditions are generally met on the Canadian Shield in south-central and southeastern Canada (Area 9 in Fig. 12-2). A smaller area dominated by Podzols exists along the western coast (area 1 in Fig. 12-2, in British Columbia, where volcanic ash is an important component of the topsoil in the mountains in the southern part of the province. Podzol soils are comparable to Spodosols in the United States.

Podzolic Great Groups

Podzols have a podzolic B at least 10 centimeters thick that has a texture coarser than clay. The great groups are differentiated on the basis of the characteristics of the podzolic B into Humic Podzol, Ferro-Humic Podzol, and Humo-Ferric Podzol, as shown in Table 12-2.

Humic Podzols have dark-colored Bh horizons high in organic carbon relative to pyrophosphate-extractable iron (see Table 12-2). The soils typically develop in wet areas where vegetation is heath, heath-forest, or sphagnum. Fluctuating shallow water tables favor the development of Bh horizons and, in some cases, the Bh is cemented, ortstein. Humic Podzols also tend to develop in low-iron parent materials and commonly occur as inclusions in larger areas of Ferro-Humic or Humo-Ferric Podzols. Humic Podzols correlate with Cryohumods in the United States.

Ferro-Humic Podzols develop on sites less moist than Humic Podzol sites and

Table 12-2 Characteristics of B Horizons of Podzolic Great Groups

Characteristics of Podzolic B	Humic	Ferro-Humic	Humo-Ferric
Type	Bh	Bhf	Bf
Percent organic carbon	Over 1 percent	Over 5 percent	0.5–5 percent
Pyrophosphate iron	Less 0.3 percent	—	—
Organic carbon pyrophosphate iron ratio	20 or more	—	—
Pyrophosphate iron + aluminum	—	0.6 percent or more (0.4 percent for sands)	0.6 percent or more (0.4 percent for sands)

have Bhf horizons high in organic carbon, but with appreciable pyrophosphate-extractable iron and aluminum. These soils occur mostly on relatively iron rich-parent materials, and their formation is favored by an ash cover that releases abundant iron. Ferro-Humic Podzols are of significant occurrence in the coastal areas of British Columbia, where a thick ground cover of moss exists (see Fig. 12-12). These soils correlate with Humic Cryorthods in the United States.

The overwhelming bulk of Podzols in Canada are Humo-Ferric Podzols with Bf horizons; they contain significant organic carbon but a relatively high content of pyrophosphate-extractable iron and aluminium compared to Podzols of the other great groups. Humo-Ferric Podzols are widely distributed and typically occur

Figure 12-12 Ferro-Humic Podzol developed in black spruce forest with thick ground cover of sphagnum moss. Note thick layer of organic matter overlying the white A2 horizon.

Figure 12-13   Typical boreal forest landscape dominated by Humo-Ferric Podzols developed from sandy glacial materials on the Canadian Shield.

under coniferous, mixed, and deciduous forest vegetation (see Fig. 12-13). A few of these soils have developed under shrubs and grass.

Land Use on Podzolic Soils

Most of the Podzols support productive boreal forests. Forest productivity, however, decreases as the tundra is approached and soils have pergelic temperature regimes. These boreal forests are an important source of paper pulp; Quebec is the leading paper-producing state or province in the world. The value of the finished forest products is equal to about one-half of those in agriculture. Other important land uses include hunting, trapping, and recreation. The use of the soils for agriculture is limited by unfavorable climate, stoniness, and shallow depth to rock. The soils are low in fertility and tend to be droughty. A small acreage of Podzols is farmed in areas of milder climate in southeastern Canada, mostly in Quebec. Pasture, forages, and small grains are the major crops. Locally, vegetables and fruits are important.

Luvisolic Soils

Luvisolic soils, Luvisols, have an illuvial horizon of silicate clay accumulation and correspond to Alfisols in the United States. In older classification systems the soils were called Gray-Brown Podzolic and Gray Wooded. Luvisols are found throughout the forested regions of Canada and occur on about 8.8 percent of the land. These soils are important in both forestry and agriculture.

Environment and Properties of Luvisols

Luvisols are well-drained to imperfectly drained mineral soils that have developed under forest vegetation in mild to cold climates. They have developed mainly from till and associated glacial deposits. Loamy textures predominate, and most parent materials were calcareous. Luvisols must have both an eluvial horizon (Ae) and an illuvial horizon of silicate clay accumulation (Bt). Luvisols intergrade with Brunisolic and Podzolic soils on coarse-textured parent materials. Luvisols intergrade with Chernozemic and Gleysolic soils on the Interior Plains. The soils are dominant in the Cordilleran region of the west (area 2 of Fig. 12-2), adjacent to and north and east of the Chernozems of the Interior Plains (area 4 of Fig. 12-2), and in southern Ontario (area 6 of Fig. 12-2).

Luvisolic Great Groups and Land Use

Luvisols are either Gray-Brown Luvisols (Gray-Brown Podzolic) or Gray Luvisols (Gray Wooded). Their counterparts in the United States are Udalfs and Boralfs, respectively. The subgroups are separated on the basis of kind and sequence of horizons and mean annual soil temperature (MAST). Gray-Brown Luvisols are restricted to areas with a MAST of 8°C or higher (mesic) and a humid or wetter soil moisture regime. As a result, Gray-Brown Luvisols are mainly confined to southern Ontario and the St. Lawrence Valley and make up only about 6 percent of the Luvisols. Undisturbed soils show evidence of high biological activity and mixing by earthworms of organic matter into the upper mineral layer, creating a mull Ah layer. Evidence of disintegration of the upper Bt may exist.

The Gray-Brown Luvisols of southern Ontario are sandwiched between similar soils (Udalfs) in southern Michigan and western New York. The northern boundary in Canada is from Georgian Bay on the west to Kingston near the eastern end of Lake Ontario (area 6 of Fig. 12-2). The soils have a mesic soil temperature regime and udic moisture regime. The original hardwood forest has been mostly removed and the soils intensively cultivated (see Fig. 12-14). Although this region represents less than 1 percent of the land area of Canada, it is one of the major crop-producing areas. A wide variety of crops is grown, including canning and market crops and fruits near Lakes Ontario and Erie. Urban enroachment is a serious problem in the fruit-growing areas because of the limited area suited for growing fruit crops. Minor areas of steeper and poorer soils are retained as natural woodlots or are being reforested.

About 94 percent of the Luvisols of Canada have a MAST less than 8°C (boreal or cryoboreal) and are Gray Luvisols. They occur mainly along the northern and eastern Chernozem border of the Interior Plains, where the climate is subhumid

Figure 12-14 Gray-Brown Luvisol landscape in southern Ontario near Brantford, where tobacco is an important crop. The soils occur mainly on nearly level glacial till plains.

(area 4 of Fig. 12-2). In eastern Canada some Gray Luvisols have humid and wetter soil moisture regimes. The leaching potential on mainly calcareous materials (some of which contain as much as 40 percent calcium carbonate equivalent) has been insufficient to develop Podzols. Base saturation of Gray Luvisols is generally high relative to most other forest soils (see Table 12-3). The cooler climate of Gray Luvisols, compared to Gray-Brown Luvisols, results in surface accumulation of slowly decomposing leaf litter under boreal forest and formation of L, F, and H layers. Ah layers are generally thin or absent. Some of the Gray Luvisols of the grassland-forest transition, however, may have quite dark and thick Ah horizons and intergrade to Dark-Gray Chernozems.

Table 12-3 Percentage Saturation of Cation Exchange Capacity of a Gray Luvisol

Depth, centimeters	Horizon	Percent Saturation				
		Calcium	Magnesium	Potassium	Sodium	Hydrogen
8–0	L-H	68.6	10.8	3.6	0.6	16.4
0–8	Ah	75.5	8.6	3.7	1.6	10.6
8–18	Ae	46.7	31.4	4.8	—	17.1
18–25	AB	68.0	16.9	3.5	0.6	11.0
25–66	Bt	71.9	19.3	2.6	0.9	5.3
66–117	BC	81.9	14.9	2.0	0.8	0.4
117+	Ck	92.4	5.6	1.3	0.7	—

From Soils of Canada, 1978.

Figure 12-15 Mixed productive forest and cultivated Gray Luvisol landscape near the Luvisol-Chernozem border west of Edmonton.

Land use on Gray Luvisols is importantly related to climate. Under subarctic climates, the land is unsuited for agriculture and most of the forests are unproductive. The main use of soils is to maintain wildlife activities. Under warmer climate regimes the soils are naturally suited for sustained growth of productive forests, and commercial forestry is the prime land use. On sites with the most favorable topography and parent materials, the soils are used for agriculture (see Fig. 12-15). Cultivated Gray Luvisols have lower organic matter content than Chernozemic soils and require considerably more fertilizer, especially nitrogen, phosphorus, and sulfur. The growing season is too short for the best quality bread wheat because the cooler and moister weather, compared to areas further south, results in wheat grain with more starch and less protein. Frost is a major problem, and large land preparation and planting equipment is used to minimize the length of time to establish crops so that the period of growth is maximized. Rape is a short-season crop

Table 12-4 Estimated Area of Cultivated and Potentially Arable Land in the Gray Luvisol Regions of Northwestern Canada in Millions of Hectares

	Alberta	Manitoba	Saskatchewan	British Columbia	Northwest Territories	Total	Percent
Total area	36.5	8.1	14.6	8.9	13.0	81.1	100
Cultivated	1.4	0.5	2.2	0.1	—	4.2	5.2
Potentially arable	3.2	0.8	1.2	1.1	0.9	7.2	8.9

Adapted from Ehrlich and Odansky, 1960.

that has had a marked increase in acreage in recent years, much of the output goes to Japan as rapeseed oil. Production is limited to coarse grains, rape, forages, and pasture. Livestock production, especially hogs and cattle, is important. Only about 5 percent of the land is now cultivated, and nearly twice this area is considered arable, as shown in Table 12-4.

Chernozemic Soils

Chernozemic soils in Canada correlate with Borolls in the United States and are the dominant soils on the Interior Plains of Alberta, Manitoba, and Saskatchewan. About one-half of Canada's agricultural land is located on Chernozems (see area 5 of Fig. 12-2). The soils occur in a climate where the soil is usually frozen in winter and dry for some time during each summer. Most of the soils formed in glacial materials. When the ice moved from the Keewatin ice cap on the Canadian Shield over the Interior Plains, it moved over land that sloped upward to the west. This caused extensive pounding of meltwater in front of the ice during glacier retreat, and extensive areas of lacustrine sediments were formed. As a result, many of the Chernozems are high in silt and clay content and occur on nearly level plains. Loess, however, is of minor importance.

Central Concept and Great Groups

The central concept of Chernozemic soils is similar to that of Mollisols—well-drained to imperfectly drained mineral soils of the steppes that have dark-colored surface horizons with well-developed structure. Specifically, Chernozemic soils must have a chernozemic A horizon (defined in Chapter 1). The soils have a MAST of 0°C or higher and a soil moisture class that is drier than humid. Lime accumulations usually occur in the lower part of the solum.

There are four great groups that reflect differences in climate and vegetation. The climate ranges from subarid to subhumid, and the vegetation ranges from xerophytic and mesophytic grasses and forbs to forest-grass parkland. The soil moisture zones form a series of concentric bands centered around the south-central part of the Interior Plains, as shown in Fig. 12-16. The soil differences are reflected in the color of the chernozemic A horizon, as shown in Table 12-5. The great groups represent a series of soils with gradually changing properties along a moisture transect, as shown in Fig. 12-17. The percent of the land area for the great groups is Brown 1.1, Dark Brown 1.2, Black 2.2, and Dark Gray 0.6; this totals 5.1 percent of Canada.

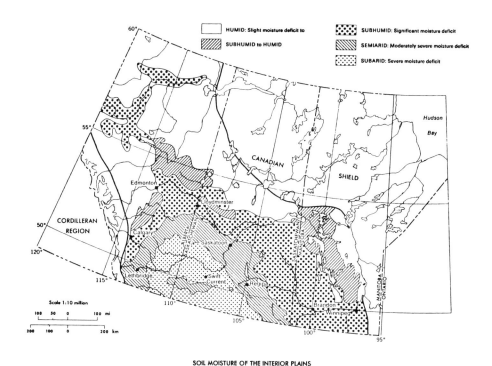

SOIL MOISTURE OF THE INTERIOR PLAINS

Figure 12-16   Soil moisture zones of the Interior Plains. (Courtesy Saskatchewan Institute of Pedology.)

Black and Dark Gray Chernozemic Soils

Black Chernozemic soils are the most extensive Chernozems and develop in a parkland landscape. There is sufficient soil moisture in the low or depressional positions of the landscape to support trees (see Fig. 12-18). Severe droughts are rare, and the soils have more organic matter and darker-colored Ah horizons than Dark Brown Chernozems do. The climate is typically cold and subhumid, and the

Table 12-5   Differentiating Characteristics of Chernozemic Great Groups

Chernozemic A Characteristics	Brown	Dark Brown	Black	Dark Gray
Color value (dry)	4.5–5.5	3.5–4.5	Less than 3.5	Less than 4.5
Chroma	More than 1.5	More than 1.5	1.5 or less	Less than 1.5
Climate	Subarid to semiarid	Semiarid	Subhumid	Subhumid

From The Canadian System of Soil Classification, *1978.*

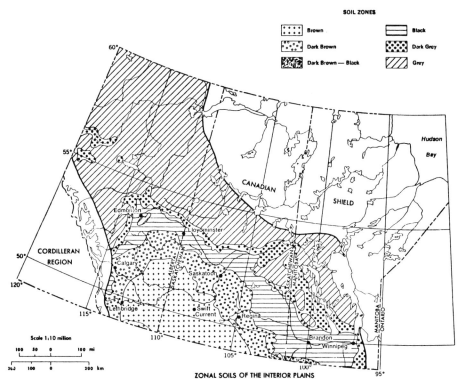

Figure 12-17 Soil zones of the interior plains, including Grey soils of the Luvisolic order. (Courtesy Saskatchewan Institute of Pedology.)

Black soils occur as a band, mainly north and east of Dark Brown Chernozems (see Fig. 12-17). About 2200 square kilometers of Black soils occur outside the Interior Plains within the southern interior of the Cordilleran region in British Columbia.

One of the surprising features of many Black soils is a relatively thin Ah horizon that barely satisfies the minimum 10-centimeter thickness criterion for a chernozemic A, as shown in Fig. 12-19. A similar situation occurs in regard to many Chernozems at similar latitude in the Soviet Union. Perhaps low soil temperature combined with more adequate soil moisture, compared to locations farther south, results in decreased rooting depth and decreased thickness of Ah horizon. In general, there is decreasing calcium carbonate content of the glacial materials from east to west that is associated with decreased abundance of calcareous subgroups and increased development of textural B horizons. The high calcium carbonate content, as high as 40 percent, has likely retarded eluviation and horizonation in many Black soils. Black Chernozemic soils developed from the clayey sediments of glacial Lake

Figure 12-18 Parkland landscape with Black Chernozemic soils developed under grass on slopes and Gleysols in the depressions occupied by aspen and willow.

Figure 12-19 Black Chernozemic soils with thin Ah horizons that increase in thickness toward gleyed soils at the bottom of the slope.

Figure 12-20   Black Chernozem in the Red River Valley. The soil cracks in dry years and dark surface soil material is moved deep into the soil. The soil is "vertic," because slickensides exist in the lower part of the soil.

Agassiz in the Red River Valley have thicker Ah horizons than Blacks on the rolling uplands (see Fig. 12-20). The Red River Valley in the southeastern part of the Interior Plains is a highly productive agricultural region due partly to the more favorable climate in southeastern Manitoba. Flooding, however, is a problem, especially after snowmelt in the spring.

Dark Gray Chernozems have a chernozemic A; however, the soils have characteristics indicative of the eluviation that is typical of soils developed under forest. Dark Gray soils occur in the forest-grassland transition and intergrade to Black Chernozems and Luvisols. The soils usually occur under mixed native vegetation of trees, shrubs, forbs, and grasses and, in some cases, the soils developed under grass but were later enroached by trees. Horizon sequence is typically Ahe, Ae, Bm or Bt, and Cca or Ck. The peds of the Ah show evidence of eluviation by occurrence of gray spots or bands in a darker matrix. The Ae is 5 centimeters or less in thickness. Increased humidity, relative to other Chernozems, results in more frequent Bt horizons. The soils are quite similar to Dark Gray Luvisols, which have thicker Ae horizons.

Under cultivation Dark Gray Chernozems are highly productive and have a fertility similar to that of Black soils. Most of the areas in the Interior Plains have been cleared and are used for cropping. Some areas are used for bushland pasture, and some soils support commercial forests.

Dark Brown and Brown Chernozemic Soils

Dark Brown and Brown Chernozemic soils typically occur in treeless landscapes where trees exist only along watercourses and farmsteads. The soils develop in areas with less annual precipitation than Black soils and have less organic matter content and lighter color. Typical horizon sequence of both Dark Brown and Brown soils is Ah, Bm, Bmk, or Btj, and Cca or Ck. The B horizon commonly has a prismatic structure. Many of the soils developed from lacustrine parent materials and have a high water-holding capacity.

The soils are used almost exclusively for agriculture. Grazing is dominant on the drier and coarser-textured soils. Less than 50 percent of the Brown soils and about 70 percent of the Dark Brown soils are cultivated with wheat and other small grains as the major crops. Frequent and severe droughts occur in the Brown soil zone. A large area of marginally arable land was cropped during early settlement days and has been abandoned. Much of this abandonment occured as a result of prolonged drought and wind erosion during the so-called "dirty thirties." Using tillage practices that leave crop residues on the soil surface to control wind erosion is virtually universal (see Fig. 12-21).

Figure 12-21   Landscape near Brown-Dark Brown border. Note crop residues left on soil surface of fallow field and dugout in foreground for water collection.

Fallowing and The Problem of Saline Seep

Fallowing is a common practice on Chernozems to increase stored water for wheat production. Forty-two percent of the cultivated land in Saskatchewan was fallowed between 1971 and 1976. The Chernozemic soils of Canada (and the Borolls of the northern plains in the United States), however, have lower evapotranspiration than soils farther south on the Great Plains. Thus, in some years of above average precipitation, fallowing results in surplus water that passes through the root zone, moves downslope above an impermeable layer, and seeps out at the surface someplace in the lower part of the landscape. The seepage water contains salts that, in time, result in soil salinity that is aggravated by the sodium content of some soil materials. As much as 16 percent of the land has been salinized in some areas by saline seep, and the problem is becoming more serious. Anything that increases infiltration of water, such as dugouts for water storage and snow accumulation behind a snow fence, is a potential contributor to saline seep. Saline seep control includes reduced frequency of fallowing, increased use of deep-rooted crops such as alfalfa, and use of fertilizers to increase plant vigor, deeper rooting, removal of water by plants.

Fallowing results in considerable increase in stored water and, in addition, an increase in available nitrogen. The combination of increased water and fertility results in "high" wheat yields and mitigates against the acceptance of a reduction in fallowing by farmers.

Land Use as Related to Soils and Climate

The Chernozemic soils of the Interior Plains support one of the world's great wheat-producing regions. Although the soils generally have favorable texture, topography, and fertility, land use is greatly affected by low soil moisture and soil temperature. The southern Brown soil zone has a boreal temperature regime with severe moisture deficit, and grazing is the dominant land use. Spring wheat production is dominant on Dark Brown soils; mixed farming (wheat plus other crops and livestock) is common on the Black and Dark Gray soils, which have more humidity and a cooler climate.

The southern boundary of the Interior Plains is 40 degrees north latitude and a short distance south of Winnipeg. An important compensating factor in crop production in the "north" is the longer days in summer. Winnipeg receives only 60 percent as much annual solar insolation as the equator; however, in the summer Winnipeg receives more insolation each day. The insolation is less intense at Winnipeg than at the equator, but plants can grow more hours per day in the summer at Winnipeg. The short nights also contribute to less heat loss and cooling during the night period. When the temperature is over 5.5°C, grain cereals can grow and do not necessarily need a specific number of days to mature. Instead, the crops must have a minimal period of time above 5.5° that is frost free. Wheat requires about 1100 degree days uninterrupted by frost (if the average of minimum and maximum temperature for a given day was 6.5, it would be equal to 1 degree day.) As one travels north on the Interior Plains, increased latitude has a small effect on wheat production until the degree days approach 1100 or less; then wheat production declines sharply. The northern limit for significant wheat production is in the Peace River area, at about the latitude of southeastern Alaska. Areas farther north can be used for forage and other adapted cool season crops.

Rape requires more moisture than wheat but fewer degree days. The wheat surplus of recent years has greatly stimulated rape production on the northerly Black and Dark Gray Chernozems as well as on the forested soil areas farther north, so that Canada is now the world's leading rape producer. The rape is grown for seed that is pressed into oil and cake or meal. The rapeseed industry was brought to Canada in 1942 to produce the oil needed for the Allied navies. Rapeseed oil was a superior lubricant because it clung to steam-washed surfaces. The development of rapeseed oil for human consumption expanded greatly after World War II. The meal is used for livestock feed.

The chinook winds of southwestern Alberta produce a unique effect on land use in that region. The mild air off the Pacific descends the eastern slopes of the mountains and is dried. The winds cause rapid melting of snow in winter, so livestock can graze fields the entire winter. The chinooks, however, cause increased soil freezing in the absence of a snow cover and greater winter killing of trees and shrubs. The dry, strong winds also increase wind erosion on cultivated land.

Solonetzic Soils

About 0.7 percent of the land in Canada has Solonetzic soils. The soils occur mainly on salinized materials in intermediate drainage positions in association with Chernozemic soils on the Interior Plains. In the United States the soils correlate with natric great groups of Borolls and Boralfs.

Genesis and Properties of Solonetzic Great Groups

The word *Solonetz* was used in Russia to indicate soils that developed under the influence of sodium salts. Solonetzic soils develop in saline parent materials after salinization ceases. The three great groups of Solonetz, Solodized Solonetz, and Solod represent an evolutionary sequence. Solonetic soils may revert to saline soils at any time along the evolution sequence if salinization reoccurs. As long as shallow groundwater allows salts to be brought to the soil surface, the soils remain Saline Regosols.

The presence of soluble salts in saline soils keeps the colloids flocculated. Any factor that causes a lowering of the groundwater table may stop salinization and allow the natural precipitation to begin the process of salt removal by leaching. Removal of salts may result in dispersion of colloids, allowing the dispersed organic and mineral particles to fill voids and reduce permeability. Continued leaching may produce a B horizon that has an exchangeable calcium/sodium ratio less than 10 and a pH as high as 10, that is characteristic of solonetzic B horizons, or Bn horizons. The peptized clay may migrate and increase the clay content of the B horizon, forming a Bnt horizon. Both Bn and Bnt horizons are solonetzic horizons found in Solonetz soils; they are produced by *solonization*. Solonetzic horizons are typically stained a dark color by dispersed organic matter, are hard when dry, and have columnar or prismatic structure.

Continued leaching of Solonetz soils eventually results in a leached and acid A horizon. Eluviation of weathering products and clay produces an Ae, and the AB boundary becomes abrupt. The soil has now evolved into a Solodized Solonetz produced by *solodization* (see Fig. 12-22). Continued solodization encourages phys-

Figure 12-22   Solodized Solonetz in the Dark Brown zone. Note light-colored Ae just at top of columnar Bnt horizon. (Scale in centimeters.)

Table 12-6   Characteristics of Solonetzic Great Groups

Solonetz	Solodized Solonetz	Solod
No continuous Ae at least 2 centimeters thick	Ae at least 2 centimeters thick	Ae at least 2 centimeters thick
Intact Bn or Bnt	Intact, columnar Bnt or Bn	Distinct AB or BA horizon (distintegrating Bnt)

From The Canadian System of Soil Classification, *1978.*

ical disintegration of the Bnt horizon, and the A horizon thickens. The soil becomes more favorable for plant growth, and the abrupt AB boundary becomes less distinct. The remnant solonetzic horizon becomes more easily broken, and the soil becomes a Solod. The characteristics of the Solonetzic great groups are given in Table 12-6.

Land Use on Solonetzic Soils

Although Solonetzic soils occupy only 0.7 percent of the land of Canada, the soils are of considerable importance. The area is about 6 to 8 million hectares, or about equal to the seeded wheatland. The major areas of solonetzic soils are shown in Fig. 12-23.

Solonetzic soils in their natural state have low productivity for crops, and many areas are bare. The soils are low in available nitrogen, have poor physical condition, and are usually wet as the solonetzic B impedes downward percolation of water. High salt content in subsoils reduces water availability and inhibits root penetration. Most of the land is used for grazing, with about 45 percent of the land used for cropping.

Deep plowing is being used experimentally to improve crop production by mixing the Ah and solonetzic B horizons with high-lime layers from the lower part of

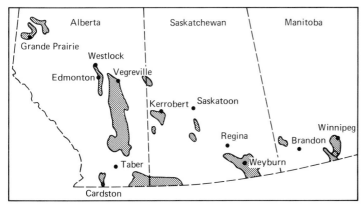

Figure 12-23   Major areas of Solonetzic soils in western Canada. (From *Solonetzic Soils and Their Management,* 1977.)

the profile. Deep plowing increases the calcium supply in the root zone, increases pH, and increases the biological mineralization of nitrogen. Increasing root depth results in increased available water. In one experiment the benefits of deep plowing were observed over a 16-year period.

Organic Soils

Organic soils are found in all provinces of Canada and comprise about 10 percent of the soils. About 60 percent of these soils have permafrost within 1 meter of the surface and are Organic Cryosols. The other organic soils without permafrost, which comprise 4.1 percent (see Table 18-1), are in the Organic order (see Fig. 12-24).

Central Concept and Great Groups

Soils of the Organic order are composed largely of organic materials; most are water saturated for prolonged periods. Vegetation is mainly mosses and sedges and other hydrophytic vegetation. Organic soils contain 17 percent or more organic carbon (30 percent organic matter) by weight. The classification of Organic soils is based on thickness and nature of the organic material, as shown in Fig. 12-25. The thickness of the surface organic layer must exceed 40 centimeters if mesic or humic

Figure 12-24   Organic soil (Mesiol) in fen in foreground supporting sedges and shrubs.

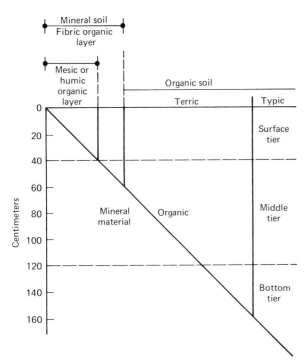

Figure 12-25 Depth relationships in classification of Organic soils. (From *The Canadian System of Soil Classification,* 1978, reproduced by permission of the Minister of Supply and Services, Canada.)

and exceed 60 centimeters if fibric in character. Fibric, mesic, and humic represent organic materials with increasing degrees of decomposition. Fibric material is least decomposed and contains large amounts of well-preserved plant fibers that can be identified as to botanical origin. Humic materials are the most decomposed and contain few recognizable fibers. Some physical properties of fibric and humic materials are given in Table 12-7. Properties of mesic materials are intermediate.

Table 12-7   Some Physical Properties of Fibric and Humic Materials

Property	Fibric	Humic
Bulk density (grams per cubic centimeter)	Less than 0.075	More than 0.195
Total porosity (percent volume)	More than 90	Less than 85
0.1 bar water content (percent volume)	Less than 48	More than 70
Hydraulic conductivity (centimeters per hour)	More than 6	Less than 0.1

From The Canadian System of Soil Classification, *1978.*

The nature of the organic matter in the middle tier (see Fig. 12-25) is used to determine whether the great group is Fibrisol, Mesisol, or Humisol. Most of the Organic soils of Canada are Fibrisols that occur extensively where the organic materials are mainly from spaghum mosses. Humisols are known to occur only as minor areas.

If a lithic contact occurs at a depth shallower than 40 centimeters the organic material must extend to a depth of at least 10 centimeters in Fibrists, Mesisols, and Humisols. Folists, however, are typically shallow or thin. They usually develop under forest vegetation, and the soil organic material is mainly leaf and other litter materials that form L, F, and H layers. These soils are not water saturated. The organic layer must be at least 10 centimeters thick over a lithic contact or fragmental material; interstices are filled or partially filled with organic materials. Folists occur on the Canadian Shield and in mountainous areas where rock materials are close to the soil surface.

Distribution and Use of Organic Soils

Organic soils in Canada occur mostly within or adjacent to forest regions. They are dominant in an area south and west of Hudson Bay and are subdominant in an area dominated by Luvisols (areas 9 and 4 of Fig. 12-2). Small areas of Organic soils are widespread, but their occurrence is less frequent in the subhumid and semiarid grasslands and in the tundra regions.

Organic soils warm and cool slowly relative to adjacent better-drained mineral soils and have less plant growth potential in cool and cold regions. In most cases forest development is restricted by poor drainage and low temperature. Wildlife habitat is the most extensive use. Some areas with milder climate have been drained and developed for agricultural use. These developments are mainly in the St. Lawrence Lowlands in Ontario and Quebec and the lower Fraser Valley in British Columbia. Engineering aspects of the use of Canadian Organic soils is discussed by Macfarlane and Williams (1974).

Gleysolic Soils

Gleysolic soils are mineral soils that have been influenced by waterlogging. The soils are water saturated all or part of the year when reducing conditions exist unless drained. Gleysols commonly have gleyed horizons with grayish color and prominent rusty mottles. The soils occur in all regions as poorly drained associates of other soils and occupy 1.3 percent of the land. Wetness and low temperature limit productivity for trees. Where extensive areas of Gleysols occur with mesic soil temperature regime they are frequently drained and used for agriculture as in the St. Lawrence Lowlands and the lower Fraser Valley.

Rockland Land Type

The Rockland Land Type includes exposed bedrock areas and areas with less than 10 centimeters of soil material above a lithic contact. Rockland occurs as the dominant land type for 15 percent of the land. About 35 percent of the land area has Rockland as subdominant, so that about 50 percent of the land is affected by rock at or near the land surface.

The major Rockland area is on the Canadian Shield (see area 12 of Fig. 12-2). The bedrock material is mainly Precambrian igneous and metamorphic rock of granitic character. The shield forms a large, saucer-shaped land surface centered on Hudson Bay. The shield extends into the United States west and south of Lake Superior. Ice accumulated on the shield during various ice ages and moved outward in all directions. Extensive areas were denuded of soil and are now bare. Most of the till derived from the shield was carried beyond it and deposited in the Interior Plains of Canada and the United States. However, significant areas on the shield have over 10 centimeters of soil material, and thick deposits of till occur in valleys and low areas.

Rockland is a significant component of most mountain landscapes (area 2 of Fig. 12-2). In the Innuitian region there is Rockland as a subdominant associate occupying much of the barren land and tundra of the northernmost Arctic islands. All provinces and territories have significant Rockland or lithic soil areas except Prince Edward Island.

References

Alberta Dryland Saline Seep Committee, *Dryland-Saline-Seep Control,* Proceedings of Subcommission on Salt Affected Soils, International Soil Science Society Congress, Edmonton, 1978.

Bliss, L. C., "Polar Climates: Their Present Agricultural Uses and Their Estimated Potential Productivity in Relation to Soils and Climate," *Proc. Intl. Soil Sci. Soc.,* 2:70–90, 1978.

Bowser, W. E., "The Soils of the Prairies," *Agr. Inst. Rev.,* 15:24–28, Ottawa, 1960.

Bridges, E. M., *World Soils,* Cambridge University Press, 1970.

Cairns, R. R., and W. E. Bowser, "Solonetzic Soils and Their Management," *Pub. 1391,* Information Division, Canada Department of Agriculture, Ottawa, 1977.

Canada Soil Survey Committee, *The Canadian System of Soil Classification,* Publ. 1646, Canada Department of Agriculture, Supply and Services Canada, Ottawa, 1978.

Canada Soil Survey Committee and The Soil Research Institute, *Soils of Canada,* Research Branch Canada Department of Agriculture, Ottawa, 1977.

Cann, D. B, and J. F. G. Millette, "Soils of the Appalachian Region," *Agr. Inst. Rev.,* 15(2):44–47, Ottawa, 1960.

Chapman, L. J., "Physiography, Climate and Natural Vegetation of Canada," *Agr. Inst. Rev., 15(2):*15–19, Ottawa, 1960.

Crown, P. H., and G. M. Greenlee, *Guidebook for A Soils and Land Use Tour in Edmonton Region, Alberta, Tours E1, E2 and E3,* 11th International Soil Science Congress, Edmonton, 1978.

Dion, H. G., "Land Use in Canada—Present and Future," *Agr. Inst. Rev., 15(2):*57–59, Ottawa, 1960.

Ehrlich, W. A., and W. Odynsky, "Soils Developed Under Forest in the Great Plains Region," *Agr. Inst. Rev., 15(2):*29–32, Ottawa, 1960.

Eilers, R. G., W. Michalyna, G. F. Mills, C. F. Shaykewich, and R. E. Smith, "Soils," in *Manitoba Soils and Their Management,* pp. 17–59, Manitoba Department of Agriculture.

FAO, *Soil Map of the World: Vol. II, North America,* Unesco, Paris, 1975.

Farstad, L., and C. A. Rowles, "Soils of the Cordilleran Region," *Agr. Inst. Rev., 15(2):*33–36, Ottawa, 1960.

Harker, D. B., G. R. Webster, and R. R. Cairns, "Factors Contributing to Crop Response on a Deep-Plowed Solonetz Soil," *Can. Jour. Sci.,* 57:279–287, 1977.

Hills, G. A., "Soils of the Canadian Shield", *Agr. Inst. Rev., 15(2):*41–43, Ottawa, 1960.

Hoyt, P. B., W. A. Rice, and A. M. F. Hennig, "Utilization of Northern Canadian Soils for Agriculture," *Proc. Intl. Soil Sci. Soc.,* 3:332–347, 1978.

Klages, K. H. W., *Ecological Crop Geography,* Macmillan, New York, 1942.

Macfarlane, I. C., and G. P. Williams, "Some Engineering Aspects of Peat Soils," in *Histosols: Their Characteristics, Classification and Use,* pp. 79–93, Soil Science Society of America Special Publication 6, Madison, Wis., 1974.

Michalyna, W., R. E. Smith, J. G. Ellis, and T. M. Macyk, *Guidebook for a Tour Across the Southern Portion of the Interior Plains of Western Canada from Winnipeg, Manitoba to Edmonton, Alberta, Tour 2,* 11th International Soil Science Congress, Edmonton, 1978.

Nowosad, F. S., and A. Leahey, "Soils of the Arctic and Sub-Arctic Regions," *Agr. Inst. Rev., 15(2):*48–50, Ottawa, 1960.

Pettapiece, W. W., "Soils of the Subarctic in the Lower Mackenzie Basin," *Arctic,* 28:35–53, 1975.

Pettapiece, W. W., C. Tarnocai, S. C. Zoltai, and E. T. Oswald, *Guidebook for Tour of Soil, Permafrost and Vegetation Relationships in the Yukon and Northwest Territories of Northwestern Canada, Tour 18,* 11th International Congress of Soil Science, Edmonton, 1978.

Rapeseed Association of Canada, "Rapeseed, Canada's "Cinderella" Crop," *Pub. 33,* Winnipeg, 1974.

Rennie, D. A., and J. G. Ellis, "The Shape of Saskatchewan," *Pub. M41,* Saskatchewan Institute of Pedology, Saskatoon, 1978.

Rennie, P. J., "Utilization of Soils of the Boreal for Forest Production," *Proc. Intl. Soil Sci. Soc., 3:*305–331, 1978.

Saskatchewan Agricultural Services Coordinating Committee, *Guide to Farm Practice in Saskatchewan in 1978,* Extension Division University of Saskatchewan, 1978.

Shaykewich, C. F., and T. R. Weir, "Geography of Manitoba," in *Manitoba Soils and Their Management,* pp. 7–16, Manitoba Department of Agriculture.

Soils Department, *This Land of Alberta,* University of Alberta, Edmonton, 1978.

Sparrow, S. D., F. J. Wooding, and E. H. Whiting, "Effects of Off-Road Vehicle Traffic on Soils and Vegetation in the Denali Highway Region of Alaska," *Jour. Soil Water Cons., 33:*20–27, 1978.

Stobbe, P. C., "The Great Soil Groups of Canada—Their Characteristics and Distribution," *Agr. Inst. Rev., 15(2):*20–23, Ottawa, 1960.

Tarnocai, C., "Distribution of Soils in Northern Canada and Parameters Affecting Their Utilization," *Proc. Intl. Soil Sci. Soc., 3:*281–303, 1978.

Tarnocai, C., and S. C. Zoltai, "Earth Hummocks of the Canadian Arctic and Subarctic," *Arct. Alp. Res., 10:*581–594, 1978.

Tarnocai, C., and S. C. Zoltai, "Earth Hummocks of the Canadian Arctic and Subarctic," *Proc. Intl. Soil Sci. Soc. 1:*2, 1978.

Tedrow, J. C. F., *Soils of the Polar Landscapes,* Rutgers University Press, New Brunswick, 1977.

Toogood, J. A., and R. R. Cairns, "Solonetzic Soils Technology and Management," *Bulletin B-73-1,* University of Alberta and Agric. Canada Res. Branch, Edmonton, 1973.

Valentine, K. W. G., P. N. Sprout, T. E. Baker, and L. M. Lavkulich, *The Soil Landscapes of British Columbia,* Resource Analysis Branch, Ministry of the Environment, Victoria, 1978.

Webber, L. R., and D. W. Hoffman, "Origin, Classification and Use of Ontario Soils," *Pub. 51,* Ontario Department of Agriculture and Food, Toronto, 1967.

Zoltai, S. C., "Soil Productivity in the Arctic Environments of Canada," *Proc. Intl. Soil Sci. Soc., 3:*348–359, 1978.

Zoltai, S. C., "Tree Ring Record of Soil Movement on Permafrost," *Arct. Alp. Res., 7:*331–340, 1975.

Zoltai, S. C., and C. Tarnocai, "Soils and Vegetation of Hummocky Terrain," *Environmental-Social Committee, Task Force on Northern Oil Development,* Report 74-5, 1974.

13

Mexico and Central America

Physiographic Environment

During most of geologic time, North America and Central America as far south as Nicaragua were separated from South America by a deep extension of the Pacific Ocean referred to as the "Caribbean Mediterranean." Volcanic activity, probably late in the Mesozoic era, linked the two continents. Since then there has been extensive volcanic activity in the area. Limestone was laid down in Tertiary time. The area in general has been geologically active over a long period of time, reflecting periods of subsidence and uplift; there are few if any Oxisols, since essentially no significant ancient land areas have survived the complex geologic changes. A map of the physiographic regions is given in Fig. 13-1.

The climate of northern Mexico is an extension of the desertic climate of the southwestern United States. Along the coasts of southern Mexico and Central America warm winds off the Pacific and Atlantic oceans bring moisture to the

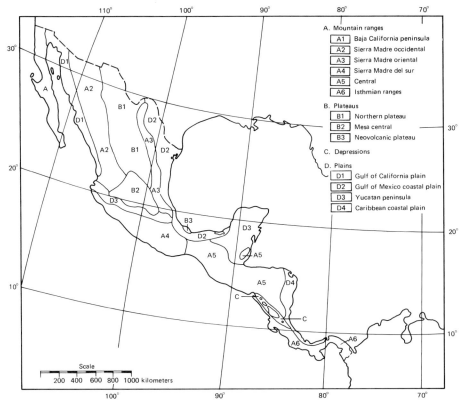

Figure 13-1   Physiographic regions of Mexico and Central America. (From *Soil Map of the World,* Vol. 3, Mexico and Central America, copyright © FAO/Unesco, 1975.)

coastal areas; there is generally greater rainfall along the eastern coast. Rainfall exceeds 500 centimeters annually near the coast by the Nicaragua–Costa Rica border. The climate is highly variable in the mountainous areas. The vegetation pattern generally follows that of climate, and vegetation types range from desert shrubs to tropical rain forest.

Soils

The dominant soils along the Mexico–United States border are Aridisols; they also dominate the northern continental portion of Mexico (see Fig. 1-11). The next most extensive general soil region is the mountainous region extending from north central Mexico to Panama. Volcanic glass has strongly influenced soils of the mountains and formation of Andepts (Andosols) in the neovolcanic uplands of Mexico and the Central American Volcanic Highlands, which extend as a continuous belt from southern Mexico down the Pacific side to Costa Rica, with isolated volcanic areas in northern Panama (see Fig. 13-2). The nonvolcanic highlands of Central America are dominated by soils developed from crystalline limestone and related calcareous rocks mixed with soils developed from lavas and tuffs of Tertiary age or older. These soils are mainly Inceptisols and Entisols. Another large mountain area, Oaxaca Uplands and Sierra Madre del Sur (area A5 of Fig. 13-2), consists of a series of complex, dissected uplands and basins dominated by Vertisols, Alfisols, Inceptisols, and Entisols. Some areas of granitic rocks in the coastal ranges give rise to Ultisols.

In Mexico and Central America the mountainous areas contrast sharply with the lowlands along the coasts. Along the Pacific Ocean the lowland areas tend to be narrow, and soils are related to the rocks of the adjoining uplands, which served as the source of soil parent materials. Some fairly broad expanses of lowlands exist along the east coast. Soils of the Yucatan developed mainly in limestone and are principally Rendolls along with Vertisols and Entisols. There are some red-colored soils on the peninsula that show evidence of influence of ash and are Ultisols. Farther south along the Caribbean coast south of British Honduras are swamps and low terraces where the principal soils are Udults and Udalfs. Udalfs are also dominant soils along the coast north of the Yucatan, and the soils gradually merge with Ustolls and Usterts farther north (see Fig. 1-11).

Land Use

Agriculture and settled areas with relatively high human population in Mexico and Central America are generally ancient compared to that of the United States and Canada. If the earliest humans in North America crossed over on a land bridge from Asia, they apparently found conditions more favorable in the highland plains and valleys of Mexico and Central America than in the areas generally farther north. Many of the soils were derived from volcanic ash and calcareous rocks and the soils are fertile. Shifting cultivation became the basic form of agriculture. The Ameridian farmers found the lithic soils of the mountains attractive because trees on steep slopes were smaller than trees on leveler, more productive land, making it easier to prepare for burning; also, the soils were certain to have adequate drainage during the rainy growing season. A diverse planting of crops provided good cover and protection from excessive erosion, as long as the fallow periods were long enough to insure renewed soil fertility. Use of coastal areas for agriculture required high installation and maintenance costs for drainage and deterred development of coastal region soils.

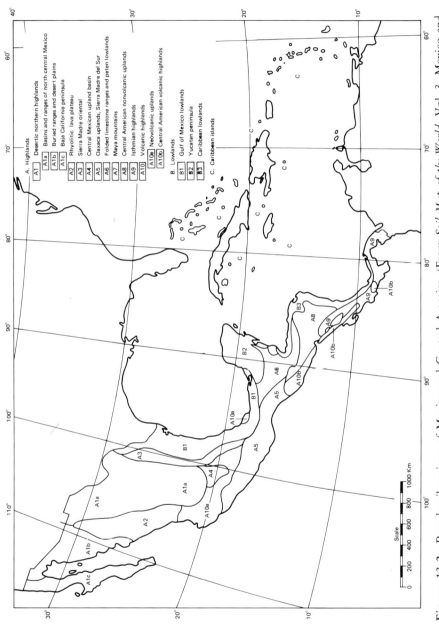

A. Highlands
- A1 Deseretic northern highlands
  - A1a Basins and ranges of north central Mexico
  - A1b Buried ranges and desert plains
  - A1c Baja California peninsula
- A2 Rhyolitic lava plateau
- A3 Sierra Madre oriental
- A4 Central Mexican upland basin
- A5 Oaxaca uplands, Sierra Madre del Sur
- A6 Folded limestone ranges and peten lowlands
- A7 Maya mountains
- A8 Central American nonvolcanic uplands
- A9 Isthmian highlands
- A10 Volcanic highlands
  - A10a Neovolcanic uplands
  - A10b Central American volcanic highlands

B. Lowlands
- B1 Gulf of Mexico lowlands
- B2 Yucatan peninsula
- B3 Caribbean lowlands

C. Caribbean islands

Figure 13-2 Broad soil regions of Mexico and Central America. (From *Soil Map of the World*, Vol. 3, Mexico and Central America, copyright © FAO/Unesco, 1975.)

The Agriculture in Mexico

Mexico is one of the nations that has been greatly influenced by the Green Revolution. In the 1930s and 1940s agricultural production in Mexico had become static, and 15 to 20 percent of the basic cereals (corn and wheat) were imported. An agricultural revolution was launched in 1943 in cooperation with the Rockefeller Foundation, and the food deficit disappeared by 1960. During this time, however, population growth and a slowing of the rate of agricultural production resulted in food imports in the early 1970s comparable to those before the agricultural revolution began. Today Mexico has some highly sophisticated farms; however, most of the farms (52.4 percent) were unaffected by the Green Revolution and remain subsistence farms. Only 7.1 percent of the farms are modern in that improved technology is used. Another 40.5 percent of the farms are semicommerical in that these farms use some recommended practices.

Mexican agriculture is strongly affected by water supply and topography. Only 9 percent of the land is cultivated, although 15 percent is considered potentially arable. The remainder (85 percent) is too dry, too wet, or too mountainous. Over one-half of the cropland is in the Central Highlands, where crops are grown by rain-fed water mostly from July to October. This area contains only 15 percent of the nation's land but slightly over one-half of its farms and people. The major crop is corn; other important crops are wheat, vegetables, strawberries, and potatoes.

One of the most productive agricultural areas in Mexico, as well as in the world, is in the dry northwestern coastal plain in the states of Sonora and Sinaloa. Crops are grown year round with the help of irrigation. The major crops are wheat, vegetables, rice, sugarcane, soybeans, cotton, safflower, and sorghum. Similar but smaller areas of irrigated agriculture are located in states of Coahuila, Chihuahua, Baja California, and Tamaulipas. An important area of intensive, year-round cropland that is not irrigated is centered in northern Veracruz and eastern San Luis Potosi. The remaining agriculture is more scattered and dispersed on narrow ridges, steep slopes, and small alluvial valleys throughout the western and eastern mountain ranges and southern highlands.

There is potential to increase the agricultural output of Mexico, but serious obstacles must be overcome. Increasing production on subsistence farms will be much more difficult than on the major commerical farms during the Green Revolution. Available water is out of balance with the location of the population. The Central Highlands, with over one-half of the people, has only 10 percent of the water resources. By contrast, approximately 40 percent of the nation's water is in the humid southeast, where only 8 percent of the people live. The major potential for new croplands is in the coastal plains of the tropical Gulf of Mexico, but conversion of these lands for agriculture is costly and they are located in an area not highly desirable for habitation.

References

FAO, *Soil Map of the World: Vol. III Mexico and Central America*, Unesco, Paris, 1975.

Martini, J. A., and L. R. Jaramillo, "Soils Derived from Volcanic Ash in Central America: 2. Soils More Developed than Andepts," *Soil Sci., 120:*376–384, 1975.

Martini, J. A., and J. A. Palencia, "Soils Derived from Volcanic Ash in Central America: 1. Andepts," *Soil Sci., 120(4):*278–287, 1975.

Watters, R. F., *Shifting Cultivation in Latin America,* FAO Forestry Development Paper 17, FAO, Rome, 1971.

Wellhausen, E. J., "The Agriculture of Mexico," *Sci. Amer., 235(3):*128–150, September 1976.

Africa

The African Environment

Africa differs from North America in several significant ways. Most of the land surfaces of North America were either covered by ice or were dramatically altered during Pleistocene times. The relationship between soils and the factors of soil formation is simpler because of the relative youthfulness of land surfaces. In Africa many of the land surfaces and the soils on them are much older. Soil properties are influenced by present environmental conditions and by a succession of several past environments. Soil studies are complicated by the necessity of determining whether present soil conditions were influenced by current environmental factors or by an earlier environment.

In addition, there is the obvious difference that much of Africa lies in the tropics and most of North America is in the temperate zone. The affect of this on African soils is due not so much to high temperatures as it is to a more or less uniformly warm temperature year round. The seasonal differences, instead of reflecting extremes in temperature, reflect extremes in rainfall, with distinct wet and dry seasons. The influence of these moisture extremes on microbial activity, leaching, and other soil processes gives rise to soils that are quite different from those in the temperate zone.

Physiography

Africa lies astride the equator. One-half of the tropics of the world, the land between the Tropic of Cancer and the Tropic of Capricorn, lies within the continent of Africa (Fig. 14-1). Because of its shape, more than two-thirds of the landmass of Africa lies north of the equator. The northern and southern tips of Africa are about the same distance from the equator.

The African continent covers an area of approximately 30 million square kilometers. About 71 percent is covered with true soils. About 20 percent is desert,

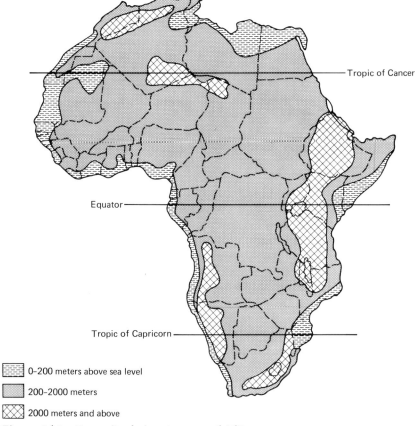

Tropic of Cancer

Equator

Tropic of Capricorn

0–200 meters above sea level

200–2000 meters

2000 meters and above

Figure 14-1   Generalized elevation map of Africa.

some of which contains Entisols or Aridisols, but much of it is not much more than raw parent material. The remaining 9 percent is occupied by bare rock, open water, or snow.

Africa can be considered a large plateau slanting upward from 200 to 500 meters above sea level in the north to 500 to 1000 meters above in the south. Much of it is extremely level. Except for narrow, coastal strips and the basins of some of the major African rivers, most of the continent is more than 200 meters above sea level. The lower-lying coastal fringe generally extends inland less than 100 kilometers from the sea. It is composed primarily of marine sediments with more recent alluvial deposits.

Behind this fringe an escarpment is frequently found. This results in rapids and falls in the rivers that hamper their use for transportation. The Tugela river in the Tugela Basin of South Africa falls over 3000 meters from its source to the sea in a straight-line distance of 260 kilometers, or 540 kilometers of actual river distance.

In general, Africa lacks the high, rugged mountain ranges found on other continents. Most of Africa is below 2000 meters except for the rather small Atlas Chain in Morocco and Algeria in the north, a small range in the Sahara, and a discontinuous "J-" shaped system in the south. The Atlas Chain is a series of folded sedimentary rock. The southern system starts in Ethiopia, extends southward through the Lake Victoria region to South Africa, and swings northwest into Angola. It is a series of high plateaus studded with lakes between high peaks.

African surfaces are among the oldest in the world. As stated earlier, relationships between soils and the five soil-forming factors are much more complex because the land surface has undergone a series of vegetative and climatic shifts. Much of the central plateau of Africa is of Precambrian age (over 600 million years old). Nearly one-third of the continental surface is an outcrop of rocks of this age. The rest of the surface is covered by sand deposits and alluvium of Pleistocene age (less than 2 million years old). There are also a few areas of volcanic activity. Recent volcanic activity occurs primarily in the west between Ethiopia and Lake Victoria. This area is also complexed by fracture systems or rift valleys.

Not only is Africa geologically old, but parts of Africa have been occupied by humans much longer than North America. Their activities in obtaining food, fiber, fuel, and shelter have often significantly altered the soil. These alterations must be recognized in order to understand the soils we find today.

Landscapes are often classified as depositional or erosional. However, parts of the African landscape are flat, exhibiting neither accumulation nor loss, and they therefore lack rejuvenation. The acidity and grain size of the parent material is very important in determining the potential productivity of the soils. The minerals contained in the parent rocks determine the natural fertility of the soil. Micronutrient deficiencies are often correlated with the nature of the parent material, especially on landscapes of advanced weathering. Parent material and topographic differences become very important in understanding soil properties.

Climate

Figure 14-2 shows a generalized precipitation map of Africa. The total annual rainfall is highest at the equator and decreases north and south to essentially zero in the desert regions. At the northern and southern extremes of the continent, rainfall increases again. In the equatorial region rainfall decreases from west to east. The wettest region is along the equator to the west, where the annual rainfall exceeds 2000 millimeters. Much of the equatorial region has an annual rainfall between 1000 and 2000 millimeters. And yet the evapotranspiration rates are so high that the soil can be dry much of the year. Evapotranspiration rates can exceed 8 millimeters per day in the dry season.

But the total annual rainfall tells only part of the story. Seasonal variation, reliability, shower intensity, and losses by evaporation are more variable in the tropics than in temperate regions. The intensity of storms can be very severe, leading to

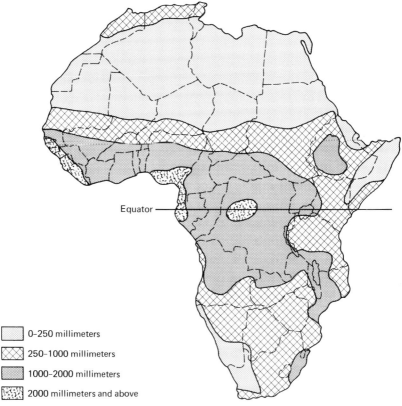

0–250 millimeters

250–1000 millimeters

1000–2000 millimeters

2000 millimeters and above

Figure 14-2 Generalized precipitation map of Africa.

runoff, erosion, and flooding, particularly in the equatorial region. The energy of a rainstorm can be tremendous. Kalpage has reported that a 1-hour storm of 7.6 centimeters with a drop velocity of 9.1 meters per second contains 29 times the energy required to plow the same area of land. Tropical storms can reach 6 millimeters of rainfall per minute, and rains of 600 millimeters per day occur occasionally.

There is no true dry season in equatorial Africa, but rainfall is heavier in March and October. To the north and south of the equator, two short dry seasons break the rainfall pattern. Farther from the equator, only one distinct dry season occurs. Except in the wettest part of equatorial Africa, most of Africa is subject to seasonal or periodic droughts because the rainfall is not always reliable. In a 10-year period the driest year may receive only one-fourth the rainfall that falls in the wettest year. This greatly influences both soil properties and soil productivity.

Approaching the desert, the rainfall decreases to 250 to 500 millimeters per year; this semiarid region is characterized by wet summers and dry winters. Parts of the desert are very dry, with rain occurring only once a decade or less. The desert region of northern Africa is much larger than the southern desert because of the shape of the continental landmass. The extreme northern and southern tips of the continent have a Mediterranean climate, with wet winters and dry summers.

Africa lies principally in the tropical and subtropical zones. It has the hottest climate of all the continents. Most of Africa has a mean annual temperature above 10°C except at high altitudes. Much of Africa has a mean monthly temperature above 20°C for at least 9 months each year. Temperature means vary with latitude; in the mountainous regions they vary with altitude. In the mountains the average temperature drops about 6°C per 1000 meters in elevation, and frosts are common above 2000 meters. Continuous snow can occur above 4500 to 5000 meters. Temperature is also influenced somewhat by ocean currents.

The only climatic factor common to the entire tropics is the uniformity of temperature throughout the year. Along the African equator the temperature averages about 28°C and varies less than 5°C from the coldest to the warmest month. Farther from the equator the average may drop slightly with more distinct seasonal variations, but the averages of the hottest and coldest months still vary less than 5 to 6°C. Desert regions lack water to absorb heat energy and, near the equator, they are under almost direct solar radiation. Here daytime temperatures are known to exceed 50°C, yet nights are cold because of high levels of reradiation to clear skies. Soil surface temperatures of over 80°C have been reported.

The generally higher temperatures of the tropics cause several important differences in soil-forming processes and soil properties. Under favorable moisture conditions, microbial activity is more intense. The rainwater in the tropics may average 15°C warmer than in temperature regions (25°C versus 10°C). Under these warmer temperatures, the ionization of water increases by a factor of four. Silicon is eight times more soluble. Bases go into solution more readily. The water is less viscous and penetrates deeper into the soil. All of these factors influence the development of soils whose properties differ significantly from their temperate region counterparts.

Vegetation

Despite our concept of Africa as being a continent of tropical jungles, only about 5 percent of the continent is covered by tropical rain forest (See Figure 14-3). The rain forests in Brazil alone cover 1.7 times the land surface covered by all of the rain forests in Africa. Most of Africa is covered by desert or savanna.

A savanna is a grassy plain in a tropical or subtropical climate with a distinct dry season. Prairies, pampas, and steppes are treeless plains containing different grass species. Extreme temperature ranges are common. A prairie is in a moister region

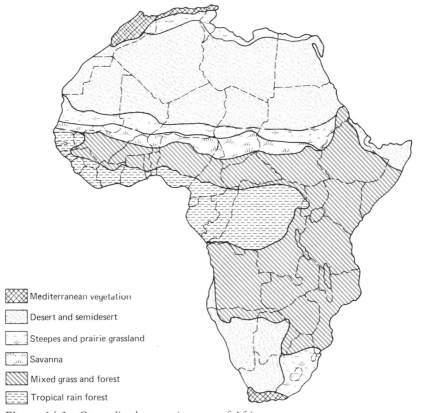

Mediterranean vegetation

Desert and semidesert

Steepes and prairie grassland

Savanna

Mixed grass and forest

Tropical rain forest

Figure 14-3   Generalized vegetation map of Africa.

Figure 14-4   A rain forest is difficult to photograph. It consists of short, medium, and tall vegetation. The dense foliage drastically reduces the light intensity on the ground surface and causes a very high humidity. Removing the vegetation alters the microclimate. (Photograph courtesy D. K. Whigham.)

typified in the upper Mississippi Valley of the United States. Steppes are drier and are typified by the Russian Plains in Europe and Asia. The *pampas* are a grass prairie in South America. The terms steppe and prairie are often used to distinguish between short and tall grass vegetation.

Much of Africa has been cut over many times by people; in many regions the present plant communities reflect this activity. Furthermore, it is much more difficult to correlate soil properties to vegetation in the tropics. In the glaciated parts of Europe and North America, there are about 16 major species of trees. In the tropics there are 18,000 species.

The forested region of tropical Africa is a luxuriant, often multistoried growth so thick that sunlight may not reach the ground (see Fig. 14-4). The native forest is in a delicate balance with soil fertility. The replacement of the native vegetation with agricultural crops quickly destroys this balance, and productivity drops drastically. Leaching is intense and weed growth is luxuriant if the shade of the forest is removed. Most of the rain forests today are secondary growth. Many temperate season crops (such as corn) do poorly in the rain forest area because of the lack of a dry period to aid in maturation.

The grass cover of the savanna grows rapidly, reaching more than 80 centimeters in height in the wet season, and scorches brown in the dry season. The heavy rains of the early wet season fall on a surface poorly protected by the parched vegetation. Erosion can be very severe. It is subject to frequent fires and is also easily destroyed by human activities. Overgrazing in the dry season or during occasional extra dry years kills back much of the grass. The natural vegetation may give way to plants less desirable to grazing animals. The loss of cover plus compaction by trampling animals can cause severe erosion problems. Erosion of shallow soils can result in a surface of weathered rock. The rock is more resistant to erosion and, in time, the savanna vegetation may return, but these soils are much less productive than the original soils.

The vegetative balance is most delicate at the rain forest-savanna boundary. In places where trees were cleared for cultivation, grass is more easily reestablished than trees. There is an increased pressure on the remaining trees to be used for fuel or lumber. The trees may eventually disappear. This loss is usually accompanied by a decrease in soil fertility and a deterioration of the physical properties of the soil.

In arid regions the surface is covered with sparse vegetation or is completely void of vegetation. The presence of any vegetation on the surface modifies the erosion potential of the climate and reduces the amplitude and variation of soil temperature and humidity. This results in soil conditions more favorable for microbial activity.

In parts of Africa large termite mounds are found. They are mostly in grassy and tilled areas and not in forests. Most of the material in these mounds has been removed from underground passages that may extend 50 meters laterally from the mound. The mounds have been reported to weigh about as much as 600 metric tons per hectare. If the soil is stratified, the mixing effect produces a more loamy surface. In stoney soils, the fines are brought to the surface and the stones are gradually buried. It has been estimated that in 2500 years termites could produce a stone-free soil in the top 50 centimeters (see Fig. 14-5).

Termite mounds in Rhodesia were shown to contain 95 percent more total calcium, 81 percent more nitrogen, 67 percent more extractable potassium, and 69 percent more available phosphorus than the surrounding soils. Thus local farmers often mix termite mounds with nearby soil or plant crops directly on undisturbed mounds.

Soil Regions of Africa

Figure 14-6 is a soil map of Africa that has been simplified from the map in Fig. 1-11. It highlights several important points about the soils of Africa. First, note that only a small region in North Africa is shown as mountainous. Throughout most of

Figure 14-5   Termite mounds can dramatically alter soil properties. Changes in soil properties over short distances due to human and animal activity make it more difficult to determine the best management practices for a field. (Photograph courtesy D. K. Whigham.)

the continent, the climate-soil relationship is not sufficiently interrupted by the altitude-soil relationship to be noted on this map. At this map scale, Africa has less area in mountainous complexes than any other continent.

Aridisols dominate the Sahara in northern Africa and are also found in southern Africa. These regions are too dry to be very productive. A relatively small area is irrigated. The productivity of these irrigated areas is high, and they are an important local source of food. There is considerable nomadic grazing in those parts of the desert that receive sufficient moisture to provide grazeable vegetation.

Equatorial Africa is dominated by Oxisols. This reflects the intensive weathering under a hot, humid climate on an old, level, unrejuvenated landscape. Without fertilizer and proper management techniques, continuous cultivation is not possible. A shifting agricultural system is more common. Where the economic return is high (e.g., the production of oil palm for an export market), a large investment in fertilizer is economically warranted, and a permanent type of agriculture is possible. But this type of agriculture is not extensive in the Oxisol region, which is primarily a subsistence agriculture area with production primarily for local consumption and little for export.

Alfisols comprise the third most extensive area and are located mainly between the Aridisols and the Oxisols. They receive sufficient rainfall to allow soil development and are frequently on rejuvenated land surfaces. Much of this region is in

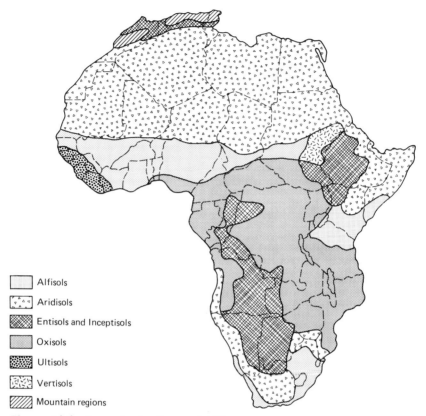

Alfisols

Aridisols

Entisols and Inceptisols

Oxisols

Ultisols

Vertisols

Mountain regions

Figure 14-6   Generalized soil map of Africa.

shifting agriculture and grazing, although there are regions of permanent agriculture. Corn and millet are the major grains grown in Africa, and much of it is produced in the Alfisol region.

Entisols and Inceptisols are also extensive in Africa. The Entisols are mostly Psamments in regions of sandy parent material. The Inceptisols are primarily Aquepts on young parent materials in flat areas with poor external drainage. An important exception to this is the Inceptisol region of North Africa, where Xerochrepts developed in a Mediterranean climate.

Despite our picture of African soils as old and weathered, Ultisols are much less widespread than Alfisols and show up only along the west coast on our simplified map of Africa. A very significant region of Vertisols is found in southern Sudan.

Mollisols, Spodosols, and Histosols are found in Africa, but nowhere do they dominate the landscape sufficiently to appear on either soil map of Africa. (Fig. 1-11 or 14-6).

Aridisols

The Aridisols of Africa are found in the Sahara, in the Ethiopian highlands, and in two areas in southern Africa. The frequent lack of biological activity in these extremely arid regions has led some to question whether or not these areas are truly covered with soils.

Aridisols of the Sahara Desert

The concept of the Sahara as a region of shifting sands is misleading. The surface of much of the Sahara is stoney. The area covered by sands is only one-half the area covered by desert detritus. Psamments dominate in some areas within the northern Aridisols (see Fig. 1-11). They will be discussed with the Entisols. The nonsandy Aridisols develop mainly from physical weathering of the parent material and the loss of the finer-sized particles by the action of wind and water. Several years have been known to pass without rain. The occasional thunderstorm in the desert is often fierce. Lacking vegetative protection, the soil surfaces are easily eroded by the flash floods and mass movements that accompany these rains. The fine particles are carried away, and the coarse particles remain. Water-transported sediment accumulates in low-lying closed areas of the desert called *wadis.*

Rainfall limits the use of Aridisols. On the desert fringes, grazing is possible, but the land frequently has no agricultural value. An important exception, particularly in the Sahara, is the presence of oases.

An oasis is a desert region in which water is available. An accumulation of soluble salts, particularly sodium, can cause problems. But where conditions are favorable, an intense garden type of agriculture is possible in an oasis. Irrigation and the use of organic fertilizers are an important part of their management. They are characterized by tall date palms; a second story of tropical fruits with vegetables grown underneath. Citrus, grapes, olives, pomegranates, and almonds are common.

Along the northern and southern fringes of the Sahara, precipitation is generally in the 200 to 600 millimeter range. The increase in rainfall compared to that in the heart of the Sahara results in soils with a low but significant level of organic matter and some movement of soluble salts within the profile. Genetic horizons can develop, but soil properties often reflect an earlier and more humid climate. Biological activity is generally very low and, in some cases, nitrogen-fixing organisms are inactive or totally absent. Under irrigation they yield well if soluble salts are not a problem.

Other African Aridisols

To the east of the Sahara in Ethiopia and Somalia is another region of Aridisols. In places the rainfall is low to nonexistent, and the parent material is weathered sandstones and limestones. Where rainfall is higher, the native vegetation tends to be subdesert and not true desert, and some soil development is expected. As with all desert regions, the availability of water limits agricultural productivity.

The Aridisols along the west coast of southern Africa lie in a narrow zone; rainfall patterns change over a relatively short distance compared to the northern desert. Most of this region receives less then 200 millimeters of rainfall, but the precipitation is generally higher and more dependable then in the deserts of the northern part of the country. Profile development is limited, and organic matter accumulation is low. The soils may exhibit a zone in which free carbonates have been leached. The agricultural use of the soils is limited generally to grazing unless they are irrigated. Cotton yields well under irrigation, but the physical properties of these soils tend to deteriorate with cultivation. Excess irrigation must be avoided, and organic matter must be added to improve structure.

Oxisols

The Oxisols of Africa lie along and to the south of the equator. They are mapped in three zones, as shown in Fig. 14-7. The largest area is along the equator and is dominated by Orthox soils. To the south, a distinct dry season leads to Ustox soils. Along the coast, the Ustox soils are associated with Alfisols while, inland, the Ustox soils are associated with Ultisols.

Characteristics

Many of the Oxisols of Africa are shallow to bedrock; they are on flat surfaces unrejuvenated by deposition or erosion, or they have developed in old eroded soils or exposed plinthite. Some of these unrejuvenated crusts date back to tertiary times. Intensive and continuous weathering, with losses of silicates by hydrolysis, results in the accumulation of sesquioxides. Chemical weathering is active and vigorous all year long. Plinthite formed, and today it caps the hills and plateau summits.

The 1938 system of soil classification classified many of these soils as Lithosols (shallow to bedrock). They are now classified as Oxisols because of the presence of the plinthite or an oxic horizon. This region corresponds approximately to the location of the African Shield. This is a large, flat area that has not been subjected to warping, folding, volcanism, mountain building, or other processes that erode and rejuvenate the landscape. The changes that have occurred here have been slow and uniform, giving a surface of great geologic age that has been deeply weathered over long periods. The soils may have developed through several climatic periods. Many of the soils developed to their current state long ago and have remained unchanged since. The adjacent Alfisol and Ultisol regions are generally on surfaces where geologic activity has exposed younger parent materials.

Orthox

Ustox with Alfisols

Ustox with Ultisols

Figure 14-7   Oxisols of Africa.

On the shield, weatherable minerals are essentially absent, and even quartz may be altered in the profile. Oxide clays often dominate the clay fraction, and the silicate clays that remain are primarily kaolinitic in nature.

The soils are often very deep. Horizons are separated by gradual transition zones with diffuse boundaries. Horizon boundary designations become very arbitrary. Large quantities of iron and aluminum oxides are present. A textural analysis, using repeated dispersion, will show the soil to be high in fine-sized particles but, in the field, these soils have a very strong structure and their properties are more similar to temperate region soils with coarser textures.

The cation-exchange capacity of the Oxisols is generally less than 20 milliequivalents per 100 grams, and the solum is generally less then 40 percent base saturated. They are very infertile because they lack weatherable minerals. Much of the fertility that is present is tied up in the organic fraction of the soil. They are generally very low in total phosphorus and often show a high phosphorus fixation capacity.

As the pH declines, aluminum and iron become more soluble. These react with phosphorus and render it unavailable. As the pH declines further, the level of soluble aluminum can become toxic. Aluminum toxicity at these pH values is the most frequent limiting factor to plant growth on tropical Oxisols.

Land Use

In general, Oxisols can be considered the poorest of African soils found in areas with adequate rainfall. Much of the agriculture of the Oxisol region is on the associated Ultisols and Alfisols and on soils of recent alluvium along rivers. The Oxisols can support lush tropical growth but usually do not produce high yields of agricultural crops once the natural vegetation is removed. In parts of Africa Oxisols are cultivated with tree crops such as rubber, coffee, or oil palm. Oil palm looks especially promising in that they do not require a rich soil. Some of the old rain forest areas that have reverted to savanna under poor management have been successfully planted to oil palms. Although oil palms can grow on these poor soils, they yield much better under a good fertilizer program.

Oxisols respond very differently to management then do adjacent Alfisols. Only in the last one-half of the twentieth century has it been recognized that Oxisols could be productive after proper management techniques were developed. Good management involves extensive use of fertilizer with particular attention to micronutrient needs. Heavy rates of lime are often required to reduce aluminum toxicity, and the lime may need to be incorporated very deeply in the soil.

Conservation practices that protect the soil from the intensity of the sun and the impact of heavy rains are important. Insects and diseases frequently flourish in the warm and humid tropics, and their control is essential to maximize yields. Sets of management principles suitable to the soils of the tropics are slowly being discovered. They frequently share the common need for chemicals, machinery, fossil fuels, and other inputs that are beyond the economic capability of farmers in their local political and economic structure. Thus the management techniques need to be further refined to fit local economics before increased productivity is possible.

Plinthite (laterite)

The term "laterite" was coined in India in the early nineteenth century to describe a material rich in secondary iron and/or aluminum, low in organic matter, and depleted of bases. It can develop on any parent material, including limestone, if there is sufficient iron present. It most commonly forms on a level to gently sloping landscape with a fluctuating water table. When there is no water table, the plinthite is presumed to be a fossil formed in an earlier time when a water table was present.

Laterally moving waters often bring in additional iron, which precipitates out at the capillary fringe of the water table.

Plinthite seems to form primarily under trees, more frequently where there is a 2- to 4-month dry season. It forms below the surface and seldom hardens under a forest. If the forest under which the plinthite developed is invaded by savanna, the plinthite may harden. This can restrict the root zone of plants and further reduce its agricultural potential. Plinthite that hardens in this manner has been known to resoften in as little as 16 years when the savanna reverts to forest. Plinthite at the surface is thought to result from erosion of the surface soil. Truncated plinthite may harden irreversibly at the surface, often resulting in a hard cap, but it is soft below. Road cuts through this material will cause the exposed soft plinthite to harden in 2 or 3 years.

Well-developed plinthite is red and white in color. The red fraction usually is higher in iron, and the white material is higher in kaolinite and quartz. The crystal growth of kaolinite during the dry period may be important in its development. Identifying soft plinthite usually is not easy. The best test is to remove a sample to the laboratory and run it through 5 to 10 wet-dry cycles. If it is friable and can be broken by hand after this treatment it is not plinthite. Plinthite will harden irreversibly when dried in sunlight, which allows it to be used for construction purposes (see Fig. 14-8). Usually the inhabitants of an area have already identified true plinthite and use it for road and building construction. Many of the ruins of buildings that have survived for centuries in the humid tropics were constructed of plinthite bricks. Angkor Wat in Cambodia, which has survived for 800 years, has a carved limestone facade over laterite.

Figure 14-8 Plinthite makes an excellent road-building material. It compacts well and provides a stable road surface.

Over the years, the use of the term laterite was broadened to include several other types of soil pans and then to describe any soil containing laterite. Finally, it was used as a general term for red soils in tropical areas, regardless of the presence or absence of laterite. *Soil Taxonomy* introduced the term "plinthite," reflecting current understanding of the pan originally defined as laterite.

Although plinthite is found in Oxisols, some Ultisols, and even a few Alfisols, its presence is not as widespread as once thought. Part of this misconception resulted from the interchangeability of the term laterite for plinthite and for red tropical soils. Plinthite is a fascinating material and has been widely studied, but its extent in the tropics and the degree to which plinthite hardening will restrict expanded food production in the tropics has been greatly exaggerated.

Shifting Agriculture

Shifting agriculture is well adapted to soils of low fertility, where most of the nutrients are tied up in the native vegetation and soil organic matter. Shifting cultivation systems are complex and varied. They all combine a period of cultivation with a fallow period in which the soil fertility is rejuvenated. Techniques range from simply buring and planting to techniques utilizing ashes, manure, and irrigation combined with a variety of crop rotations, intercropping, and double cropping.

Usually, the land is cleared of vegetation in the dry season. Trees may be partially or completely girdled as much as a year ahead of time. Toward the end of the dry season, the cut material is burned off (see Fig. 14-9). The primary purpose of

Figure 14-9   Burning the vegetation in preparation for planting under a shifting agriculture system. (Photograph courtesy D. K. Whigham.)

burning is to get rid of excess vegetation. The burning increases the pH and the base level of the surface soil. Carbon, nitrogen, and sulfur are lost from the litter but are essentially unaltered in the soil humus. The heat seems to improve structure and may have some sterilization affects. The crops are then planted. A variety of crops are often planted as a hedge against the weather. Yields are generally good for the first 2 to 3 years but decline rapidly after that. Rice, corn, pulses, or vegetables are commonly planted the first year or two. Then perennials and root crops are grown.

Soil structure deteriorates rapidly when bare soils are exposed to direct solar radiation and raindrop impact. Some rotations recommend grain or vegetable crops in the first year or two of cultivation, followed by bananas or a crop that either shades the ground or is tolerant of the shade of the regrowing trees.

Most fields are abandoned within 3 to 5 years for the following reasons.

1. Grassy weeds take over. With hand tools and fire it is easier to clear trees and undergrowth than to control grassy weeds in the humid tropics.
2. Soil fertility is depleted as nutrients are leached and harvested.
3. Physical properties deteriorate due to raindrop impact on bare soil, and the heat of the sun destroys organic matter.
4. Insects or disease problems become severe, since there is no period of freezing temperatures to break life cycles.
5. The village moves on to another area because of tradition or for reasons unrelated to agriculture.

Once a field is abandoned, the native vegetation quickly takes over and the fallow period lasts 8 to 20 years. The purposes of the fallow period are to reverse the detrimental effects of cultivation mentioned in points 1 to 4. Under tropical conditions, organic matter levels are quickly reestablished. The organic matter level can usually be brought back within 5 to 8 years. Five years after corn production, a forest can accumulate 94 percent of its maximum leaf production, 80 percent of its maximum litter, and 64 percent of its maximum root growth. The single most critical factor seems to be the level of available phosphorus. It seems that as soon as the available phosphorus level is restored, the soil is ready for another production cycle.

The Influence of Trees and Grasses on Tropical Soils

Contrary to our experiences in temperate climates, trees and not grasses are most efficient in restoring soil physical properties and nutrient levels in the soil. Tropical grasslands often provide less organic matter than temperate region grasslands. A tropical forest will add up to two to five times as much fresh organic matter as a tropical grassland. While the potential for growth may be higher in tropical regions, it is often severely limited by poor soil conditions, especially aluminum toxicity. Since grass provides less litter, it supports a reduced population of beneficial soil insects and small animals. Thus the restoration of structure is slower under grass. The deeper-rooted trees seem better able to extract phosphorus from deep in the profile and bring it to the surface. In one 20-year fallow experiment, the phosphorus of the surface layer of the soil under trees was 30 percent higher than the soil under a savanna fallow. Likewise, the nutrient content of the leaves deposited on the soil surface by trees helped to quickly build up the nutrient level of the soil.

A series of short fallow periods favor a savanna regrowth instead of trees. Thus formerly forested areas can be converted to savanna. Compared to a forest fallow, a savanna fallow requires more tillage to eliminate grasses in preparation for cultivation. This extra tillage is detrimental to structure, so it is desirable to use techniques that encourage a tree fallow. One method is not to kill back all the trees completely. A few trees are left or allowed to reestablish from the old stump during the growing period. This gives them a head start for the fallow period.

Another interesting difference between trees and savanna is the available nitrogen level in the soil during the first year or two of cultivation. The available nitrogen level is high after the clearing of forest, and a moderate level of nitrates may be maintained in the soil for several years. Burning of the savanna for cultivation may result in an inhibition of the nitrification process. The exact mechanism is unclear, but it causes the nitrogen to remain tied up in the organic matter. The nitrate level in the soil remains low for some time after the savanna has been destroyed. Under natural conditions, this phenomenon would suppress leaching of nitrates while natural vegetation was reestablished. This would be very desirable after the vegetative cover was destroyed by a fire or a severe drought. A second benefit of the nitrate suppression is the corresponding suppression of leaching of cations. Nitrate is an important anion accompanying the leaching cations. Its suppression limits excessive cation leaching after a loss of vegetation when water movement through the unprotected soil would be high. But in a shifting agricultural system this inhibition of nitrification suppresses yields.

Future Potential

As populations increase, the pressure for land to produce food intensifies. There is a tendency to lengthen the cultivation period and decrease the length of fallow. The land is returned to production before it is fully restored, and in the next production cycle it yields less and is productive for a shorter period. These lower yields in an area with an expanding population force even shorter fallow periods, which simply intensifies the problem. Shifting agriculture is only feasible in areas of low population densities.

Research continues for methods to develop better management practices to obtain higher, long-term productivity from the land under shifting agriculture.

Studies have shown that the use of manure and fertilizers, especially nitrogen and phosphorus, can give good yields with continuous cropping.

In parts of Africa economics and politics allow large-scale cultivation. But problems occur with this system, too. Erosion is much more serious under continuous cropping, where the fields are large and the soil surface smooth. Under shifting agriculture, the fields tend to be random and the soil surface irregular, trapping potentially erosive water. Since the natural fertility balance is fragile, the less the soil is disturbed, the less likely the soil will deteriorate. In some areas perennial crops such as oil palm, coffee, and rubber are a better land use than shifting agriculture.

Perhaps it should be pointed out that shifting agriculture is not unique to African Oxisols. The basic principles developed, probably independently, in many parts of the world where natural fertility limits yields. Shifting agriculture is about as old as agriculture and was common in the Middle Ages in Europe on Alfisols. It was known as the three-field system, a 3-year rotation that included 1 year fallow. A type of shifting agriculture was common in the Ultisol regions of North America during colonial times. Land was cultivated until yields dropped to unprofitable levels. Then the land was abandoned and the settlers moved west. Extensive erosion occurred during the latter parts of the cultivation period and during the first few years after abandonment before a good vegetative cover was established. Today many of these eroded rolling landscapes are not economical for crop production and are forested. Others are being brought under production again. The development of new management techniques involving the use of chemicals and heavy machinery allows these lands to be productive once again.

Alfisols

The widespread distribution of Alfisols on the African continent has caused some heated debates among soil scientists. Figure 14-10 shows the currently accepted distribution of Alfisols in Africa. Figure 14-11 is a soil map of Africa published by Kellogg in 1941. It has been simplified to highlight the most significant kinds of African soils. Kellogg's Sierozens and related soils can be roughly translated as Aridisols. The northern Sierozems in Fig. 14-11 approximate the northern Aridisols in Fig. 14-6. The southern African Sierozems include much of the Aridisols and part of the Entisol area of Fig. 14-6. The grassland soils were subdivided by Kellogg into: (1) Prairie soils and degraded Chernozems, (2) Chernozems and Reddish-Chestnut soils, and (3) Chestnut, Brown, and Reddish-Brown soils. These have been frequently translated as Udolls and Ustolls. These grassland soils were mapped here primarily on the widespread distribution of savanna vegetation, but further studies of the soil of this region have shown that they do not have the properties necessary for Mollisols. In fact, Mollisols do not appear as the primary soil in any part of Africa on our soils map (Figure 14-6).

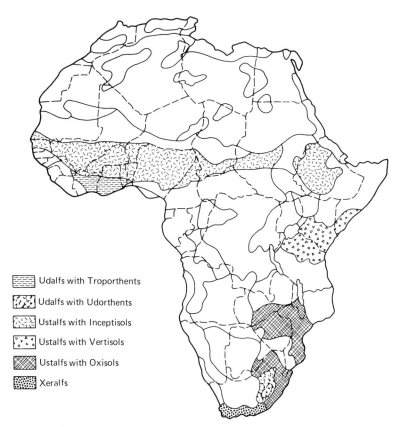

Udalfs with Troporthents

Udalfs with Udorthents

Ustalfs with Inceptisols

Ustalfs with Vertisols

Ustalfs with Oxisols

Xeralfs

Figure 14-10 Alfisols of Africa.

Early Classification

The Lateritic soils of Kellogg's map are often translated as Oxisols. The latest map of Africa shows much of this area as Alfisols. Alfisols have been stereotyped as soils of the cool, forested, temperate zone. Alfisols are found in the northeast part of the United States. Some soil scientists feel they do not belong in the warm tropical parts of Africa. The basic problem lies in the changing criteria for classification.

Early classification of Lateritic soils emphasized their color (usually red or reddish yellow), their low productivity, and the presumed presence of laterite (plinthite). American soil classification wherever possible, ignores color and considers productivity as an incidental property in classification and the presence or absence of plinthite of secondary importance. Primary emphasis is placed on the presence of an oxic horizon (Oxisols) or an argillic horizon with either a high base status (Alfisols) or a low base status (Ultisols). While it is helpful to draw parallels

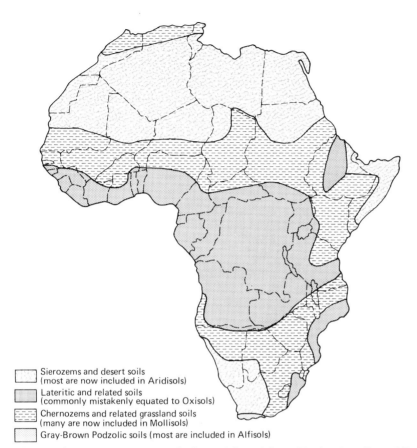

Sierozems and desert soils
(most are now included in Aridisols)
Lateritic and related soils
(commonly mistakenly equated to Oxisols)
Chernozems and related grassland soils
(many are now included in Mollisols)
Gray-Brown Podzolic soils (most are included in Alfisols)

Figure 14-11   Early soil map of Africa. (Adapted from Charles E. Kellogg, *The Soils That Support Us,* The Macmillan Company, copyright © 1941 and used by permission of The Macmillan Company.)

between terms in various systems for comparison purposes, it is dangerous to make comparisons on soil maps. In theory, most Chestnut soils would be included as Ustolls. But the area in central Africa once mapped for Chestnut soils contains mostly Alfisols.

Certainly it is inaccurate to imply that the tropical African Alfisols are identical in properties and potential productivity to the Alfisols of the U.S. grain belt. On the other hand, many of the early classification systems lumped together soils of widely divergent properties and referred to them simply as "tropical soils." This masked important distinctions between them and has delayed our understanding. Because of this emphasis on different soil properties in newer classification systems, more recent detailed studies have lead to soil maps that differ considerably from earlier soil maps.

The Northern Ustalfs

The largest area of Alfisols lie in the area earlier mapped as grassland soils in a belt 10 degrees north of the equator. The area contains Ustalfs and receives an average of 1000 to 1800 millimeters of rainfall per year, with a distinct dry season. The parent materials tend to be crystalline acid rocks higher in quartz and lower in iron than those of other parts of tropical Africa. The resulting soils are lighter in texture and, because of the absence of iron, are less likely to have plinthite. The Ustalfs are associated with Inceptisols, which are generally on the more recently exposed slopes.

The Ustalfs have a distinct textural B horizon. Kaolinite clays dominate, with 2:1 clays often present in small amounts. This gives a low cation-exchange capacity, but the base saturation is sufficiently high to classify them as Alfisols. Free iron has been leached and may form iron concretions within the profile. The reserve of weatherable minerals is often appreciable, which is characteristic of Alfisols. These soils frequently have bedrock within 100 centimeters of the soil surface that provides a source of weatherable nutrients within the root zone of many of the tropical plants.

The agricultural potential is good, but it is not without problems. Alfisols tend to be moderately low in fertility but respond well to fertilization. The pounding effect of heavy monsoon rainstorms on unprotected surfaces shallow to bedrock can lead to severe erosion problems. Management systems incorporating rotations, fertilization, surface water control devices, and protection of the bare surface during the rainy season must be practiced.

Pullan describes a region in northern Nigeria in this soil region. Tribes have traditionally lived on the high terraces and farmed the lower terraces. As population pressures increased, they began farming the uplands. In some cases, upland fields were as much as 20 kilometers from the village. The lack of erosion practices on cultivated land, combined with overgrazing by cattle, has caused devastating erosion. Some of the soils are beyond reclamation.

Even though this is an old, flat, shield area, many soils are young because of natural erosion or erosion induced by humans. In parts of Africa, the bedrock is within 2 meters of the surface. Where this rock is high in weatherable minerals and within the reach of plant roots, crop yields can be good.

Encrouching Deserts

In western Africa along the Aridisol-Alfisol boundary is a region known as the Sahel. This region, just south of the Sahara, receives sufficient rainfall for some drought-resistant grains. Good yields of cotton and millet are obtained where water is available for irrigation. However, much of this area is restricted to grazing. During normal to wet years, the natural vegetation easily supports the nomadic herdsmen who roam the area with their livestock and the subsistence farmers (see Fig. 14-12).

Figure 14-12   A herd of cattle being driven to better grazing land by nomadic herders. Dry years at the fringes of the desert are devastating to cattle numbers. (Photograph courtesy D. K. Whigham.)

As populations increase, there is an increased pressure on the land. This is causing deserts to spread and new deserts to appear. It has been estimated that one-third of the world's arable land could easily be lost by this process of *desertification.* Shifts in rainfall patterns in Africa in the early 1970s brought a severe drought to the Sahel and Ethiopia. A scarcity of water reduced plant growth, and the pressure on the land resulted in overgrazing by livestock, overcutting of forests, and improper tillage of cropland. With the loss of many watering holes, nomads and their livestock concentrated near the remaining water resources. Vegetation was stripped from the land, preventing natural regeneration. Feet and hooves trampled the ground, compacting it to the point that seedling development and root growth were almost impossible.

Ironically, western agricultural techniques introduced into the area in the last few decades seem to have aided the desertification processes. Religious practices and traditions dictated periods of fallow, which encouraged rejuvenation. Fertilizer, pesticides, and irrigation pumps have allowed more intensive land use. Improvements in human and veterinary medicine have resulted in more people and animals. In a normal year there is no problem, but a series of dry years can lead to permanent losses of productive land.

Udalfs Southwest of the Sahel

To the southwest of the Sahel lies an area of Udalfs. The boundary between the Ustalfs and Udalfs corresponds roughly to the boundary between interior savanna and the coastal tropical forests, which reflects the increase in rainfall to the south. The forested Udalfs to the south are closer to the equatory and to the coast.

The increase in temperature and rainfall would suggest that the Udalfs would be more weathered and less fertile than the Ustalfs south of the Sahara. In fact, to some extent, just the opposite is true. The effect of erosion has been more serious in the humid region and has the effect of keeping the soils younger. This is partly suggested by the shift in the secondary soils. Entisols are associated with the Udalfs of the more humid coastal region, while Inceptisols are found in the Ustalf region south of the Sahara. The youthfulness of the Udalf region results in generally more fertile soils with a higher potential agricultural productivity. This potential is aided by the greater and more dependable rainfall. However, past erosion has lead to a rougher terrain, and conservation practices are essential. This area is often more suited to perennial industrial crops such as coffee, tea, or tung oil, which do well on rolling landscapes when the surface is protected.

The Southeastern Alfisols

To the east of Lake Victoria is another region of Ustalf soils. Here the secondary soils are Vertisols (see Fig. 14-10). This region of Africa is drier than the Alfisol regions to the west and has a more distinct dry season. Much of this area receives 200 to 600 millimeters of precipitation annually. Frequently, the climatic pattern alternates between excessive drought and inundating floods. In addition, the area is mountainous. The Alfisols are found in the uplands, with the Vertisols formed in lake clays. In some parts of Kenya, erosion off the hills has all but eliminated the productivity of the upland soils, but the deposition of this eroded material in valleys has resulted in several very productive alluvial areas.

Traditionally, much of this region was composed of cattle-dependent subsistence farms of low productivity. Sorghum, millet, and corn are grown widely. Cotton and sugarcane are good cash crops but are basically confined to the deeper soils on stream banks, deltas, and colluvial slopes. Many of these soils are Vertisols. Fertility levels of the Alfisols are low. Aluminum and manganese toxicities are common. The relatively low cation-exchange capacity requires frequent, light fertilizer application to avoid saturation, imbalance, and leaching. Heavy applications of nitrogen, phosphorus, and potassium often interfere with the uptake of calcium, magnesium, zinc, and occasionally copper.

Some of the better upland soils of the region are occupied by plantations, and coffee and tea are produced.

Further south along the east coast are two areas of Ustalfs with Ustox. Still further south is an area of Udalfs with Entisols. Along the southern coast of Africa is an area of Xeralfs. The Xeralf region is in the southern Mediterranean climate zone and receives its rainfall primarily in the cool season.

The southern Alfisols are just east of a desert and in places are adjacent to Aridisols. Grain, including corn, are produced in this region, but rainfall is not always dependable. Management of these soils therefore emphasizes erosion control and moisture conservation. The success of these techniques has resulted in good productivity. The Alfisols of southern Africa are included in the area known as the "maize triangle." This is one of the few parts of Africa where agricultural systems approach the "family farm" concepts of the United States.

Entisols

The soil map of Africa shows five major Entisol regions (Fig. 14-13). Two are located in desert regions, and three lie south of the equator in areas of higher rainfall. As mentioned earlier, the sand dunes of the Sahara are not widespread. The sandiest areas are Psamments with shifting sands. They are surrounded by Aridisols of moderate to low sand content. The climate is hot, with very low, erratic rainfall. Unprotected by vegetation, the sand particles are subject to continuous wind movement. The climate, lack of vegetation, and continued mechanical movement prevent profile development. A second region of shifting sands is in the

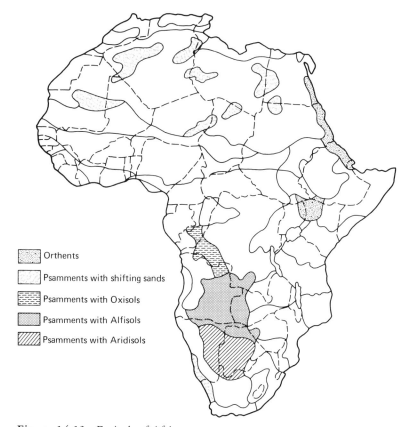

Orthents
Psamments with shifting sands
Psamments with Oxisols
Psamments with Alfisols
Psamments with Aridisols

Figure 14-13 Entisols of Africa.

Namib Desert along the southwest coast of Africa, in both areas agriculture is restricted to a few isolated areas where water is available.

In Kenya around Lake Turkana (formerly Lake Rudolf) and on the western shore of the Red Sea are two regions of Orthents. The Kenya area lies along the equator but at the eastern or drier end of the equatorial wetlands. As Fig. 14-2 indicates, the precipitation along the equator declines from west to east. The soils around Lake Turkana tend to be rocky and stoney, and this restricts soil profile development. It is also a more hilly region with steeper slopes. Continued erosion prevents extensive soil development. Recent volcanic activity has lead to youthful soils in local areas. Organic matter levels of the soils are low, and biological activity is very low. The overall agriculture potential of the region is low. Where irrigation is possible, yields may be good, but soluble salt problems often occur. Where local climate is sufficiently humid, forests grow well. In some areas, trade in resins, incense, gum arabic, or other forest products has been developed.

The region of Orthents along the Red Sea in northern Africa is generally drier and not as hilly. Because of the low rainfall, agricultural production is even more limited in this area than in the similar soils of the Lake Turkana area.

The more humid Entisols are located in southern Africa. These are primarily Psamments. They are divided into three subregions from north to south based on the common secondary soil present. This generally reflects a decrease in rainfall from the equator toward the Kalahari Desert.

The precipitation in the northern region often exceeds 1000 millimeters. Here the Psamments are associated with the more intensively weathered Oxisols. The Oxisols may, in fact, be a remnant of an earlier climatic condition instead of having been formed in the present dry savanna.

The precipitation decreases in the central Entisol region. Here the Psamments are associated with Ustalf soils. The Ustalfs are generally found on older surfaces or where local conditions are favorable for soil development.

The soils in the driest region are formed on aeolian deposits in areas where the precipitation seldom exceeds 500 millimeters. Aridisols are the associated soil. Grazing is the best use of the land on this steep landscape. Where irrigation is possible, millet and cotton are produced.

Inceptisols

Throughout the world Inceptisols and Entisols are frequently found in river valleys. Most of these valleys are too narrow to be shown on a generalized soil map except along the major rivers, in this case, the Nile, Niger, and Zaire rivers.

The upper regions of the Zaire and the Nile rivers in central Africa are in large, flat plains near the equator (areas 12h of Fig. 1-11). The tropical climate and the poor drainage of these plains give rise to Tropaquepts. The frequent flooding and low relief result in a complex soil pattern. The soils and their agricultural potential range from very good to very poor in a complex pattern.

In theory, these two headwater regions should have good potential for expanded agriculture, but such expansion is not without problems. First, an extensive soil survey to identify the location and extent of soils with good potential productivity is needed. Second, a system to control flooding and to drain excess water must be developed. Third, an agricultural system that will work in a continuously warm (and in the western area, a continuously humid) climate is needed. Here there is no cold period to help control crop pests, so suitable management systems using rotations, chemicals, and other practices must be developed.

The Zaire (formerly the Congo) River delta is relatively small and does not show up on our map. The Nile delta is larger and is mapped as Haplaquepts. For the last 400 kilometers or so, the alluvial valley often exceeds 25 kilometers in width. Further south, the valley is perhaps 2 kilometers or less across and nearly disappears when the river cuts through rocky parent material. This is one of the oldest agricultural regions of the world. Early in human civilization people learned to work with the floodwaters of the Nile, which provided an annual application of nutrients. Drainage and irrigation systems were designed that were able to support a large population. Unfortunately, for most of its length the river flows through a desert, and a region of productive soils is confined to a narrow valley.

The Niger River flows into the Pacific in Nigeria just north of the equator. These delta soils are also classified as Tropaquepts. The region along the mouth of the Niger River has a very different character from the soils at the mouth of the Nile. This character is shared by many of the river deltas along the west coast of Africa. Mangrove swamps have developed in fine-textured, marine, and alluvial sediments along the coast. In their present location they are subject to periodic flooding at high tides by brackish water or saltwater. They accumulate sulfur, which is not a problem as long as poor drainage keeps the soil in a reduced state.

Drainage and oxidation of these soils lead to the formation of large concentrations of sulfates. The pH of the soil drops to a value of 3 or less, and the availability of aluminum, manganese, and iron may increase to levels toxic to plants. Reclamation is a substantial undertaking requiring dikes and other devices to prevent salt intrusion. Large quantities of organic matter and other amendments to neutralize the acidity are also needed. Because of the expense and difficulty of reclaiming these soils, they are frequently used for lowland rice or simply are wasteland. Lowland rice does well where the continuous water cover maintains the sulfur in a reduced, nontoxic form. These high sulfur soils are called Sulfaquepts.

Inceptisols are also found along the northern coast of Africa. Here the soils are very different from the Inceptisols of the alluvial valleys. These soils are Xerochrepts and are associated with Xerolls. They formed in a Mediterranean climate, principally in Morocco and Algeria. This area is on the fringe of the Sahara, and rainfall often does not exceed 500 millimeters, most of it falling in the winter season. Leaching of the carbonates is not extensive in these soils, giving them a high percent base saturation and pH value. The soils may contain up to 3 percent organic matter.

The agricultural potential of this region of North Africa is good, but it is confined to a very narrow band between the sea, the desert, and the Atlas Mountains. The rainfall is not always dependable, and yields in unirrigated regions can be very poor in dry years.

Ultisols

Ultisols dominate in three areas in equatorial Africa. Two adjacent areas are along the west coast and a third is west of Lake Victoria (see Fig. areas U4e, U3k, and U3h Fig. 1-11).

Ultisols of the west coast are divided into Ustult and Udult regions that primarily reflect the length of the dry season. The southern Udult region receives more precipitation, and the dry season is shorter than in the northern region. The FAO/UNESCO system maps this area as ferrallitic soils. In the FAO/UNESCO system the ferrallitic soils are the most extensive soils found in Africa (ignoring dry, soilless areas of the desert). These soils are considered to be in the final stages of weathering, and even kaolinite and quartz have been altered. Many are polygenetic and reflect previous as well as present environmental conditions. Ferrallitic soils frequently lack an oxic horizon but contain an argillic horizon and so are included as Ultisols and Alfisols as well as Oxisols.

Ustults and Udults

The Ustults on the west coast often lack sufficient moisture in the growing season for top productivity. This is especially true on sandy and coarse-textured soils with a low water-holding capacity. The Udults adjacent to them lack the distinct dry season. Both receive sufficient moisture at some time each year to leach the profile and prevent any signficant nutrient buildup. Generally, most of the bases are in the organic matter or the surface mineral horizon. Fertility and percent base saturation decrease with depth. Clearing the forest vegetation and planting crops quickly exhaust the native fertility. These soils must be either heavily fertilized or managed by shifting agricultural methods. Heavy fertilization is often economically impossible, although this undoubtedly will change some in the future.

The third region of Ultisols is located north and west of Lake Victoria. In the immediate vicinity of the lake, the soil patterns are very complex because of historic changes in the character of the lake during different geologic periods. In addition, it is a more mountainous region. The predominant soils are Udults associated with Oxisols rather than with Alfisols, as in the west. As a result, the agricultural potential of this area is lower.

Potential Productivity

In general, the potential productivity of Ultisols depends on the use of shifting agriculture or chemical fertilizers. Nitrogen and phosphorus fertilizers are almost always beneficial in combination. Lime and other macronutrients are usually beneficial, and the periodic application of one or more micronutrient is frequently necessary. Because of the generally low fertility levels, organic matter applications are often more profitable than the use of only one or two nutrients in mineral form. But with the proper use of fertilizers, many Ultisols in Africa can give good sustained yields.

Vertisols

Africa has nearly 40 percent of the world's Vertisols. They occupy about 3 percent of the African continent, but few are cultivated, despite the fact that they include some of Africa's best soils. They are found in at least 33 countries in Africa, but only in the upper regions of the Nile is the area sufficiently large to appear on the maps in Fig. 1-11 or Fig. 14-6. This region is primarily Usterts—Vertisols with a pronounced dry season. This is probably the largest expanse of Vertisols in the world that are formed from by base enrichment. Water brings in bases leached from the surrounding uplands. Africa has nearly 700,000 square kilometers of this type of Vertisol.

The Vertisols of the upper Nile often have poor internal and external drainage in a climate with a marked dry season. The mean annual precipitation is seldom over 1000 millimeters, and the dry season lasts at least 3 months. Since the base enrichment is primarily from drainage water leaching from the surrounding upland areas, excess salts are sometimes a problem.

In addition, there are another 250,000 square kilometers of Vertisols that are shallow to bedrock. These shallower soils tend to be found in small acreages scattered across the continent. They develop in basic rocks rich in ferromagnesium minerals or calcium-containing rocks. They tend to be more fertile and have a higher productivity potential and generally have fewer problems with soluble salts. These are some of the best soils in Africa. They are capable of producing a wide variety of crops, including sugarcane, cotton, grain, tobacco, and fodder. But many of these Vertisols are uncultivated or are used extensively for pasture and grazing.

The water-holding characteristics and tilth of Vertisols makes them difficult to manage. When they are wet, they are sticky and puddle easily. When they are dry, they are often hard, difficult to till, and susceptible to shattering. The resulting fine powder is very susceptible to wind erosion. The moisture range for ideal tillage is relatively narrow. They have been referred to as "noon hour soils," implying that in the morning they are too wet and in the afternoon they are too dry. They move quickly from one extreme to the other through the ideal tilth zone. Because of this, they are used more extensively in areas of the world where a high degree of mechanization is possible. This allows quick tillage of large acreages of land during the short ideal tilth period.

Many of Africa's Vertisols are in less developed areas where animal and human energy are the principal source of power for tillage. Such tillage operations are too slow, and thus Vertisols are primarily in pasture or are grazed. They do have a good potential for future development but only after social and economic changes are made. Most of the African managers of Vertisols are unable to accumulate the kind and size of power equipment and the trained manpower to maintain and operate the equipment. This would be needed for only a few ideal days each year when the soil must be tilled at top speed. During the rest of the year, much of this equipment and manpower could not be used on Vertisol soils. Farmers cannot afford this investment on small holdings that exist at, or not far above, subsistence levels.

Trees that provide a canopy and protect Vertisols from erosion are desirable. Cocoa and rubber are suited for Vertisols. Coconuts and citrus do not give a complete cover and are less suitable. Erosion control on slopes, including contouring, is beneficial. Because of the cracking nature of the clay, it is often impossible to construct earth terraces that will hold water.

Histosols, Spodosols, and Mollisols

The soil map of the world (Fig. 1-11) shows that Histosols do not dominate in any part of Africa; they are found scattered among the other soils, in swampy areas, particularly in areas of very high rainfall, and in areas of subhumid to semiarid conditions, where local depressions cause water accumulation. In Uganda reclamation of papyrus (L; *Cyperus Papayrus*) swamps is hindered by the high level of sulfur, which oxidizes when the soil is drained. The oxidized sulfur lowers soil pH and renders the soil essentially worthless for agricultural production. Phragnite swamps *(Phragmites communis Trin.)* in Mozambique do not have a high sulfur content and can be reclaimed more easily.

The main problem with their use, particularly in a dry region, is that the soil rapidly oxidizes upon drainage and hardens, often irreversibly. This alters the physical properties of the soil, reducing both wetability and water-holding capacity. The ease with which dry organic matter burns, particularly in an area with a dry season where shifting agriculture techniques are practiced, is a serious problem. When properly reclaimed, they are suitable for very intensive cultivation of vegetables and bananas.

Other Histosols develop at very high altitudes. These are not extensive but are found in a few of the highest mountain regions. Often they are associated with bare rock, glaciers, and snow packs and therefore have no agricultural value.

No extensive region of Spodosols is mapped in Africa. However, Spodosols are found on the continent, principally scattered along the western coastal fringe. Most are Aquods found on sandy deltas and in alluvial deposits in equatorial or subequatorial climates.

There are no large areas where Mollisols dominate, so no area of Mollisols is shown on the map. They are, however, found throughout Africa as a secondary soil, particularly along the Alfisol-Aridisol boundary. Xerolls are found in the Mediterranean climatic zone in North Africa.

The Soils of Africa and Their Food Production Potential

It has been estimated that the cultivated areas in the world could be doubled. One-half of this new area would be in the tropics. Africa contains more tropical land area than any other continent. Therefore there should be considerable room for expansion of agricultural land in Africa. The problem is to identify the suitable land and develop a technology to maximize its use.

Africa is a large, old, dry continent where the rejuvenation processes of glaciation, erosion, and deposition have not played such an important role. Thus the relatively youthful and productive soils found in the Northern Hemisphere are not as extensive in Africa. The soils map of Africa would indicate that Aridisols are the most widespread soil order. With a few exceptions they are too dry to be of agricultural value and are primarily restricted to extensive grazing. The potential for developing these areas by irrigation is severly limited by the lack of water.

Surrounding the desert regions are the Entisols, Alfisols, and Inceptisols. In wet years they are productive, but they go through frequent cycles of restricted rainfall in which crops fail and grazing pressure causes overgrazing. This results in the destruction of surface vegetation and, ultimately, the deterioration of the soil itself. Thus the region of Aridisols is currently enlarging into surrounding areas as vegetation and soils are being destroyed.

Vertisols hold considerable promise for expanded productivity. However, the physical limitations require a highly mechanized agricultural system in order to utilize this potential to the fullest. In much of Africa, the present social, political, and economic atmosphere precludes that rapid development.

Current Problems

The more humid areas are dominated by Alfisols, Ultisols, and Oxisols. The potential of these soils is uncertain. It may be true that nearly one-half of the potentially arable land in the world is in the tropics, much of it in the African tropics. But this statement is very misleading. On the one hand, the statement is based on current technology. Much of this technology is not currently applicable to Africa because of political, social, and economic constraints. In many African countries these constraints must change dramatically before the potential of the soils can be realized. This will be a slow and gradual process, full of frustration for both the developing countries and the developed countries that are trying to help.

Second, agricultural technology is complicated. Progress involves an understanding of local climatic and biological factors (see Figs. 14-14 and 14-15). Those can change dramatically over short distances, and technology must be adapted to each unique set of conditions. This hinders or prevents the mass transfer of technical knowledge across large areas. As an example, the timing of our tillage, fertilizing, spraying, and other field operations in the U.S. midwest is geared to a planting season that follows a cold period. During this time, pest populations die back and low evaporation allows rain and snow to recharge the soil moisture to near field

Figure 14-14 Tea on a rolling and eroded land surface. The plant canopy essentially stops erosion and allows a high economic return from poor land. (Photograph courtesy D. K. Whigham.)

Figure 14-15 Soybean production on the African savanna. The beans are chlorotic because they were not inoculated. Note the termite mound at the edge of the field. (Photograph courtesy D. K. Whigham.)

capacity. In the tropics the planting season follows a dry period. There is no stored soil water. Plants that are dormant during the dry period spring to life with the first rains. Simple things such as timing and sequencing of cultural practices must be adapted to local conditions. Transplanting temperate regions practices frequently will not work.

A third handicap is that soils have not been inventoried over much of the continent. Where they have been inventoried, much of the work is preliminary, not systematic and not correlated with the efforts of other workers. We do not know as much about the soils in the field nor do we have extensive laboratory studies to back up the fieldwork as we do in North America. We do know a lot, but much remains to be learned.

Building Soil Productivity

In general, we know that soils of the humid tropics are frequently very low in fertility. With fertilization and proper management, productivity can be very good. Shifting agricultural systems have been developed by trial and error. More recently, scientific studies have been undertaken to understand better precisely how and why these systems work and to identify their limiting factors. From this research has come new ideas that can be combined with a low level of technology that can be justified in the local economy. This will allow a higher level of productivity and raise the standard of living above the present subsistence level. Hopefully, this can cycle upward, lifting agricultural productivity along with the economic and social standards. But improving economics is not always simple. It is one thing to provide a village storekeeper with adequate supplies of low-cost fertilizer. It is something else to convince a farmer to use it when he has 2 hectares of land 10 kilometers from the village on a trail too narrow and steep for a bicycle.

Much emphasis has been placed on the fertility aspect of improving yields. Erosion problems in Africa also need emphasis. The combination of long, dry periods followed by heavy, high-energy rains causes severe soil erosion problems. Residues help protect soil structure, but they often interfere with soil preparation when hand labor or animal power is used. Second, this residue material is valuable for thatched roofs, animal feed, and bedding. Contour tillage and the construction of ridges to control water movement is beneficial, but they take effort to build and represent one more field operation that is in itself destructive of structure.

Plantation agriculture amid subsistence farmers is possible only because productivity can be very high when technology is implemented and markets are available. Part of the low productivity of the tropics as a whole is due to poor management and failure to implement known technology. But the main factor limiting food production is the farmer. His health, education, goals, and social status have a significant impact on potential. Improving the financial reward for his effort while reducing the amount of hard physical labor are necessary to make progress.

The potential for increasing food production of Africa is real, but it depends on social, economic, and political growth and stability. The potential has been realized in some areas and can be realized in others. The agricultural potential is probably not as great as some proponents would like to believe, and it will not be realized without extensive research and hard work.

References

Buringh, P., *Introduction to the Study of Soils in Tropical and Subtropical Regions,* Center for Agricultural Publishing and Documentation, Wageningen, 1968.

D'Costa, V., and S. H. Ominde, *Soil and Land Use Survey of Kono Plain,* University of Nairobi, Department of Geography, Occasional Memoir 2, 1973.

D'Hoore, J. L., *Soil Map of Africa,* Publication 93, Commission for Technical Cooperation in Africa, Lagos, 1964.

Drosdoff, M., Ed., *Soils of the Humid Tropics,* National Academy of Science, Washington, D.C., 1972.

Kalpage, F. S. C. P., *Tropical Soils,* MacMillan of India, Delhi, 1974.

Kellogg, C. E., *The Soils That Support Us,* MacMillan, New York, 1941.

Moberg, J. P., *Formation, Mineralogy and Fertility of Some Tropical Highland Soils,* Copenhagen, 1974.

Moss, R. P., ed., *The Soil Resources of Tropical Africa,* Cambridge University Press, Cambridge, 1968.

Papadakis, J., *Soils of the World,* Elsevier, Amsterdam, 1969.

Pomeroy, P. E., *Some Effects of Mound-Building Termites on Soils in Uganda, Jour. Soil Sci., 27:*377–394, 1976.

Pullan, R. A., *The Soils, Soil Landscapes and Geomorphological Evolution of a Meta-sedimentary Area in Northern Nigeria,* Department of Geography, University of Liverpool Research paper No. 6, 1970.

Sanchez, P. A., "Soil Management Under Shifting Cultivation," in *A Review of Soils Research in Tropical Latin America,* P. A. Sanchez, Ed., *N.C. Agric. Exp. Stat. Tech. Bull., 219:*46–67, 1973.

Sivarajasingham, S., L. T. Alexander, J. G. Cady, and M. G. Cline, "Laterite" in *Advances in Agronomy,* Vol. 14, New York, Academic, 1962.

Van DerEyk, J. J., C. N. Macvicar, and J. M. DeVillers, *Soils of the Tugela Basin,* Town and Regional Planning Commission, Natal, South Africa, 1969.

Watson, J. P., "The Use of Mounds of the Termite *Macrotermes Falciger* as a Soil Amendment," *Jour. Soil Sci., 28:*664–672, 1977.

15

Asia

The Asian Environment

Asia is longer and wider than any of the other continents. It is a continent of extremes. Asia contains both the highest (8848 meters above sea level) and the lowest (396 meters below sea level) land surfaces on the earth. It includes some of the hottest, coldest, wettest, and driest areas found anywhere. Some of the most densely populated areas of the world are found in Asia along with uninhabited desert, jungle, tundra, and mountains. Asia is the most rugged of the continents. It is dominated by a mountain chain (the Himalyas and associated ranges) that runs east and west across the continent rather than north and south, as is the case in North and South America. The mountain system was formed by the folding, uplifting, and colliding of ancient tectonic plates. Asia, like Africa, has extensive desert zones. Most of them are near the mountains. To the north of the mountain chain, the land is cold and to the south it is strongly influenced by the monsoonal climate. About 30 percent of the continent has no drainage to the sea. The rivers and streams disappear into the earth to fill aquifers that provide water hundreds of kilometers away. One spot in Asia is nearly 3000 kilometers from the ocean in any direction. Thus Asia is a vast and varied area, and the soils reflect this variability.

Physiography

Asia lies on the eastern portion of the Eurasian continent. It extends from well north of the Arctic Circle to more than 10 degrees south of the equator (see Fig. 15-1) and from the Ural Mountains and the Caspian, Black, Mediterranean and Red seas on the west to the Pacific coast on the east. It includes a long chain of Pacific islands, including Japan, the Philippines, and Indonesia. It contains 30 percent of the landmass of the earth and approximately 60 percent of its people. It includes a higher proportion of inhospitable land than any other continent with the exception of Australia, and yet many believe that civilization had its roots in the fertile valleys of the Tigris and Euphrates rivers in western Asia.

The central mountain range is over 1000 meters above sea level. More than one-quarter of this range is above 5000 meters, and much of that has a permanent snow cover. Compared to Africa, Asia has a much higher percentage of its land below 200 meters. In Southeast Asia, China, central India, Saudi Arabia, and Russia are extensive regions of plateaus between 200 and 1000 meters in elevation. The Asiatic islands have a large proportion of their land surface either above 1000 meters or below 200 meters, with little land in between.

The ruggedness of much of the landscape combined with the large areas of desert result in the population density of less than 10 people per square kilometer over most of the continent. And yet Asia contains some of the most populated areas of the world.

The most densely populated areas include the Ganges River valley and coastal plains of India, the North China Plain and Szechwan Basin in China, the southern parts of Japan, and the island of Java in Indonesia. In much of this area the population density is over 200 persons per square kilometer.

In North America the population tends to be concentrated on the two coasts. The soils tend to be arid in the west and of lower fertility on the east coast. The fertile, deep soils of the prairies lie in the midlands. This region of lower population density is devoted primarily to the production of crops and livestock. In Asia the people live in the best agriculture regions. There is strong competition between urban and agricultural needs for land.

Climate

Figure 15-2 show the precipitation distribution in Asia. Much of Asia receives less than 250 millimeters of precipitation annually. At the other extreme, rainfall in parts of southern Asia exceed 3000 millimeters annually. Temperatures vary from the continuously cold areas of the Arctic Circle to the heat of the tropical jungles along the equator.

The climatic pattern of Asia is dominated by the vast mountain ranges of central Asia and the monsoon winds. To the north of the mountains, the climate is cold,

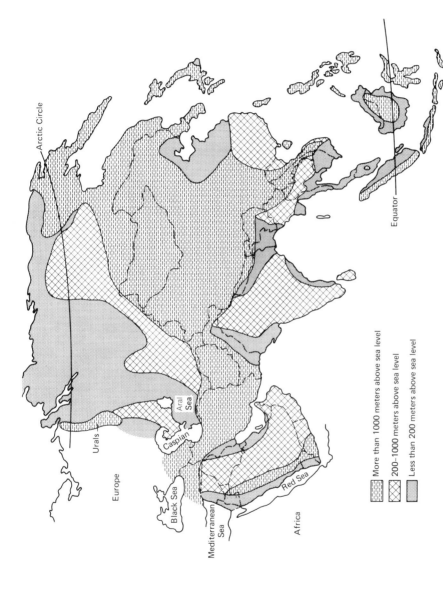

Figure 15-1 Generalized elevation map of Asia.

Figure 15-2   Generalized precipitation map of Asia.

comparable to that of Canada and Alaska. The precipitation is generally low but increases slightly to the west and north across a region of vast plains and low mountains in Asiatic Russia. The altitudes of the vast mountain ranges keep the temperatures cool in the central region. South of the Himalayas, the temperature is warm to hot to very hot, with precipitation influenced by the monsoon season.

   The precipitation of Asia is influenced by a high-pressure system that sits over the Gobi Desert for most of the winter. Winds radiate outward and rotate in a general clockwise direction around this high-pressure system. The winds move across the cold desert landmass toward the somewhat warmer ocean waters. These cold winter winds, moving initially across dry land, are low in moisture. As they leave the continent and move out across the ocean, they pick up moisture. The system then drops winter rains in a clockwise arc along the coast of Japan, the Philippines, Indonesia, and the coast of India.

In the summer there is a shift in the stratospheric jet streams, and the land is dominated by a low-pressure system centered in Pakistan and extending northeast to the Gobi Desert. This low-pressure system causes winds to blow inland and in a general counterclockwise direction. The winds first move across the warm ocean waters, picking up moisture. Then, moving inland, the moisture is dropped on the land between the ocean and the mountains from China and Japan to India.

In parts of Asia 80 to 90 percent of rain falls in the 3 to 4 month summer monsoon season. Much of Southeast Asia has two seasons—wet and dry. Other areas have three seasons—warm, cool, and wet. Still other regions have two distinct wet seasons, receiving both summer and winter rains. The pattern varies considerably from one locality to another, causing local variations in soil properties. Local patterns of soils and climate strongly influence the use of the soil.

Monsoon storms can contain tremendous amounts of energy and cause some fantastic climatic extremes. One of the wettest places in the world is in the plains north of the Bay of Bengal in India. The precipitation averages over 10,000 millimeters annually and has been known to reach 23,000 millimeters. China, in the wet year of 1935, reported one location receiving 2790 millimeters of rain in 24 hours and 13,200 millimeters in one 6-day period.

The summers are warm all over the continent but, in the winter, there is a wide range of temperatures. Near the equator, the monthly temperature mean from the hottest to the coldest month varies less than 5°C. In Siberia the January to July mean temperature may vary by as much as 65°C.

Vegetation

Like Africa, much of the inhabited parts of Asia have been cut over many times, so the influence of vegetation on soil properties has been altered. In general, the vegetation north of the central mountain chain shifts from tundra at the northern fringe across a vast area of coniferous forests to a region of steppe vegetation north of the mountains (see Fig. 15-3.) The steppe vegetation is most extensive in western Asiatic Russia. The central mountainous region is generally desert and high steppes, except where it gives way to alpine and snow-covered surfaces at the highest altitudes. Southern Asia is largely covered by rain forests, temperate forests, and savanna, reflecting local precipitation patterns.

The effect of large animals on the soil is usually considered to be less important than the influence of plants. There are, however, a few notable exceptions of great local significance. Prairie dogs in some parts of the steppes have reworked much of the soil volume. In other places the activity of earthworms is so important that vermic great groups of Mollisols have been defined in *Soil Taxonomy*. In the tropics the activities of termites can be very significant. It has been estimated that over 1 percent of the humid tropic uplands and over 2 percent of the subhumid tropical uplands have been reworked by termites. In some places over 50 percent of the surface soil has been reworked. The exact influence of this reworking on the soil is

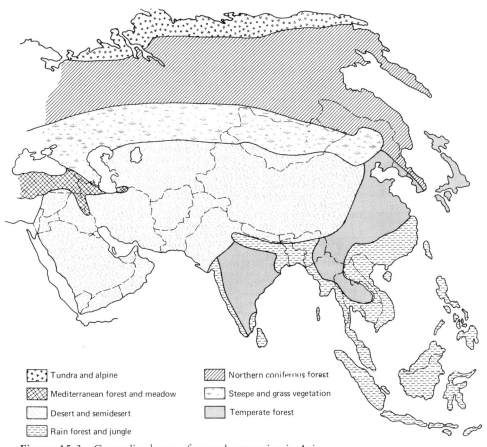

Figure 15-3   Generalized map of natural vegetation in Asia.

Legend:
- Tundra and alpine
- Mediterranean forest and meadow
- Desert and semidesert
- Rain forest and jungle
- Northern coniferous forest
- Steepe and grass vegetation
- Temperate forest

not clear. It apparently varies with the original soil properties and the species of termites. The effect is generally considered beneficial, although some detrimental affects have been reported.

Soil Regions of Asia

As we did with Africa, we begin our study of the soils of Asia by looking at a generalized soil map, Figure 15-4. The first striking feature of the generalized map is the dominance of mountains and deserts in central Asia. The mountains extend nearly the full length of the continent from east to west and occupy much of the northeastern part of the continent. Vast desert regions are found north and southwest of the central mountains.

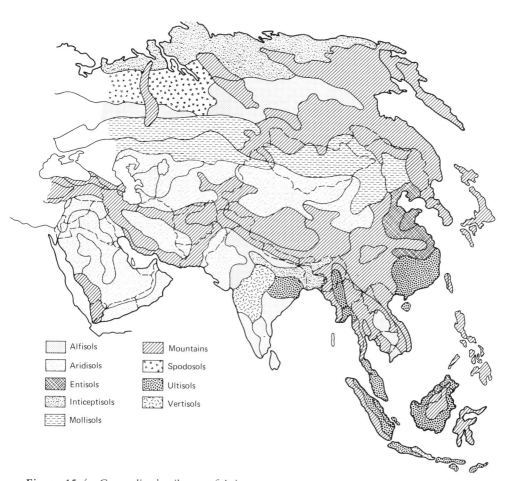

Figure 15-4 Generalized soil map of Asia.

This range of mountains and deserts splits Asia into two sections. North of this region in Asiatic Russia are four broad soil belts that range from the Inceptisols of the arctic tundra through Spodosols, Alfisols, and Mollisols. Figures 15-5 and 15-6 show the temperature and rainfall distribution pattern of Russia. This is in contrast to the United States, where temperature increases north and south and rainfall increases west to east. In Russia the temperature increases north to south and the rainfall increases south to north. This is superimposed over a landscape that rises in elevation from west to east.

Thus, where the climate is warm enough for crops, it is too dry, and where the climate is humid enough, it is too cold. At the lower elevations to the west, good crops are possible in years when the weather is good. To the east, as the elevation

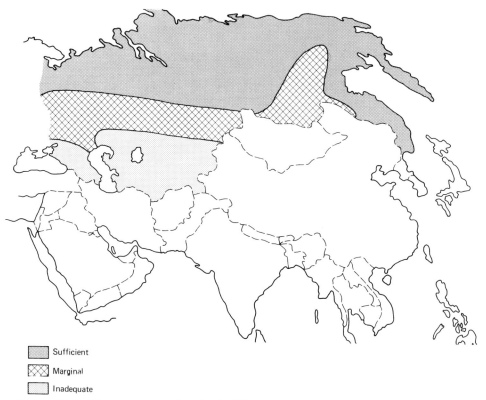

Sufficient

Marginal

Inadequate

Figure 15-5   Moisture zones of the Soviet Union.

rises, farming gives way to vast forests. This whole region is essentially north of the U.S.-Canadian border, and the mountains to the south and east prevent the penetration of warm air masses from the ocean.

Because of the lay of the land, most of the northern Asian rivers flow north from the central mountain range to the Arctic Ocean. In the spring the headwaters of these rivers melt first, while snow and ice still block the middle and lower reaches of these rivers. This causes extensive spring flooding over the low-lying areas of northern Russia. The flooding delays spring fieldwork, shortens the effective growing season, and further hinders agricultural production.

South of the midmountain range are large regions where the mountains and deserts reach to the sea. Alfisols, Ultisols, and Vertisols dominate in the uplands along with Inceptisols and Entisols in the river valleys. Because of the age of the continent, the uplands near the equator tend to be old and weathered. Alluvial soils are frequently the most productive. In some areas, particularly in the Pacific

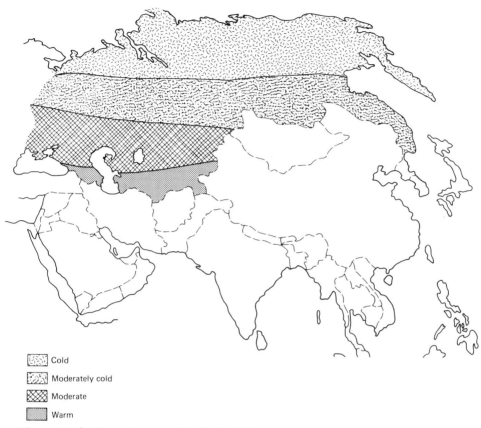

Cold

Moderately cold

Moderate

Warm

Figure 15-6 Thermal zones of the Soviet Union.

islands, recent volcanic activity has rejuvenated the landscape. The agriculture of the region is controlled by the monsoon season.

Mountain Region Soils

A large mountain range sprawls from east to west across central Asia. Changes in altitude influence both moisture and temperature regimes over relatively short distances, giving a wide range of soils. The ruggedness of the terrain and the extent of bare rock surfaces confine the soils of these regions largely to glacier-shaped mountain meadows or plateaus and to alluvium along rivers. From a soils point of view, the mountains of Asia can be divided into six zones. One is so rugged that it is essentially devoid of soils, and the soils of the other five regions are influenced by local climatic conditions (see the X areas and Z area of Fig 1-11).

Rugged Mountains

The highest and most rugged of the mountain ranges is the central Himalayas stretching from northwestern China across Tibet into Russia. This region comprises the largest area of high-altitude peaks in the world and is essentially devoid of soils. Much of the area is perpetually covered by snow or glaciers. The climatic regime is polar, and the rainfall generally is less than 250 millimeters. Much of the area is inhabited only by occasional isolated villages that depend heavily on the meat and products of cold, tolerant livestock (yaks, sheep, and goats) and the production of vegetables, tubers, and quickly maturing grains such as barley.

Cold Mountains

The coldest of the mountain ranges containing soils are the cryic great groups. These soils are found just north of the crescent-shaped rugged mountain region of the Himalayas, and they also stretch over a vast area in northern Russia. The Tibetan highlands include some of the western parts of the cryic mountain regions of central Asia and the adjoining rugged mountains. The Tibetan Highlands comprise more than one-fourth of the area of China, yet less than 1 percent of China's population lives here. Much of this region is greater than 4000 meters above sea level, and many areas have no drainage to the sea, yet many of the major rivers of Asia—the Yellow, Yangtze, Mekong, Irrawady, Indus, Ganges, and others—have their headwaters in this highland system, Barley, corn, hardy vegetables, and fruits are produced primarily on outwash and alluvial soils on local mountain valleys.

The cryic mountain region of Russia and northern China is likewise very sparsely populated. The unfavorable climate caused by both high altitudes and latitudes is not conducive to a comfortable life or the production of crops. Most of the region receives less than 500 millimeters of rainfall annually. This is partially compensated for by low evaporation rates in this cool climate. Only in a few isolated areas is agriculture possible. Here, too, the agriculture is mainly along the large rivers where the landscape is reasonably level, soils are fairly fertile, and the water supply is dependable. But the general ruggedness of the area and its isolation from other populated areas result in primarily subsistence type of agriculture.

The Ural mountains region separating Europe from Asiatic Russia has similar soils. But this mountain range is narrow, less extreme in elevation, and closer to population centers. Even though suitable areas for villages and agriculture are not abundant, a slightly higher population density can be maintained.

An island of mountains between the Urals and the northeastern mountain region contains Boralfs. This region in central Asiatic Russia is also severely limited by the general ruggedness of the area and the cold climate at these latitudes.

Udic Mountains

The mountain ranges with udic soils start in China and Korea in eastern Asia and extend southward from Japan to the tropical islands of the Philippines and Indonesia. The udic moisture regime results from a monsoonal rainfall pattern. The rainfall in this area is frequently very high, and the islands may receive both winter and summer monsoon rains. The temperature varies considerably from the north to the south end of this region. Korea and northern Japan have cold, snowy winters. To the south, warm tropical conditions prevail.

The southern coastal regions, except where salinity problems are serious, are intensively cultivated to rice, coconuts, tropical fruits, and vegetables. The uplands tend to have Alfisols and, where they are flat enough to be cultivated, a wide range of adaptable crops including corn, cotton, and rubber are grown. In the most densely populated areas many of the steeper mountain slopes are terraced, and rice is grown entensively up and down the mountain sides (see Fig. 15-7). Most of the population live in small valleys and the river deltas.

Figure 15-7 Rice paddies in the mountain regions of Java, Indonesia.

The southern mainland regions extend from central China to the Indochina peninsula. This region differs in one important aspect from the northern and the island regions—its shape. It is generally wider and not so close to the sea. Thus it contains areas that tend to be more isolated both physically and politically. Parts of this area are inhabited by tribes of various ethnic groups that are fiercely independent and resist outside efforts to develop the areas uniformly. Only about 10 percent of this area is actually in cultivation. There may be considerable room for expansion, but it will still be in small, isolated areas and development will be slow. The lowlands of this region are mainly in rice and receive an annual supply of fertility from yearly floods. The uplands are largely uncultivated; much of the area is still in native forest and is used by tribes for hunting native game. In places the uplands have been cultivated for a long time using shifting agricultural methods. More recently, a permanent type of agriculture has been developed in some areas in this region. With irrigation, better varieties, fertilizer, and erosion control, much of this area can be intensively used. Corn is becoming much more important along with upland rice, cotton, and wheat. Rubber, tea, tobacco, and other specialty crops are also grown extensively in places on the upland soils, which are primarily Ultisols and Alfisols.

A small area of these soils is also found in the Caucasus Mountains between the Black and the Caspian seas. These mountains serve as the boundary between Europe and Asia.

Other Mountain Regions

The mountain region with ustic soils is found in northern Thailand and the surrounding countries. The ustic region indicates an area with a more distinct dry season. Like the udic regions, much of the area is occupied by isolated villages. The more extensive dry season is a limiting factor. Most of the agriculture is in the alluvial area where rice, sometimes double cropped, is produced. Tobacco, peanuts, garlic, soybeans, and cotton are grown in the uplands, where solum thickness, slope, and soil conditions permit.

The final two mountain regions contain xeric and torric great groups. In addition to rugged landscapes and warm temperatures, the area lacks adequate rainfall. Agriculture is limited largely to where irrigation water is available.

Aridisols

The Aridisols of Asia are found primarily north and southwest of the central mountain range and occupy a considerable part of the continent (Fig. 15-8). They are mapped in two units, Aridisols with Orthents and Aridisols with Psamments. As in Africa, this reflects primarily the extent of sands in each zone. Much of the area is

With Orthents

With Psamments

Figure 15-8 Aridisols of Asia.

uninhabited. Settlements are mainly along watercourses or where water can be brought in from nearby mountains. There is some grazing by nomadic herdsmen on this land of rather low carrying capacity.

In areas where the desert extends to the seacoast, serious salt problems can arise. The windblown salt spray can eventually result in serious accumulations of salt. Salinity problems can also occur in areas affected by tides.

The Gulf of Cambay in western India is of special interest. It is narrow, and the water level rises considerably in the monsoon season when the floodwaters of the rivers flow into the bay.

The area surrounding the gulf is low-lying and flat. It is subject to local flooding at high tide. At high tide, saltwater from the gulf rushes in over the land. At the turn of the tide, there is a short period of equilibrium in which silt and salts are deposited in a thin layer on the soil surface. This creates a local saline condition. Later in the dry season, when the floodwaters are gone, winds blow across this area, picking up these sands and carrying them onto adjacent lands. This causes more extensive salt problems.

Some of the Earliest Known Agriculture

Some of the earliest human agricultural activities are found in the Euphrates River valley. The headwaters of this valley is in the desert regions of Syria. Geographically, this was once part of the Sahara, but it has become separated by rift valleys and the Red Sea. As early as 6000 B.C., this region was inhabited by grain farmers during a humid period. Irrigation became extensive between 4000 and 3000 B.C., when the region became arid. Later, conditions improved and, from 1000 B.C. to the year 0 B.C., the Romans built many aqueducts. This was again a very fertile and productive area. Conditions deteriorated after that, both agriculturally and politically. Today, agriculture is often limited to the irrigation of a monoculture of cotton on the upper terraces of the valley, but the uplands are occupied by nomadic herdsmen. Like many dry areas, the air is often full of dust that frequently shields the sun from view.

The soil climate is a nonflushing regime with the groundwater too deep to influence the soil. The high evaporation of water at the soil surface reverses the influence of leaching. Chemical weathering of the soil is the highest in the spring, when the maximum soil moisture occurs. This moisture, combined with the higher temperatures, results in relatively high biological activity. The summer is too dry for biological activity or chemical weathering, but organic matter can be destroyed.

When sodium is present in the parent material, sodic soils can develop. In some places $CaCl_2$ and $MgCl_2$ accumulated near the surface. These salts are hygroscopic, attract water, and form dew, which appears as dark patches on the soil surface.

Despite all these problems, where water is available, production can be very high. Near Baghdad, Iraq, along the Tigris River, a three-tier system of agriculture is practiced. Tall date trees tower over citrus. Below that three crops of vegetables per year are produced. The transformations of the desert in parts of Israel is another example of what people can do with water in the desert (see Fig. 15-9). The problem is that the most common feature of deserts is the lack of water.

Deserts of Russia and China

The desert region in southern Russia along the Caspian Sea is hot, dry, and subject to windstorms called *sukhovey*. These windstorms can last from several hours to several days. During the *sukhovey*, the relative humidity can drop below 30 percent

Figure 15-9 Irrigation of an Aridisol in Israel allows crop production in the desert.

and the winds range from 8 to 30 kilometers per hour. The temperatures range from 25 to 40°C. The storms frequently will last all night and the plants have no chance to recover during the night from the day's moisture stress. In this region of Russia, the probability of damage to winter wheat each year from a *sukhovey* is about 70 percent. Despite this problem, agriculture is expanding in this area. Water from the rivers that flow through this region is available for irrigation, and additional water is supplied from river reversal projects and from water brought down from local mountain ranges to the east. A wide variety of crops can be grown, including cotton, corn, grains, tea, tobacco, and tropical fruits and vegetables.

To the east, the desert reaches into northern China. Again, the area is one of extremes. It contains land more than 150 meters below sea level and is surrounded by high mountain peaks. Exceptionally high and low temperatures have been recorded here. The area lies primarily in the Sinkiang region, which contains one-sixth of the land area of China but less then 1 percent of its population. Most of this population is concentrated in the Mollisol belt, which extends into this area. Precipitation in the deserts of the area averages less than 5 millimeters annually, but local irrigation is possible where streams, fed from the snowmelts off the surrounding mountains, are dependable. In some places rivers simply disappear into the sands of the desert to reappear elsewhere as a spring-fed oasis. Intensive agriculture is practiced in these oases, and a wide variety of crops are grown. Sinkiang is especially famous for its melons, seedless grapes, and dried fruits.

The Major Northern Soils

Inceptisols

Inceptisols are found widely scattered throught Asia. The northernmost area is in Russia, almost all of it north of the Arctic Circle (area I2a of Fig. 1-11). These Cryaquepts are in a region that generally lacks sufficient thermal energy for crop production. The growing season is never more than 60 to 90 days, and the warmest months average only 10 to 15°C. Only quick-growing, cold-tolerant vegetables and some adapted varieties of barley and oats will grow. Almost all of the area receives less than 500 millimeters of rainfall annually, and about one-half of the area receives less than 250 millimeters. And yet excess water is more serious here than nearly anywhere else in Russia. Although the precipitation is not excessive, evaporation is low and the soils of much of this area are excessively wet at least 2 years in three.

As mentioned earlier, the rivers in this part of Asia flow north from the mountains to the Arctic Ocean. The deltas of these rivers are in this Inceptisol area. Extensive spring flooding further shortens the effective growing season. Thus, land use is restricted mostly to forestry and fur trapping. More recently, it seems that there may be considerable mineral wealth in the tundra of Asia and North America.

Spodosols

The only region of Spodolsols shown on the Asian map is in the northwestern part of the continent in Russia (area S1b. of Fig. 1-11). It is an extension of the larger region of European Spodosols, which will be discussed in more detail in the chapter on Europe. Like the Inceptisols of northern Russia, this is a region of cold climates and low evaporation that gives rise to wet soils much of the year. The Spodosols in this area are largely cryic, and they are associated with Histosols. The area is primarily forested; agriculture is limited almost exclusively to the major river valleys. The Ob and the Yenisey are the most significant of the river valleys of this region. The Ob river is 5400 kilometers long and over 58 percent of its length the fall is less than 3.2 centimeters per kilometer. This compounds the flooding problem mentioned under the heading "Inceptisols".

Crop production in this area just south of the Arctic Circle is limited to hardy grains, potatoes, green fodder, and vegetables along with cattle and pigs. The production is generally used for local consumption. Part of the Ob River will be involved in river reversal projects. These projects dam the river and transport the water south through canals into warmer and drier regions. Most of these projects are south of this Spodosol region but could influence this region by alleviating some of the spring flood problems.

Alfisols

Just south of the Spodosol region is an extension of the European Boralf zone (see areas A1a and A1b of Fig. 1-11). This region is discussed in more detail in the chapter on Europe, where the bulk of this soil region lies.

The region of Asiatic Alfisols is warmer than the Spodosol region to the north, but much of it is still in the cryic temperature region. In addition, it is drier than the western European Alfisols. Thus the agricultural production from the land is much less. Only around 10 percent of the total area is in crops, primarily wheat. The remainder is largely in forest. The local drainage is poor, giving rise to many Histosols. Most of the good agricultural land is along the alluvial plains of the major rivers.

Farther east in central Asiatic Russia, the Boralfs and Spodolsols merge to form a large area of Boralfs with Spodosols in a cryic temperature regime. The elevation of this area is higher, and the landscape is rolling to hilly. It is better drained than the region to the west. Agricultural use of the land occupies less than 5 percent of the total area, primarily in the river valleys. The climate is cold and the winters are harsh. There is snow on the ground 6 months or more per year, and agriculture is limited to short-season, cold-resistant crops. Wheat and barley along with sheep and timber are the most important products.

There is another area of these soils along the China-Russian border in a mountain valley landscape. Here, too, the landscape is limited by the cold and steepness of the area.

Mollisols

Most of the Asiatic Mollisols lie in a large belt across central Asiatic Russia, northern Mongolia, and into the Manchurian Plain of China. They, too, are an extension of soils from European Russia. The northern and eastern Mollisols are Borolls indicating that the temperature regime is still cool (areas M2a and M2b of Fig. 1-11). The region is a large plain somewhat comparable to the prairie provinces of Canada and, like Canda, this area is an important spring wheat area. The extensive grain production is hampered by three problems. The cool temperatures give a short growing season, and the production of many crops is frequently limited by early and late frosts. Second, the moisture of the area is adequate for grains most years, but the total precipitation and its distribution reduce yields other years.

Rains during harvest often reduce yields. Finally, this region shares a problem associated with the low-lying Alfisols and Spodosols of the north. The drainage is not generally good, and a high proportion of Aquolls are present.

Borolls or Chernozems

It is in this region that the Chernozem soils (largely included in the Borolls) were first explored. This was one of the soils that fascinated V. V. Dokuchaev and his Russian associates over 100 years ago. Their studies of these and other soils of Russia resulted in the development of some of the basic concepts that are still fundamental to modern soil science. The Russian Chernozem has a thick, dark, organic, rich A horizon of excellent structure. The A gradually decreases in organic matter and grades into a B horizon, with a minimal amount, if any, of illuviated clay. The structure of this B horizon promotes good drainage and aeration. The lower parts of the B and the C horizons are usually calcareous. An accumulation of carbonates is common in the lower parts of the solum. The rainfall is high enough to support a good vegetative cover, which provides organic matter to the soil. And yet the precipitation is low enough to preclude extensive leaching of bases. Furthermore, the relationship between rainfall and evaporation results in a dry soil for extended periods of time. This pattern reduces the effectiveness of microbes in decomposing organic matter. Thus the soils are rich in organic matter and nutrients.

Russian maps show 2 million square kilometers of Chernozems, or about one-half of those known in the world. The Chernozems have been extensively mapped in the United States along the corn belt-wheat belt border from the Dakotas to Kansas. There is some indication that the Russian and U.S. Chernozems are not identical but are closely related soils. The U.S. Chernozems almost always develop under higher rainfall than the Russian Chernozems. Many of the Russian Chernozems have thin dark-colored surface horizons similar to many Candaian Chernozems.

This same mapping unit, Borolls, is found in the Manchurian Plain of China. This is an area of significant food production. While still densely populated by world standards, the density is perhaps one-quarter that of the North China Plain. (This area will be discussed later under Entisols.) In the North China Plain most of the area is organized into communes, which were formed by grouping many small farms into larger production units. In the Manchurian Plain, large state farms are found. This, combined with the lower population density and the character of the Mollisol landscape, permits the area to be laid out in larger fields, and crop production is more heavily mechanized. In addition, the precipitation in China is slightly higher and its distribution is more favorable than in the comparable region of Russia. In China the climate is monsoonal; a very large proportion of the precipitation occurs as rain during the growing season.

It is an important area for wheat, corn, and other grain production. Pulses and sugar beets are also grown widely. This is one of the first areas in the world where soybeans were cultivated, and they are widely cultivated there today.

Ustoll Regions

South of the Boroll region in Russia is an Ustoll belt (area M 5d of Fig. 1-11). As we move south in Russia, the temperature becomes more favorable for agriculture, but the precipitation becomes more restrictive. This is a very important wheat-and grain-growing region of Russia. The area is on a high plateau, higher in elevation than the northern Mollisol region. The elevation increases toward the east and becomes rugged and mountainous at the eastern boundaries of this zone. The Borolls and Ustolls contain the "New Lands" of Russia. In the past few decades much research and technology has been poured into the development of this area. In addition to the obvious agricultural goal of obtaining more food for its people, there is also a political goal. The Russians are developing a stable political and social system in this heretofore sparsely populated prairie of Asiatic Russia with strong ties to European Russia (two-thirds of the people of Asiatic Russia are ethnically European). The political stability and increased population reduces the temptation for neighboring countries to invade and conquer this area in order to add agricultural land to their own country.

The Ustolls are important for the production of wheat, oats, barley, rye, sunflowers, and flax. In the 1950s there was a push to expand corn production in this area. The goal was to make corn as important here as in the midwest of the United States. The lower moisture in Russia prevented the crop from maturing, and the plan was abandoned. Corn still is important where conditions are favorable, but the potential is limited.

The most limiting factor to food production generally is the unpredictability of rainfall. Projects designated to reverse the north-flowing rivers will bring additional irrigation water to this region. The Ob River project, for example, will add some 50 cubic kilometers of water to this region each year. High costs and environmental disruption are likely to hinder development of these projects.

Other Mollisol Regions

North of the Gobi Desert in Mongolia, the land becomes more rugged and the Mollisols are mapped as Borolls with Orthids. The elevation is higher here (1000 to 3000 meters), resulting in a cooler climate. In the winter the outflow of air from the high-pressure system in the Gobi Desert to the south keeps this area dry. In the summer much of the moisture in the incoming winds is removed by the sur- rounding mountains. Thus this area is generally cool and dry. Agriculture is limited to flat areas along the river valleys, especially where irrigation is possible. The uplands are idle or are grazed largely by nomadic herdsmen.

A small region of Xerolls is located between the Mediterranean and the Black seas. It will be discussed with the Xerolls in the chapter on Europe.

Entisols

As stated before, Entisols are extensive in river valleys throughout the world.

Considering the world as a whole, alluvial soils occupy less than 5 percent of the land surface. But in tropical Asia less then 5 percent of the land surface is alluvial. Looking only at arable land, we find that over 10 percent of the world's arable land is alluvium, and a full 30 percent of this is in tropical Asia. Or, to put it another way, Asia, with one-sixth of the world's area, has one-third of its arable alluvial land.

It should be remembered in the following discussion that only about one-half of these alluvial soils are actually Entisols. The remainder are Inceptisols, Vertisols, or Mollisols. Most of the Entisol regions are two small to appear on our map. The one important exception to this is the North China Plain.

With proper drainage and flood control, they are capable of supporting an intense, high-yielding agriculture. The major river valleys in Asia support a high proportion of her population. Because alluvial areas are so heavily populated, it has been said that one-fourth of the world's population is fed from alluvial soils.

North China Plain

This plain is roughly bracketed by Peking in the north and Shanghai in the south (see Fig. 15-10). It is one of the most intensely cultivated areas of the world and, at the same time, there is probably no other part of the world that is both as large and as highly populated as this plain. The plain is a valley fed by the Yellow, Yangtze, and several other rivers flowing from the mountains to the west. It is in the mon- soon area and receives its rainfall primarily in the summer. Much of the plain is essentially dry in the winter.

The predominating soils (Aquents and Fluvents) reflect the character and prob- lems of the area. It is a vast floodplain, flat, poorly drained, and historically subject to periodic, devastating floods. Several of the rivers flow out of what is reported to be the deepest loess area in the world. The loess is believed to originate from wind

Figure 15-10 Entisols of Asia.

action in the Gobi Desert further west. Erosion off this loess area, especially where it is cultivated without adequate erosion control, provides a heavy sediment load for the major rivers in the upper reaches. This is carried into the plain, where the sediment is deposited in the riverbed between the natural levees. This deposition has raised the bed of the Yellow River 12 to 15 meters above the present flood-plain. Even today the elevation of the riverbed rises 5 centimeters each year. When the river occasionally breaks out of the levee and spills over this vast plain, the results are devastating. The People's Republic of China has put tremendous efforts into flood control of the rivers, drainage in the floodplain, and terracing of the loess slopes to help contain the sediment.

The crops of the area are primarily winter wheat combined with corn to the north and rice to the south. Large areas are also devoted to cotton, soybeans, and other grains and to fruits and vegetables to feed the population. Silk is also produced. Grazing land is almost nonexistent because most of this flat area is in crop production. Only a limited number of dairy and beef cattle, sheep, and goats are produced. Large number of hogs are fed on kitchen scraps and wastes from food processing. Poultry are also common. Excellent yields are possible with extensive use of both organic and chemical fertilizers. The extensive use of hand labor has also helped in producing extremely high yields to feed the population.

The cooler temperatures north of Shanghai limits double cropping but, to the south, double cropping is common. Because of the need to feed such a large population, experiments with double cropping are being attempted further and further north, combining winter wheat with a short-season summer crop. By hand transplanting and hand harvesting, it is possible to double crop further north, often combining rice and winter wheat. Transplanting allows the germination and early stages of growth to take place in a nursery. Hand harvesting allows harvesting earlier and at a higher moisture level than possible by machine. Each crop gets a start in a nursery while the preceding crop is maturing and is then transplanted immediately after harvest.

Figure 15-11 shows another example of double cropping. This region along the Mekong River in Cambodia was planted to quick-growing sweet corn at the beginning of the rainy season. Here we see the crop being harvested as the floodwaters

Figure 15-11 Sweet corn being harvested along the Mekong River in Cambodia. The short-season crop is planted at the beginning of the rainy season and harvested before the floodplain is inundated with water as the rainy season progresses.

of the Mekong inundate the area. Later the area will be planted to floating rice, a plant that can survive the deep floodwaters along this stretch of the Mekong.

Other Entisol Areas

A second area of Entisols is at the western edge of Asia in southern Russia (see Fig. 15-10). It is on a plateau between the Caspian and Ural seas. The soils are classified as Torriorthents and Aridisols. The region is surrounded by desert except for the shorelines of the two seas. The agriculture of the region is restricted almost exclusively to a few kilometers along the Caspian Sea coast. There local climate permits sufficient rainfall for a wide variety of the small grains and corn. Otherwise, most of this area is limited to grazing, except for a few irrigation projects. The annual rainfall is generally less than 200 millimeters with a possiblity of a drought occuring between May and July at least twice every 3 years. It is one of the warmest regions of Russia.

A third region of Entisols is composed of Psamments and shifting sand. It covers most of Saudi Arabia and can be considered an extension of the Sahara Desert, which lies just to the west in Africa. The abundance of oil and not agriculture raises this area above subsistence level. Agriculture is largely confined to the gulf coast and to oases. Some grazing is possible in the area of steppe vegetation.

Inceptisols of the South

In contrast to the northern Inceptisols, which are found principally in the tundra of the Arctic Circle, the southern Inceptisols are found primarily in Asia's river valleys (see I2a areas of Fig. 1-11).

The Tigris and Euphrates rivers have their headwaters in the desert regions north of the Persian Gulf. The lower reaches of these rivers contain archeologic evidence of settlement before 4000 B.C. Many believe that this region may be the biblical site of the Garden of Eden, and that humanity originated in this area. In any case, this Inceptisol region reflects human activity, with evidence of ancient irrigation systems and siltation problems. Emmer wheat, six-row barley, and flax were grown here about 6000 years ago.

About 3300 B.C. wheat and millet appeared. Between 2400 and 2000 B.C. a salinity problem began to show up; this was followed by a gradual shift in favor of the more salt-tolerant, two-row barley. By 2000 B.C. wheat had completely disappeared and, within 300 years, the center of civilization shifted north to Babylon in the central part of the valley above the reaches of the salinity from the sea. While the northern part of this area remains fairly productive today, the farmers in the southern part continue to battle the salt from the sea. Aquepts and Salorthids dominate this region.

Major Aquept Regions

The Ganges River in India is another large Aquept valley. Its origin dates back to ancient times when the ancient continent called Gondwana land broke up. It drifted apart, eventually forming South America, Africa, Australia and Antarctica. A small part of this continent drifted northward, colliding with southern Asia. This collision helped push up the Himalayas and India lies primarily on this ancient piece of Gondwana land. The Ganges River flows along the south edge of this mountain range, and the river valley contains Haplaquepts and Humaquepts. This is one of the richest agricultural areas of India. The annual rainfall over much of this valley is between 2000 and 3000 millimeters, as much as 80 percent falls in the monsoon season between June and September. In some areas meltwaters from the Himalayas provide irrigation water in the dry season. The alluvium varies in texture and acidity, and excess wetness is a widespread problem in the rainy season. Where physical and chemical properties permit, good yields of paddy rice, sugarcane, wheat, corn, jute, oil crops, vegetables, and tropical fruits are possible. In some areas, three crops per year are produced. On the northern fringe of this valley mulberries are grown to support a silk industry.

The deltas of the Irrawady, Chao Phraya, and Mekong rivers in Burma, Thailand, and Viet Nam also contain Inceptisols (see Fig. 15-11). The soils here classified as Tropaquepts and Hydraquepts. This area represents some of the most extensive paddy rice areas of the world. Evidence of rice cultivation goes back to 4000 B.C. Much of the area is essentially a big lake in the rainy season. The precipitation in this area is well over 2000 millimeters per year, coming in a monsoon pattern. The soils of these areas are generally very complex. In the delta region, the rivers tend to break up into many streams, each spreading across the delta each with its own natural levee. Over time the location of these drainageways shifts, giving rise to a complex pattern of old streambeds, levees, and settlement basins. This is further complexed by a series of terraces that relect ancient changes in sea level. The intrusion of saltwater from the ocean gives rise to acid sulphate soils (Sulphaquepts). In the lower part of the central valley of Thailand, salt intrusion is a very serious problem because the land is so flat. The slope is about 4 centimeters per kilometer. Even in the upper region of the valley, the slope is still only 30 centimeters per kilometer.

Sulfaquepts

Sulfaquepts (acid sulfate soils) are found along the seacoasts from Burma to Viet Na a. It has been estimated that nearly 10 percent of the soils in South Viet Nam have some degree of acid sulfate properties. They form in depressions that pond brackish water rich in sulfates. Iron sulfides are oxidized to ferrous sulfate ($FeSO_4$) and ferric sulfate ($Fe_2(SO_4)_3$), which are, respectively, light blue and yellow. The soils are grayish blue with yellow streaks or nodules and are often locally referred to as *cat clays*. In the dry season the iron and aluminum compunds are carried to the surface by capillary water from the water table. In the rainy season aluminum levels become toxic as pH levels drop to the 3.8 to 4.5 range. Water leaching through soils can carry the toxic compounds to nearby water sources and to cropland. Pineapples, sugarcane, and oil palm are grown where the soils are moderately acid. Strongly acid areas are wastelands.

The low pH leads to toxicity of aluminum and manganese as well as to toxicity from H_2S. In some cases, toxic organic acids are also formed. Phosphorus and copper deficiencies are frequent.

Reclamation requires levees to cut off water that brings in acids or salts. The dikes are built using material removed during the contruction of drainage ditches within the reclaimed area. This drainage lowers the water table, allowing rainwater or introduced river water to leach the toxic aluminum compounds and the sulfuric acid. High-tide water is preferable to low-tide water for leaching. The pH of the water can drop from 7 at high tide to 5 at low tide. Lime is often beneficial, but it may be expensive and not locally available. Furthermore, roads are often poor in areas where Sulfaquepts are found. Since the subsoil is often more acid than the top soil, even reclaimed land may have shallow root zones. Fertilizer, especially phosphorus, is an important part of the management of these soils. Pineapple and cassava are often grown first, since the are more tolerant of the acidity. Later corn, sorghum, and vegetables are grown in the dry season and rice in the wet season. But it is a slow process. It may take 10 years to remove 50 percent of the toxic aluminum.

Where reclamation is impractical, paddy rice can often be grown on these soils. The continuous water cover prevents the oxidization of the iron sulfides and the formation of sulfuric acid. This is practiced only where the water level can be controlled in the wet season and the land is beyond the reaches of salt intrusion. On the higher, sandier, natural levees and the younger terraces that are not highly weathered, a wide variety of crops including corn, sorghum, tobacco, mung beans, cotton, sugarcane, pineapple, and many tropical fruits and vegetables are grown. The warm and moist climate of the rainy season is ideal for their growth.

The better-drained upland soils contain Alfisols and Ultisols, generally of low fertility. Deep-rooted trees such as rubber, jackfruit, mango, and cashew that can extract nutrients from deep in the soil do well.

Upland Inceptisols

Another significant area of Inceptisols is located along the west coast of India (area I4c of Fig. 1-11). It includes the coastal strip, deltas of local rivers, and much of the western Ghat mountain range. The mountains generally begin less than 50 kilometers from the coast of the Arabian Sea. This area receives both summer and winter monsoons broken by dry periods. Total rainfall varies from a low of 400 millimeters to over 7600 millimeters annually. Under high rainfall, severe erosion problems can occur on the steeper slopes. It is this past erosion that is responsible for the extensive Inceptisols in this region. One-quarter of the area is in luxuriant tropical forests, with another one-quarter of the land suitable for crop production. The presence of the very old and infertile Ustox soils on the stable surfaces and the presence of steep land reduce the general agricultural potential of the area. Rubber and tea are sometimes grown on the moderately sloping areas when the forests are cleared. Coconut, rice, tobacco, bananas, grains, vegetables, and pulses are grown where climate and soils permit.

Alfisols of the South

Alfisols of India

The uplands of southern Asia, unless too dry or too steep, generally contain either Alfisols or Ultisols. A large, T-shaped area of Ustalfs with Ustochrepts is found primarily in India; it extend generally northward from the Arabian Sea and lies west and north of the Ganges River Inceptisols. The southern part of this region includes considerable areas of alluvium. To the north, the landscape is more rolling as the altitude increases. The region is strongly influenced by the summer monsoons, which usually bring very heavy rains to this region. The winter monsoons bring additional light showers that decrease in significance toward the north. Total precipitation in this region ranges from 500 millimeters to well over 2000 millimeters annually. A wide variety of crops is grown in the alluvial area where the Ustochrepts are extensive. The soils in the foothills of the Himalayas tend to be acid and low in natural fertility. The most extensive crops in the area are wheat, potatoes, and tea.

Along the eastern coast of India are two more zones of Alfisols. To the north are Plinthustalfs and to the south Ustalfs with Orthents(see Fig. 1-11).

The northern section lies primarily in the state of Andhra Pradesh. It is generally hilly, sloping from an elevation of 650 meters in the west to the seacoast. It is a hot area, with summer temperatures reaching nearly 50°C in May just before the onset of the summer monsoon rains. Total precipitation ranges from 500 to 1000 millimeters to the south and west to 750 to 1700 millimeters along the coast. Most of the rainfall is in the summer monsoons; some additional lighter rains come with the winter monsoons. The region is dominated by what are locally called red-earth soils, which are subdivided by texture. They are well drained, and the available water-holding capacity decreases as the clay content decreases. They are generally low in organic matter and slighly acid to neutral in reaction. They respond well to irrigation and have been intensively cultivated for a long time. Many of the

Orthents in this area developed because cultivation induced eroision, which removed most of the original soil. About two-fifths of the area is cultivated, and one-fifth is in forestry. The rest is barren, idle, grazed, in tree or orchard crops, or in urban areas.

The Alfisol area to the south lies primarily in the state of Tamil Nadu and is mapped Ustalfs with Orthents. It extends from the western Ghats to the eastern seacoast. Paddy rice is grown extensively wherever water can be ponded. On the better-drained sites, a wide variety of grains along with peanuts, tobacco, fruit, and vegetables are grown. This soil mapping unit extends into Sri Lanka.

Alfisols of Eastern Asia

North of the Mekong Delta is another area of Alfisols that lies primarily in the eastern half of Cambodia and the Khorat Plateau of eastern Thailand (area A3c of Fig. 1-11). This area is mapped Ustalfs and Ustults. The inclusion of Ultisols in this mapping unit reflects an older, more stable land surface. The precipitation is dominated by the monsoon season, with 80 percent of the rain falling from May to October. The total varies from 1000 to 2000 millimeters, but the amount and timing of the rainfall are variable and agriculture suffers severely in some years. A wide variety of crops is grown. Rice is the dominant crop grown wherever ponding of water is possible, principally on alluvium. Corn, kenaf, cassava, sugarcane, cotton, bananas, and pulse crops are found on the uplands and the high terraces above the floodplains. The higher the terrace, the older and less productive the soil. Historically, the best nonrice soils are those on the lowest terrace just above the floodplain. More recently there has been a considerable expansion of agriculture into the uplands on both Ustalfs and Ustults. Improved management techniques emphasizing improved fertility and pest control are responsible for this expansion.

Another area of Alfisols is found in the Szechwan Basin of China along the middle reaches of the Yangtze River (see Fig. 1-11). It is completely surrounded by mountains and contains Udalfs with Udults. Rainfall ranges from 750 to 1250 millimeters. This is an important rice-growing area; rice is grown continuously or in rotation with wheat or peas. Corn and sweet potatoes are also grown. Double cropping is generally not possible here.

Ultisols

As would be expected, Ultisols are extensive in southern Asia and the Asiatic islands (see Fig. 15-12). They are found in Java and the other islands of Indonesia, in Borneo and the Malay Peninsula. The climate is monsoonal, rainfall is adequate, and irrigation water is plentiful. The topography is generally rolling to somewhat rugged but, in this densely populated area, rice terraces go right up the mountain-

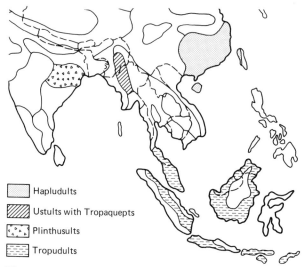

Haplududs

Ustults with Tropaquepts

Plinthusults

Tropududs

Figure 15-12 Ultisols of Asia.

side. This area was once known as the East Indies; the soils and the climate here are ideal for the production of a wide variety of spices and drugs. Over three-fourths of the island of Java is under cultivation. Altough soils are naturally infertile, the use of fertilizers permits year round cropping in southeast Asia (see Fig. 15-13).

Ultisols of Southern China

The largest region of Ultisols lies in southern China. It is a rolling to mountainous region and is heavily populated in the less steep areas. Rice is the dominant crop. In the southern parts of this soil region rice is double cropped, there is triple cropping in some local areas. Other grains and tropical fruits are also grown. The central part of this soil region is more rolling. Double cropping of rice is prevalent in some areas, but double cropping of rice with wheat or peas is more common. Tea is raised on the steeper slopes. The northern part of this soil region is transitional between China's southern rice-and northern wheat-growing areas. Both grains are grown extensively in this region and sometimes are double cropped. Cotton is also grown extensively.

The soils are largely Udults on the stable surfaces and Dystrochrepts on steep slopes and in alluvium. Locally, the dominant uplands soils of this region are also classified as Red or Yellow Earths. This reflects differences in the degree of weathering, specifically the amount of oxidation of the iron compounds. It generally reflects the extent of soil saturation during the year. In the uplands the Yellow

Figure 15-13 Fertilizer combined with the longer growing season on Ultisols in the Cambodian highlands allows staggered planting dates of corn. In an area where machinery is not plentiful the hand labor is spread out, allowing more production by the local farmers.

Earths are moister either because of climate or topography. They are preferred and are more extensively used than the Red Earths. The Red Earths are less fertile, more weathered, and have less desirable physical properties. Most of the Yellow Earths on more level landscapes are cultivated. The steeper slopes are terraced for rice or are in forest. Tung oil and varnish are important forestry products.

The Red Earths frequently are drier during some part of the year. Nearly 90 percent of the Red Earths are not cultivated, although many once were. Over 30 percent of the area is severely eroded. Some tea, bamboo, and tung oil are produced on the steeper slopes. Under rice, Red Earths can slowly gain characteristics of the Yellow Earths.

Other Ultisols

The upper Irrawady River in Burma also contains Ultisols. Away from the river the landscape is rolling to rugged and is surrounded by steep mountains, except to the south where the river flows to the sea. It is not densely populated, and much of the area is in jungle. Ustults dominate in the uplands with Tropaquepts in the valleys. Rice is the main crop. Despite its low population density and its rolling landscape, Burma is the worlds fifth largest producer of rice.

The Ultisols of India are along the eastern part of the Deccan Plateau. Much of the region is gently rolling, about 1500 meters above sea level. The precipitation for the region normally ranges from 700 to 2000 millimeters per year. Summers are hot, and winters are warm. Like the Ultisols of southern China, red and yellow soils are recognized locally. The dominance of Plinthustults in India reflects an older, more stable land surface. In some places the plinthite is quarried for building materials or is used in road construction.

Vertisols

Central Africa was reported to be the largest area of depressional Vertisols in the world. India probably contains the largest area of upland Vertisols in the world (see Fig. 1-11). They include some of the most productive soils in the country. The soils are primarily on the basaltic rocks of the Deccan Plateau, which is a large, rolling plain in central India. There are 60 million hectares of Vertisols on the Deccan Plateau. These soils are found on parent materials containing 45 to 55 percent silica and do not occur where the silica content is higher. The most strongly developed Vertisols reportedly develop cracks up to 25 centimeters wide in the dry season.

The cation-exchange capacity of these soils varies from 30 to 70 milliequivalents per 100 grams. The clay content may reach 80 percent. The soils are very productive where the moisture is adequate. Because of their sticky nature, they are not easy to work in the monsoon season. They are often underlined by bedrock at a depth of a few centimeters to over 6 meters. The shallower varieties often have rock within the top meter. The deeper ones may develop from rock or from alluvial material washed from the higher regions. These deeper Vertisols may have a gypsic horizon in the second meter, and they often have a salt accumulation below that. Others have calcic horizons, and still others lack any zone of accumulation. Despite problems with too much or too little water at times during the year, these are important soils for wheat production in India. Rice is also a very important crop, particularly on the alluvial Vertisols.

The Vertisols are often developed side by side on the landscape with Alfisols, primarily Ustalfs. The Alfisols develop where the parent material contains more than 55 percent silicates. Here the clay fraction is dominated by kaolinite and the hydrous oxides of iron and aluminum. They have an argillic horizon between 15 and 45 centimeters from the surface. They are well suited for irrigation and respond well to fertilizer and management.

A second area of Vertisols is located in eastern Java in Indonesia. Java exists in an area of favorable climate in which soils have been frequently rejuvenated by volcanic activity. It is a densely populated island, and over three-fourths of the area is cultivated. Rice is the most extensive crop grown. Sugar, tea, coffee, rubber, and palm oil are grown on plantations. Coconut, tobacco and various fibers, spices, and drug crops are also important.

Oxisols and Histosols

There are no regions in Asia where Oxisols or Histosols dominate sufficiently to appear on our soil map. Oxisols are included as an important secondary soil in the Tropept region of southwest India. Histosols are an important secondary soil in the Spodosol region of Russia and the Inceptisol regions of the river valleys of southeast Asia.

Soils Disturbed by Humans

Human activity in some cases has had just as dramatic an influence on the proper-ties of the soils as the five soil-forming factors. Evidence of human activities in southwestern Asia has been traced back continuously to 4000 B.C.. and sporadically before that. In some areas, people have drastically disturbed an area by building villages or fortifications. In many cases, these areas were abandoned and then later reinhabited. The reason for the departure of people from an area is not always clear but, in some cases, there is evidence of salinity problems. In other cases, poor management on the slopes allowed sediment to move down into the lower areas, plugging irrigation and drainage systems and occasionally overriding whole villages.

Erosion of Soils

The neglect of erosion control and its effect in the ancient past has at times been very dramatic. Lowdermilk reported that to enter a small church in Cyprus, one had to walk down 2.5 meters of stairs into a courtyard and then descend another meter to the floor level of the church. This floor is 60 centimeters above the level of the original floor. At some time this church was higher than the surrounding area, or at least level with it. Erosion from surrounding hillsides has filled the area in and, after one disastrous period, it was decided to build a second floor on top of the accumulated sediments instead of removing them. The hills were probably originally forest covered. Many of the forests were removed by the slave labor of King Solomon to build temples in Jerusalem 3000 years ago. Without adequate protection the soils on the slopes were completely eroded away. Today, instead of moderately deep productive soils, the slopes are covered by stones and rocks of little agricultural potential. The valley has been silted in and is less productive than it was in the past. The soils would generally be classified in various suborders of Entisols, but an understanding of the historical development of these soils is important in understanding their current potential.

Paddy Rice Soils

Another unique soil is found extensively throughout tropical Asia in rice paddies. As yet there is no place in the American classification system for this type of soil despite the fact that it undoubtedly covers hundreds of thousands of square kilometers. In places the soil properties have been strongly influenced by the cultural practices of paddy rice production, which keeps the soil submerged 3 to 4 months each year during the growing season.

Paddy rice is found extensively on poorly drained sites, but much is found on well-drained sites. The structure of the soil is destroyed to reduce the movement of water through the profile. The area is diked to pond natural or irrigation water-for extended periods of time. This puddling and ponding should not completely eliminate drainage. A small amount of downward movement may be desirable to leach butyric acid and other toxic compounds formed by the decomposition of organic matter under anaerobic conditions.

The ponding alters the chemical, physical, and biological properties of the soil. Below the water surface (usually between 15 and 30 centimeters deep) is an Ap horizon overlying a plow pan. The top centimeter or so is often in an aerobic state. Oxygen diffuses in from the water, results from the activity of blue green alage, or is exuded by the roots of the rice plant. However, the bulk of the A horizon is in a reduced condition (see Fig. 15-14). Organic matter decomposes slowly. Iron, manganese, and sulfur are all reduced. The reduction of iron often increases the level of phosphorus available to plants. Nitrogen from the decomposition of organic matter, plus that fixed by blue-green algae, remains in the water in the ammonium form.

Much of the water above the soil is lost by evaporation. Any nutrients or sediments carried by the water are deposited on the soil and are available to the plant. The pH of the soil is usually between 5.5 and 6.5 but, in some cases, it is basic. Sometimes the soil is saline. A few varieties of rice are salt tolerant and will grow in brackish water or in areas that are occasionally flooded by seawater. If a soil contains excessive amounts of sulfur, ferrous sulphate forms and coats the soil particles with a black coating.

Beneath this reduced A horizon is an oxidized zone if the soil is well drained. The iron and manganese are reduced in the A. Being mobile, they move below the plow pan and precipitate in the oxidized horizon. There they remain, forming a Bir horizon, a Bmn horizon, or both. Below the B horizon may be an oxidized C horizon, with a reduced C horizon occuring at the natural water table level. If the original soil was in a poorly drained site, the oxidized zone of the B and C may be

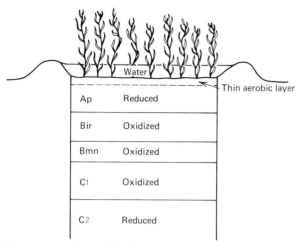

Figure 15-14   Profile of a rice paddy soil.

missing. Probably the most frequent horizon sequence is Ap (reduced), Bir (oxidized), Bmn (oxidized), Cl (oxidized), and C2 (reduced) (see Fig. 15-14). This type of profile can develop in as little as 10 years but, more frequently, takes 40 to 50 years to develop.

In many areas this type of soil development does not occur. It can be prevented by constant siltation, a high water table, which eliminates the oxidized zone below the A, or the churning action that results from the presence of montmorillonitic clay. Where this type of development does not occur, the soils resemble the non-paddy soils of the surrounding floodplains or uplands. But where these unique soils do occur, they are very distinct. They are worthy of further study and deserve a special category within the American soil system classification.

Paddy Rice Cultivation

The profile described in the preceding section develops because of the unique management system for paddy rice. The purpose of the flooding system for rice production seems to be threefold: first, it allows the utilization of wet lands that are unsuitable for any other crop; second, flooding seems to be an effective means of controlling most of the native weeds; and third, rice is very sensitive to moisture stress.

The water level is maintained on most rice soils by a series of dams and dikes to control lateral water movement and by puddling the soil to restrict downward movement. The initial flooding of the soil at the end of the dry season often traps air in the peds. These then explode like popcorn, destroying soil structure. The process can be reversed by proper tillage and several wet-dry cycles. Silicate clays are easier to puddle and more difficult to restore than are oxide clays.

Flooding eliminates water stress in rice. It takes perhaps twice as much water to produce rice as it does to produce corn, but much of this extra water is evaporated. Rice can grow under water because it can oxidize its own rhizosphere by transferring oxygen through the stems from the leaves. As the floodwaters rise in the paddy, the soil is soon depleted of much of its oxygen, and the pH shifts into the 5 to 6.5 range almost regardless of its aerobic pH. With this shifting in pH the solubility of phosphorus often increases. Under anaerobic conditions, the nitrogen remains in the NH_4^+ form instead of being oxidized to NO_3^-.

The shift in pH is almost always an improvement. Under high basic conditions the water absorbs CO_2 from the air that combines with soil water to form carbonic acid.

$$H_2O + CO_2 \rightleftharpoons H_2CO_3$$

This acid tends to reduce the pH of highly basic soils. If the original conditions were acid and the soil was rich in iron, the iron is reduced.

$$Fe(OH)_3 \rightleftharpoons Fe(OH)_2 + OH^-$$

This causes an increase in pH and decreases the toxic potential of aluminum.

Subsistence yields less than 1.4 metric tons per hectare of continuous rice have held steady for thousands of years in some areas. Nitrogen fixed by algae supplies the needed nitrogen, and the remaining nutrients are supplied by the soil or the sediments contained in the floodwaters. But proper fertilization practices can often dramatically increase yields (3 to 4 metric tons per hectare). Rice almost always responds to nitrogen, although too much nitrogen can lead to lodging in long-stem varieties. The ammonium form of nitrogen fertilizer generally gives better results, since nitrate nitrogen is readily denitrified and lost from the soil. Phosphorus is deficient in many rice soils, and phosphorus and nitrogen in combination often give a bigger boost in yield. One problem not yet fully solved is the development of a soil test for phosphorus that is accurate under paddy conditions.

Double cropping of rice has been practiced in some areas for a long time, there is much interest in expanding the practice to provide more food for an expanding population. The rainy season generally must be over 6 months long before double cropping is feasible. Double cropping of rice is practiced, but double cropping of rice with a dry land crop is more common.

The ameolerating influence of water on pH, nitrogen fixation by algae, availability of phosphorus, sulfur, aluminum, and other nutrients and toxicants is, of course, diminished when the paddy is drained for the upland crop. Thus double cropping is further restricted to soils with higher fertility or where fertilizers and/or lime are available and economical. Furthermore, the puddling effects must be at least partially reversed in order to have good aeration and drainage in the root zone of the nonrice crop. Structural deterioration is more quickly reversed in soils higher in oxide clays. The structure of silicate clays is not as quickly restored after puddling and may take several wet-dry cycles. Double cropping is more practical on oxide clay soils or in floodplains, where a high water table and not puddling is responsible for maintainng a high water table.

Soil Building in the Peoples' Republic of China

Two types of soil building now being carried out in the People's Republic of China are worthy of mention. The first involves reclaiming stoney and sandy alluvial land. Walls are first constructed to contain floodwaters in the central part of the floodplain. The land behind the walls is then covered by loess brought in from a nearby eroding loess hill (see Fig. 15-15). The final result is a flat loess area where there was once a steep hill and a loess area where there was once a stoney floodplain.

A second type of project is typified by the Tachai Production Brigade in central Shansi Province. Originally the brigade farmed seven gullies and eight ridges in rolling limestone countryside covered in places by deep loess. Curved terrace walls were constructed across the gullies. Tunnels beneath the terraces remove excess rainfall. The ridges were lopped off to provide fill for the gully terraces, and the

Figure 15-15 Chinese construct a productive soil. Loess from nearby hills is dumped behind a levee over stoney alluvium in the North China Plain.

Figure 15-16   Terracing of a mountain at the Tachai Brigade in the udic mountain zone of China.

ridge slopes were themselves terraced (see Fig. 15-16). Efforts were then made to incorporate organic matter and other organic and inorganic amendments and fertilizers.

Studies have shown that their efforts were rewarded with greatly increased yields, improved percolation rates and water-holding capacity, and increased microbial activity in the soil.

Much of this effort was the result of handwork by large work parties. Throughout this text book we refer to the need for modern technology, with associated machinery and chemicals to improve soils and increase yields. But the story of Tachai is one example where people with a sufficient social commitment and a strong, organized political system can make great improvements by combining hard work with a lower level of technology.

Floating Gardens

Another example of manufactured soil is the floating gardens of Kashmir. Mud from the lake bottom is mixed with plant residues, aquatic weeds, and mineral wastes. The resulting mound floats and can support the weight of people (see Fig. 15-17). The high fertility of the mixture combined with the subirrigation from the lake allow three to five crops per year of transplanted vegetables. This procedure

Figure 15-17 Artificial soils floating on a lake near Srinagar, Kashmir, India. Cucumbers, melons, and other vegetables are produced on these mounds, which are floating on water 2 to 4 meters deep.

creates productive soils on water where the alternative is to try to improve rocky, shallow soils on steep mountainside slopes.

Agricultural Potential

Predicting the future agricultural potential of Asia is hazardous at best. Asia contains some of the most populated areas of the world, and its growth rate is still relatively high. Much of the area is too cold, steep, or dry for more intensive use. At the same time, much of the area is yet to be seriously explored to determine its potential. The North China Plain and the southern tip of the Indochina peninsula support a dense population but, in between these areas, is a region that is mostly underdeveloped. Potential resources are yet to be determined and tapped. Likewise Japan, parts of the Philippines, and Java in Indonesia are heavily populated and produce large amounts of agricultural products. But other parts of the Philippines, Indonesia, and Malaysia contain soils of the same suborder and are sparsely populated. There is every reason to believe that the agricultural production of this area could be expanded. In southwestern Asia, irrigation projects, funded by oil profits, are under consideration in arid and semiarid regions.

Certainly there is not enough money or water to irrigate all of the desert, but significant production increases should be possible. China is putting a high priority on mechanization, and India is trying hard to take advantage of modern technology to increase its production. As in Africa, much of Asia's potential is held back by political, social, and economic conditions that have yet to produce the education and resources needed at the villiage level. China plans to feed its population from its own production in the year 2000, when they estimate their population will be approximately 1 billion people. There is every reason to believe that they can make it. In some other parts of Asia, the potential for food production is just as high.

References

Alexander, T. M., Ed., *Soils of India*, The Fertilizer Association of India, New Delhi, 1972.

Bloomfield, C., and J. K. Coulter, *Genesis and Management of Acid Sulfate Soils*, in *Advances in Agronomy*, Vol. 25, Academic, New York, 1973.

Bradfield, R., "*The Future of Soils and Crops Research in the Tropics*," *Soil and Crop Sci Soc Fl. 26:* 299-305, 1966.

Bridges, E. M, *World Soils*, Cambridge University Press, London, 1970.

Buol, S. W., F. D. Hole, and R. J. McCracken, *Soil Genesis and Classification*, Iowa State University Press, Ames, 1973.

Central Intelligence Agency, *People's Republic of China Atlas*, U.S. Government Printing Office, Washington D.C., 1971.

Central Intelligence Agency, *USSR Agricultural Atlas*, U.S. Government Printing Office, Washington D.C., 1974.

Glinka, K. D., *The Great Soil Groups of the World and Their Development*, Translated by C. F. Marbut, Edwards Brothers, Ann Arbor, Mich., 1927.

Kalpage, F. S. C. P., *Tropical Soils*, MacMillan of India, 1974.

Kawaguchi, K., and K. Kyuma, *Lowland Rice Soils in Thailand*, Center for Southeast Asian Studies, Kyoto University, Kyoto, Japan, 1969.

Kawaguchi, K., and K. Kyuma, *Paddy Soils in Tropical Asia*, University Press of Hawaii, Honolulu, 1977.

Lowdermilk, W. C. Conquest of the land through 7000 years" USDA Inf. Bull. 99 Revised, 1975.

Mulders, M. A., *The Arid Soils of the Balikh Basin (Syria)*, Drukkerij Bronder Offset N. V., Rotterdam, 1969.

Raychaudhuri, S. P. R. R. Agarwal, N. R. Datta Biswas, S. P. Gupta, and P. K. Thomas, *Soils of India.*, Indian Council on Agricultural Research, New Delhi, 1963.

Sanchez, P. A., *Properties and Management of Soils in the Tropics*, Wiley, New York, 1976.

The scientific Research Group of the Tachai Production Brigade, *Tachai Field—Its Reconstruction and its Fertility Characteristics," Scientia Sincica, 19:7*-20, 1976.

Thai-Cong-Tung, *Major Soil Groups in South Viet Nam and Their Management*, Institute of Agricultural Research, Republic of Viet Nam, 1972.

Thorp, J. *Geography of the Soils of China*, National Geographic Survey of China, Nanking, 1936.

Australia and New Zealand

The Australian and New Zealand Environment

Australia is the smallest and least densely populated of the continents. It is slightly smaller than the United States and about three-fourths the size of Europe. It rivals Africa in its high proportion of old land surfaces, and it is the most arid of the continents. It is the only continent (ignoring Antarctica, which has no agriculture) that lies totally south of the equator, and about two-thirds of it lies south of the tropics. Long periods of erosion without periods of mountain building or volcanic acitivity gives a landscape mostly below 500 meters. Only 7 percent of the continent is above 2000 meters, which is the smallest proportion of high altitudes of all of the continents. Its highest peak (2230 meters) is barely one-fourth of the height of Asia's highest peak (8840 meters).

New Zealand lies 1900 kilometers east and south of Australia. It consists of two main islands and hundreds of smaller islands. Its total area is about the size of the state of Nevada. The landscape is much more rugged than that of Australia. The vast ocean surrounding the long, narrow islands serves as a source of rainfall and moderates temperature fluctuations. Volcanism and mountain building have produced a much younger landscape.

Physiography of Australia

Australia can be divided into three areas (Fig. 16-1). In the west is a large, flat shield area covering more than one-half of the continent. It is primarily an area of low rainfall, and much of it is nearly level. Over most of the area there are no permanent rivers.

The eastern highlands is a low-lying mountainous chain parallel to the east coast. The coastal fringe is seldom more than 15 kilometers wide except in a few alluvial valleys. A few scattered plateaus in the mountains are of particular agricultural significance.

Between these two areas is the central highlands. It consists of a series of sedimentary basins, mostly below 150 meters. It includes Lake Eyre, a 1 million square kilometer dry salt flat without external drainage. It has filled once since settlement (in 1950). Much of the central highlands is so flat that rivers simply spread out over the countryside and disappear into the earth. They form underground aquifers that are the lifeblood of much of the agricultural irrigation of this region. The Murray-Darling River Basin in the south is the largest river system in the continent, but it is small in comparison to the great rivers of the other continents. Navigaton is haphazard; boats are occasionally trapped by low water for 2 years or more. Many

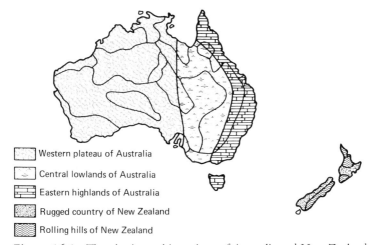

Western plateau of Australia
Central lowlands of Australia
Eastern highlands of Australia
Rugged country of New Zealand
Rolling hills of New Zealand

Figure 16-1   The physiographic regions of Australia and New Zealand.

of Australia's rivers are frequently no more then a series of elongated ponds. Only on the island of Tasmania (technically part of the eastern highlands) is stream flow dependable. Here, hydroelectric power generation is possible. In most of the country storing water behind dams is a problem because of the variable flow of rivers. Large storage areas are required to compensate for this variability. This results in a large surface area for evaporation. The continent has no true alpine country, and the snow on all mountaintops completely melts nearly every summer.

Australia was influenced only minimally by active glaciation during Pleistocene times. The activity was limited to Tasmania and the southeastern corner of the continent. Features such as moraines and loess deposits are rare. Changes in climate during Pleistocene times did influence landscape development on older surfaces.

Physiography of New Zealand

In contrast, much of New Zealand has a youthful landscape that is hilly to rugged. Few surfaces are older than Pleistocene, and many are very late Pleistocene. Nearly three-fourths of the country is mountains or hills. Only 10 percent of the country has low relief, much of it on the North Island. The North Island has a long history of volcanic activity occurring up to the present. The South Island is more rugged, with older rock near the surface. Nearly three-fourths of the South Island is covered by glacially eroded materials.

The ruggedness of the countryside produces rapid runoff that feeds many small rivers. Very few are navigable, and the largest is only 420 kilometers long. Much of the water in the rivers of the South Island are fed from snowmelt in the mountains. This gives a large fluctuation in waterflow. Winds blowing across these streams during low water flow often carry sediments into the surrounding uplands, rejuventing these soils. The Canterbury Plain of the South Island is the largest continuous flat area of the country. The soils of this plain developed in about 3 meters of alluvial sediments. It is the largest crop-producing area of the country.

Climate

Figure 16-2 shows the rainfall distribution for Australia and New Zealand. Australia consists of a central desert with higher rainfall areas around it, particularly to the east. Approximately 42 percent of the continent is classified as arid, only 9 percent is humid, and an additional 5 percent is seasonally humid. The climate today is both more arid and warmer then in the past, and many of Australia's soils reflect weathering under earlier climatic conditions. The eastern coast mountain range contains the only mountains on the continent that can extract water from the air.

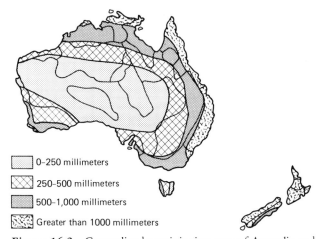

Figure 16-2 Generalized precipitation map of Australia and New Zealand.

Like Asia, Australia has an inland winter high-pressure system. From April to September, dry winds blow clockwise and outward around the continent. In the summer a low-pressure system in northwestern Australia brings monsoonal rains to the north coast from the Asiatic monsoonal system. In the south the winds tend to blow parallel to the coast instead of moving inland. Thus the rainfall along much of the coast is minimal. The south and west coasts of the continent are among the longest dry coasts of the world. The southwest tip and the southeast section of the continent receive winter rainfall, giving a Mediterranean climate. As with many areas of marginal rainfall, the most serious problem is variability. Too frequently Australia fluctuates between flood and drought.

About 40 percent of the continent receives less then 250 millimeters and 70 percent receives less then 500 millimeters of precipitation per year. Generally, irrigation is considered essential when the precipitation is less than 500 millimeters. By contrast, one spot along the northeast coast (near Tully) receives about 4500 millimeters of rainfall annually as water-laden winds from the ocean move up the moutain range.

New Zealand has a more complex but higher rainfall pattern. The precipitation is more evenly distributed among summer, winter, and fall. The total is much higher, with much of New Zealand receiving over 1000 millimeters of precipitation annually. The mountains have a significant influence on local climate. Milford Sound in the southwest corner of the South Island receives over 6300 millimeters of precipitation annually. Alexandria, 140 kilometers over the mountains to the east, receives only 325 millimeters. The prevailing winds cause the west side of the islands to receive more precipitation then the east.

The temperature in New Zealand is generally cooler due to their more southern latitude and the moderating influence of the sea. The temperatures increase south to north but are often influenced by local elevation, particularly in the south.

Vegetation

The isolation of Australia and New Zealand, the absence of large native grazing and browsing animals, and the absence of humans until very recently have produced thousands of species of plants that are unique to this part of the world. Some types of vegetation have had a profound local influence on the soil.

Agricultural areas are all less then 1000 years old, and much of the agricultural land has been cultivated less then 100 years. But cultivation and grazing has destroyed much of the native vegetation.

Most of Australia is covered by desert shrubs and grassland vegetation. Forests are found only along the coast, primarily in the east. In areas of higher rainfall agricultural crops and improved pastures have replaced much of the native vegetation. In the drier areas cultivation has had much less influence on local vegetation. However, the dominant species of grasses has shifted because of the grazing pressures of cattle and sheep.

New Zealand was originally covered with forest and tussock grass. The grass was converted to pasture about 100 years ago. Today, 70 percent of the agricultural land is in pasture, primarily for sheep and cattle. They have the highest livestock-to-human being ratio of any nation. The production of cereal crops, horticultural crops, hops, and vines occupy less than 2 percent of the landscape.

Soil Regions

Australia is composed of a cental Aridisol area that reaches to the coast in the south (see Fig. 16-3). Within this region and to the north, large areas of Entisols are found. On the fringes of this dry area precipitation increases.

There are three small mountainous areas in the west, central, and eastern parts of the country. The island of Tasmania is also mapped as mountainous. Like most mountain regions, local areas of agriculture are important. Alfisols are found in the southeast, southwest, and north-central part of the country. This is more productive agricultural land, particularly where the precipitation is higher.

Ultisols are found only in the northeast, where the climate is both warmer and humid. Tropical agriculture is possible here. A large, crescent-shaped region of Vertisol soils is found on the western slopes of the eastern mountain range, and it swings around the north edge of the desert. In this region there is extensive grazing and dry land agriculture. Irrigation is of some local importance.

New Zealand's North Island has a region of productive Alfisols in the north. The central Inceptisols and southern mountain regions of the North Island have

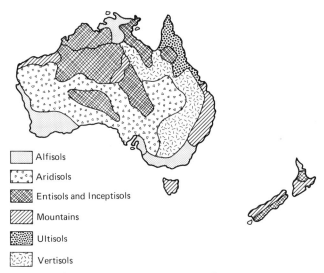

Alfisols

Aridisols

Entisols and Inceptisols

Mountains

Ultisols

Vertisols

Figure 16-3   Generalized soil map of Australia and New Zealand.

resulted from recent volcanic activity. The South Island consists of a rugged mountain chain on the west coast that has little agricultural value. Alluvial plains and deltas on the east coast contain Inceptisols, which are very productive where the land is level enough for cultivation.

Aridisols

New Zealand lacks a true desert region (Fig. 16-4). By contrast, Australia has a vast, central desert. Like the Sahara, the desert contains sandy areas of Entisols. The Aridisols frequently contain remnants of past weathering sequences. In some places plinthite gravels are a significant proportion of the solum. Some soils show evidence of having been salty at one time, but they have since been acidified. There are also areas of acid soils that have been recently recalcified from windblown sediments. Land use is severely limited by the lack of rainfall. And yet Australia has no extreme deserts. Most desert regions have a rainy period of less than 100 millimeters to over 750 millimeters that breaks the dry period, allowing some seasonal vegetative growth for some low-density grazing.

Land Use

The Australian Merino sheep industry is based largely on native pasture and shrub browse. The beef industry is heavily dependent on rangeland pastures. Most of it is in arid and semiarid areas. In some areas the dingo, a wild dog, is a more serious problem in sheep production than the lack of water. A dog-proof fence is maintained along the New South Wales-Queensland border and the New South Wales-South Australia border. Beyond the fence sheep must be tended by herders. Because of the dingo problem, cattle are frequently raised, even though sheep are more adapted to the grazing lands.

Since cattle normally graze no more than 8 kilometers from water, water supplies must be maintained. As these dry up during periodic droughts, loss of cattle from dehydration increases rapidly. Most grazing is controlled. Cattle are moved from one area to another, depending on the condition of the animal, not the condition of the grasses or the soil. At times this leads to serious degradation of the vegetative cover and the soil.

Soils of the Aridisol Region

Both the Australians and New Zealanders have developed their own soil classification systems. The correlations between these systems and the American system has not yet been developed in detail. Five unusual soils in the Australian system that are found on Aridisol landscapes are of particular interest. They are the Solodized Brown, Solonchak, Solonetz, Solodized-Solonetz, and Soloth. Similar soils are found in other arid regions of the world but, in Australia, there seems to be more evidence that their present properties cannot be explained by current weathering conditions. This again emphasizes the complexity of soil development on older land surfaces.

Near Adelaide, in south-central Australia, is a large area of Solodized Brown soils. They contain great amounts of free carbonates in both the fine and coarse fractions of the soil. They contain a calcic horizon below the argillic horizon. Some are classed as Alfisols, usually as Xeralfs. But because they are found in areas of low rainfall many are classified today as Argids. Calcium dominates the exchange on the surface, but they are also high in magnesium. A unique feature of many

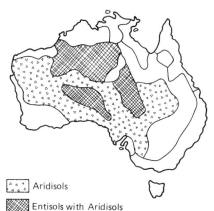

Aridisols

Entisols with Aridisols

Figure 16-4 Aridisols and dry sandy Entisols of Australia.

Figure 16-5 Irrigation in the Murray River valley allows fruit production in a region of low rainfall in southern Australia.

Australian soils is the high magnesium levels in udic and aridic areas. The pH of the surface usually is near neutral but may reach 9 or 10 in the subsoil. Solodized Browns can form on highly calcareous parent materials, but often the carbonates seem to have come from wind or sometimes water-carried sediments. The landscape was once covered by a Eucalyptus shrub called malee. This shrub is somewhat unique to these soils, which are known locally as malee soils.

The Solonized Brown soils are found where the rainfall varies from less than 250 millimeters to 400 millimeters. As the rainfall increases within this range, wheat or barley are grown in a summer fallow system. Sheep grazing is common in the middle range. In areas of lowest rainfall soils are unused unless irrigated. In the Murray River Valley a wide variety of crops can be grown under irrigation (see Fig. 16-5). Fruits, vines, and vegetables are common. Wind erosion is a problem, and segregation by particle size can cause dunes to form. Local depressions are susceptible to frost damage and are often left unplanted.

High Salt Content Soils

Four kinds of soils contain or may have previously contained high concentrations of salt. They usually develop in regions of low precipitation combined with evaporation of saline waters from a water table near the soil surface. In humid regions they result from salt enrichment from wind-carried ocean sprays or from flooding by saltwater.

The Solonchak soils are the least developed of these salty soils. They often contain a white salt crust on the surface or have a powdery surface structure containing free salt crystals. In some parts of the world they are referred to as white alkali or saline soils. In *Soil Taxonomy* many would be classified as Salorthids. Below the top 30 centimeters the salt content is low, and soil properties reflect the properties of the parent material, which are little altered in the arid climate. The aridity combined with salinity results in most of these soils being unused or lightly grazed.

Solonetz soils have a textural B that may have two to three times the clay content of the A. They are neutral to alkaline in pH. Sodium and magnesium dominate the exchange surface, and the B horizon is prismatic to columnar in structure. The soils are extremely hard when dry and not much better when moist.

Early theories explained the Solonetz profile as the result of translocation of clay from the A to the B. Dispersion of the clay in the A by sodium occurred after the leaching of excess salts. The clay was then flocculated by the excess concentration of salts in the B. More recent studies in Australia have raised some doubts about this theory. Some believe that the textural contrast may be due to lateral clay movement in the A or to differential weathering that caused total destruction of the surface soil clay. The structure is explained as a result of successive wetting and drying of the sodium-rich B. In any case, the soils are largely grazed at a very low carrying capacity. Some produce cereal grains where they occur as small inclusions within other soils. These soils are similar to the sodic or Natrargid soils described in North America, but the Australian soils seem to be higher in magnesium and clay.

A third type of soil influenced by salt is Solodized Solonetz. The Solonetz soils have a thin A (about 10 centimeters). The Solodized Solonetz soils have an A horizon as thick as 30 to 45 centimeters, including a very distinct A2. Both the A and B horizons of the Solodized Solonetz are acid, but the pH of the C horizon is often above 9. The B horizon has strongly developed columnar structure. The structural units may be 10 to 25 cemtimeters across the top and average about 30 centimeters tall. Some peds as large as 45 by 45 centimeters have been reported. Many consider the Solodized Solonetz to be a Solonetz of more advanced weathering. They believe that leaching has removed excess salts, leaving the exchange capacity dominated by magnesium and sodium.

The high sodium and magnesium content results in their infertility. Deficiencies of nitrogen, phosphorus, calcium, and micronutrients are common. They have poor physical properties and a low water-holding capacity. Traditionally, they have been used for unimproved pasture. Recent research has suggested management techniques that allow a variety of crops to be grown successfully. Alfalfa seems especially suited, and sugarcane also responds well.

The Soloth soils are the last member of this sequence. They have even more strongly developed A1 and A2 horizons and are acid throughout the solum. The B horizon contains up to three times as much clay as the A and is columnar to blocky

in structure. In the Soloths hydrogen and magnesium dominate the exchange surface, although sodium may still account for 10 to 30 percent of the exchangeable
cations. Soloths are often found in association with Alfisols on the landscape. As
with Solonetz and Solodized Solonetz, the clay content differential between the A
and B horizons cannot be fully explained by illuviation processes and probably
results partly from differential weathering, layering of the parent material, or some
other process. The Soloths are low in fertility and primarily grazed. Recent
research, however, has expanded their value for alfalfa, sugarcane, and cereal crop
production.

How the salt content changes during the development of these soils is not clear.
It could represent the influence of leaching of the whole profile, an enrichment of
salts following leaching in an earlier more humid climate, or a sequence of periods
of leaching and enrichment. It is also possible that these soils once contained a
perched water table and that they developed during a period of alternate wetting
and drying. In any event, while there does seem to be a sequential relationship
among these soils, there is little evidence that the current soil has developed
directly from others. Instead, they apparently developed concurrently under
slightly different weathering conditions. The genesis of the Soloths and that of
many other soils found throughout the world remains a partial mystery to be unrav-
eled by future soil scientists.

Before we leave this topic, we should repeat an earlier observation of the tendency of soil scientists to concentrate on unique and unusual soils. The dominant
soils of arid regions are dry and weakly developed. The soils reflecting high salt
concentration either currently or in an earlier development stage are not the most
widespread. It is their uniqueness, not their extent, that draws our attention to
them.

Entisols

Entisols appear as three different mapping units on our soil map (areas E2d, E3a,
and E3c of Fig. 1-11). Orthents are found along the northern coast. To the south,
Psamments with Aridisols are found in the sandy regions of the desert. In between
is the Great Victorian Desert. Here Psamments with Torriorthents developed on
siliceous sands.

The soils of the southern region lack horizon development except for an A1
resulting from a small accumulation of organic matter. The A1 horizon is most
distinct where the rainfall is higher and where the surface is stabilized against wind
erosion by vegetation. Most of these sand deposits have resulted from the removal
of fine material by wind action. They are generally acid in nature except for local
salt problems. Lake Eyre is located along the south edge of the central Psamment
area. It is a large, dry, salt flat without any outlet. The land in this area is essentially

useless. A few minimg operations are scattered across the desert, and there is some sparse grazing where this soil unit grades into better soil regions. However, the area is primarily an uninhabited sandy desert.

The central Entisol region is the largest area of Entisols and consists of Psamments with Torriorthents. The Psamments are similar to those just described. The Torriorthents are known locally as earthy sands. These soils contain some clay and have some structural development resulting from the presence of iron oxides. Although excess salts and calcium are generally leached, there is little evidence of clay movement. They are located on large plateaus without great relief. Since they are associated with the desert, they are generally unused.

A third Entisol mapping unit is located along the northern coast in three separate areas. These regions contain Torriorthents with Ustalfs. The area is hot, dry, and often shallow to bedrock. Rainfall is low but generally increases eastward. Land use is still minimal because of the low rainfall. The rainfall is highest and its distribution more monsoonal in the easternmost of these three regions. Here forest vegetation is possible, and forestry has some local significance.

Apatite deposits have been discovered and are being mined in this region of northeastern Australia. It has been estimated that the total phosphorus level of Australian soils is about one-half the level of U.S. soils. But the level of available phosphorus is much lower. Thus phosphorus fertilizer is essential to maintain good production. During World War II, the Germans cut off the source of Australia's phosphorus, and yields dropped dramatically. This newly developed source of phosphorus reduces Australian dependence on outside sources to sustain its agriculture.

In addition, Entisols are found locally throughout Australia and New Zealand. Many are found on floodplains and deltas. These soils are often very valuable for grain and cash crops, fruit crops, and hay and pasture. When irrigated, they are used for specialty crops such as vegetables and sugarcane and, in Tasmania, hops are common.

Vertisols

The Vertisol region of Australia rivals the Vertisols of Africa and India in extent and importance (see Fig. 1-11 or 16-3). Both residual and depositional Vertisols are found here. Those that develop directly on residual rock high in bases are generally 1 meter or less to rock. Others are formed in the sediments from these basic rocks and are generally greater then 1 meter in depth. The profile is weakly developed, contains 50 to 80 percent clay, primarily montmorillonite, and is generally alkaline in the profile. Carbonate nodules are common in the lower parts of the solum. The precipitation ranges from 500 to 1000 millimeters with a distinct dry season.

In some cases the lower part of the profile is gypsic. The top 2 to 5 centimeters are often very powdery or granular, and they are referred to as "self-mulching" soils. If the structure is destroyed by rain or humans, it quickly develops again in the next wet-dry cycle. Below this is a rapid change to a coarse, blocky structure with slickensides. The smooth slickenside ped surface results from compression and slippage of peds past each other during swelling.

The term *gilgai* is an Australian aboriginal term used to describe the land surface that results on many Vertisol landscapes. It was described earlier in the section of North American Vertisols. The *gilgai* feature can be destroyed by tillage but will reappear in 1 to 11 years. It can appear very suddenly and dramatically in certain Vertisols when heavy rains follow a severe drought.

Vertisols are difficult to work because they are frequently either too wet or too dry, so many Vertisol areas are devoted exclusively to grazing by cattle and sheep. New management techniques combined with mechanization allow more extensive cultivation of these soils, particularly where the temperature is warmer and the rainfall higher. Wheat and sorghum are the two most important crops grown on Australia's Vertisols. Rice, linseed, safflower, sunflowers, millet, corn, barley, and oats are also important. Where irrigation is available, cotton and alfalfa are produced.

As in the other parts of the world, Vertisols (Black Earths) are often associated with Alfisols (Red Earths), depending upon the silicate content of the parent material. In Australia Ustalfs are common in the Vertisol area. The Ustalfs are generally less fertile but often have better physical properties. The local Alfisols tend to be in pasture. However, many are cultivated along with the adjacent Vertisol soils.

Vertisols are not significant in New Zealand.

Alfisols

Three suborders of Alfisols appear on our Australia map: Ustalfs in the north, Xeralfs in the south and southwest, and Udalfs in the southeast. Udalfs also appear on the New Zealand soil map (Fig. 1-11).

Ustalfs

Australia's northern Alfisols are on a very old land surface in a monsoonal climate. Over much of the area the total precipitation is marginal for crops (see Fig. 16-6). It is highest on the coast and decreases rapidly inland. In some places the flat landscape is subject to widespread flash flooding during rainstorms. The monsoonal rains provide a fairly constant rainfall from year to year. Seldom is there a year exceptionally drier than average. But the start and end of the monsoon are quite variable. The rainy season lasts for about 3 months but occurs anywhere from November to April.

Figure 16-6   The Ustalfs of northern Australia can be used only for grazing at a low carrying capacity, since there is no good source of irrigation water.

Many of the soils are shallow to rock or contain ironstone gravels that are probably the remnants of old plinthic horizins. The soils are often sandy to loamy at the surface, with the B horizon higher in clay by one to three textural classes. Soils are frequently red in color because of their content of oxidized iron. The surface horizon may be moderately acid, but the lower parts of the solum approach neutrality or may be alkaline. There is evidence that many of these soils reflect past weathering environments. Many were once strongly leached but have been more recently recalcified from wind-carried dust.

Like much of Australia, the surface has been fairly stable since precambrian times. Because the uplands are often shallow and stoney and the terraces are subject to flooding, the levee soils are best for agriculture.

Xeralfs

The southwest corner of Australia lies in a Mediterranean climate, and here Xeralfs are found. The low total rainfall and its undependability limits land use primarily to the coastal areas. A number of soils are found on well-weathered landscapes. Again, these soils probably reflect a previous climatic weathering sequence.

As in the arid regions, a very high proportion of the soils of the Xeralf region are calcareous, somewhat saline, or show evidence of having once had high concentrations of salts and were later acidified. Xeralfs may be high in calcium, magnesium, potassium, or sodium. They often have a strong argillic horizon. Another feature common in this area is the presence of ironstone gravels which, like those of the northern Alfisols, probably reflect a previous weathering cycle. Sand dunes are common along the coast and some of the dunes are calcareous. The calcareous coastal sands often show cobalt or copper deficiencies that lead to serious "diseases" in livestock. Wind erosion is always a problem; good Alfisols can be buried by advancing dunes.

Population density of this region, while higher than most of the regions studied so far, still is relatively low, and much of the population has been here a relatively short time. Therefore the properties of these soils have not been extensively studied. Recent fertility research has shown how to increase productivity considerably by the application of lime and both macro- and micronutrients. Irrigation is not extensive, and precipitation limits much of the land to pasture and dry land grains. A small area of Xeralfs is also found in south-central Australia near Adelaide.

Udalfs

The third Alfisol region is along the southeast coast of Australia. It stretches from Sydney in the east almost to Adelaide in the west. It includes several of the middle- and larger-sized towns of Australia; this soil region has the highest population density of any soil region in the country. The rainfall is higher and more dependable here, and more of it falls during the growing season. The southeast corner of Australia and Tasmania are the only parts of Australia that have been glaciated. Thus the Udalf soils are on some of the youngest land surfaces in Australia.

The same mapping unit appears in the northern part of New Zealand's North Island. This is the warmest part of New Zealand, and frosts are rare. Much of it receives 1250 to 2000 millimeters of rain annually. The landscape is rolling to steep. It is interspersed with river flats and dunes along the west coast. The warmer climate results in more intensive weathering here than in the other soil areas of New Zealand. Thus there is considerable clay movement and some destruction of 2:1 clays in favor of 1:1 silicate clays and oxide clays. Compared to the rest of New Zealand, more of the fertility of these soils is tied up in the organic matter and less in the mineral fraction of the soil. Many soils contain some materials of volcanic origin. In general, these soils are more weathered (compared to the rest of New Zealand soils), and grazing is the primary land use.

Mountain Regions

The mountains of Australia and New Zealand are primarily udic or torric (see X areas of Fig. 1-11). The torric soils are primarily Entisols and Aridisols and are located in central Australia and along the northwest coast. These two regions are hot, dry, largely uninhabited and of little agricultural value. In western Australia some mountains are over 65 percent hemitite. These mountains are being systematically cut down as a source of iron ore.

The other mapping unit is largely udic great groups of Alfisols, Entisols, Utisols, and Inceptisols. The rainfall is adequate for agriculture, and the temperatures are moderate. The ruggedness of the landscape is the primary limiting factor.

Australian Mountains

The mountains of eastern Australia extend essentially to the coast. As a soil region, the population density is second only to the Alfisol region of the south. Most of the population and agriculture are confined to the narrow coastal plain and river valleys. Wherever a flat plateau or a river valley with suitable soils exists in the mountains, it is in production. The mountains extract rain from winds coming off the ocean. This part of Australia has a good rainfall pattern for crop production. The western slopes of the mountains are drier but still productive.

Much of the landscape is stoney, gravelly, or shallow to bedrock and is occupied by Entisols. Like the parent materials of many Australian soils, they are often very low in phosphorus, and frequent deficiencies of molybdenum and other micronutrients occur. Because of this, much good research on micronutrients has been conducted in Australia. In the more rugged and less easily accessible parts of Australia, subterranean clover is seeded in improved pastures by airplane. Usually applications of superphosphate, molybdenum, and other micronutrients are essential for a good stand. These, too, are frequently applied by air.

Tasmania is further from the tropics. The temperature extremes are moderated by the surrounding ocean. The central part of the island is a high plateau of shallow, stoney soils, and its use is restricted largely to grazing. The low-lying coastal areas are very productive and grain production is extensive. Grazing of cattle and sheep produces good export commodities. Fruit, especially apples, are also exported. The rainfall here is more dependable than any other part of Australia, so large hydroelectric plants have been built to take advantage of the energy in the rainwater as it makes its way from the high central plateau to the coast (see Fig. 16-7).

New Zealand Mountains

New Zealand has two regions of mountains. The mountains of the South Island are more rugged. They result from glacial sculpturing of tilted and uplifted ancient rocks. The landscape of the west coast, particularly the southwest, is compared to the fjord regions of Scandinavia. Rugged mountains plunge to the sea without any coastal plains. Much of this region lacks good roads and can be reached only by air. Active glaciers are found. The central ridge of this system is referred to as the Southern Alps. The high altitude and rugged terrain limit the agricultural use of this area (see Fig. 16-8).

Figure 16-7   Hydroelectric power plant in Tasmania. Tasmania is one of the few areas in Australia where there is sufficient dependable rainfall at high altitudes to support the generation of hydroelectric power.

The mountain region of the North Island is volcanic in origin, some of it rather recent. The landscape is less rugged, with extensive grazing on the sloping areas and crop production on the flat lowlands (see Fig. 16-9).

The soils of this region are closely correlated to the age of ash. Some ash deposits were low in cobalt and have lead to "bush sickness" in grazing livestock. Soil maps have been helpful in identifying cobalt-deficient soils where livestock need supplemental cobalt. Except very close to the vent of the volcano, the ash layers are thin and rejuvenate the previous soil. Close to the vent, the layer is thicker, and completely new soils have formed. The ash beds are porous and weather readily, releasing nutrients and producing fertile soils except for an occasional micronutrient deficiency. An A horizon incorporating organic matter into the surface of the ash becomes evident within 2 to 3 years after ash deposition. But nitrogen is essential for crop production on younger soils.

The clays are primarily allophane, which cause the soils to be slippery without being sticky. They are generally well aggregated, but the aggregates are not stable under continuous cutivation. These soils are more suited for orchards, forests, or pastures. The bulk density is generally less than 0.6 gram per cubic centimeter.

Figure 16-8 Agriculture in the southern Alps on the South Island is limited to small valleys.

Figure 16-9 Sheep grazing on Andept soils on the rolling volcanic hills on New Zealand's North Island.

This gives these soils a high water-holding capacity, which is beneficial for plants but causes engineering problems. Building foundations can fail if they are improperly designed.

Inceptisols

No Inceptisol regions are mapped in Australia.

Two Inceptisol regions appear in New Zealand (see Fig. 1-11). The region in the North Island is mapped as Dystrandepts, relecting the volcanic nature of the parent material. It is a region of subdued relief, generally below 300 meters. Wide valleys are broken by low, rolling to steep hills. The region was covered by volcanic ash probably between 5000 and 50,000 years old. The more hilly regions have been severely eroded and often expose the mudstones, sandstones, or limestones beneath the volcanic deposit. The area receives 1000 to 2000 millimeters of rainfall per year and has a warm, humid climate. The original vegetation was forest and shrubs. The younger soils have a well-developed structure and a high water-holding capacity. They are high in allophane and do not swell. The older soils show some movement of silicate clays into the B horizon. The presence of kaolinite along with the allophane gives these soils some stickiness. Summer dryness can be a problem, and irrigation is very beneficial. A wide range of crops can be grown, but most of the area is grazed by dairy cattle and fat lambs because of the slopes. Some local production of grapes and vegetables is important.

The South Island has a region of Dystrochrepts. In contrast with the Inceptisols of the North Island, these soils develop primarily on alluvial sediments orginating in the western mountain range. The soil map published by the New Zealand government classifies the two most widespread soils of the region as Southern Yellow-Gray-Earths along the coast and the High Country Yellow-Brown Earths along the foothills of the mountains to the west.

The Yellow-Gray Earths developed under tussock grass in the lowlands. The gentler slopes are loess covered. They have a weakly developed structural B horizon. The clay content increases with rainfall. Fragipans are common. This leads to waterlogged conditions and shallow rooting zones that reinforce the summer drought problems. Many of the soils are subject to erosion and molybdenum deficiencies are usual. Mixed farming and fat lamb production are widespread on the drier soils. The more humid soils are in hay for dairy cattle. A wide variety of grains and other crops are common (see Fig. 16-10).

Approaching the mountains the land becomes more rolling; and the High Country Yellow-Brown Earths dominate. The countryside is dominated by mountains and hills of sedimentary rocks with alluvial-colluvial basins, terraces, and valleys between. Rainfall is lower in this region. At the time of settlement, the area was covered mostly with tussock grass or forest. There is frequently some indication of

Figure 16-10 The Canterbury Plain, a part of the Inceptisol region of South Island, New Zealand. A rich agricultural area.

clay movement giving a well-structured B horizon, but the fragipans common in the lowlands are absent. Erosion on this more rolling topography can be a very severe problem. Summer moisture deficiencies require irrigation for cultivation and, in places, the altitude restricts crop growth. Sheep grazing on improved pastures is the most common land use. Soils are often deficient in molybdenum and phosphorus, and responses to sulfur are usual.

Ultisols

Ultisols are mapped in the more humid, monsoonal northeast part of Australia as Ustults (Fig. 16-3). On the steeper slopes of the subdued mountain range, lithic soils shallow to bedrock are common. In the alluvial valleys, Solodized Solonetz soils are common. On the intermediate landscapes are the Red-Yellow Earths (mostly Ustults). As noted in the discussion of China's Ultisols, the red and yellow color is thought to reflect the degree of weathering and periodic saturation by water. Both soils have a wide range of properties, and not all of them would qualify as Ustults. Their primary agricultural use in Australia is for grazing, although forestry is common where precipitation permits. In some areas the climate is conducive to the production of tropical fruits, sugarcane, citrus, tobacco, and peanuts.

The coastal region of Australia is becoming an important region for finishing cattle. This industry has developed in the southern part of the coastal Ultisols and extends into the valleys of the coastal mountain region to the south. Feedlots are not found often in Australia. The cattle are raised on the range, usually west of the Great Dividing Range. They are then brought to the coast, where they are finished on improved tropical grass pastures.

A unique type of soil known locally as Krasnozems is found in this part of Australia. Krasnozems are also found scattered in adjoining mountain and Alfisol regions to the south. They generally have little development beyond a rich accumulation of organic matter. The parent materials are frequently basalt or recent volcanic deposits. They are always high in iron and not siliceous. They are high in kaolinitic clay. The climate is warm enough to keep the soil well supplied with iron from iron-rich parent material. The iron strongly flocculates the clays. This results in physical properties more like a loam soil despite the fact that they often have a clay texture. They are widely cultivated and very fertile. They produce excellent yields of sugarcane and peanuts in the north and potatoes, peas, and cereal crops in the south.

Ultisols are not known in New Zealand, which is considerable south of the Australian Ultisol region.

Spodosols

Spodosols do not appear on our soil map as either a major or secondary soil in either Australia or New Zealand. They are, however, found in both countries and are of local significance.

In Australia Spodolsols are found primarily along the coastal fringe in wind- and water-sorted sands where there is sufficient moisture for their development. A wide variety of Spodosols exist. They vary in the proportion of humus, iron, and aluminum in the spodic horizon. However, they generally lack an O horizon, which is common to the Spodosols of North America and Europe. This is generally attributed to the influence of vegetation. Many of Australia's Spodosols are under eucalyptus forests. Some profiles are very thick (see Fig. 16-11). An A horizon extending to 4.5 meters has been recorded on some sites, and solums 9 meters deep have been studied. They are so deep that they would be excluded from Spodosols by current definitions and would most likely be classified as Psamments in *Soil Taxonomy*. Because of their limited extent, they are often left idle or grazed along with nearby less sandy soils. Heavy applications of superphosphate, lime, and potassium and a variety of micronutrients permits subterranean and white clover pastures on soils without pans and without prolonged dry periods. Some apples and other orchard crops are also found on these soils. They are subject to wind erosion, and extensive conservation is practiced particularly along coastal dune areas. The Humods are especially poor in fertility with a wide range of deficiencies, including cobalt, copper, zinc, and molybdenum.

Figure 16-11 A thick Spodosol soil in southern Tasmania, Australia. The bleached A2 horizon is over 100 centimeters thick.

Spodosols are also found at higher altitudes in Tasmania and the eastern mountain range.

In New Zealand Spodolsols are scattered across the country, many of them developing in low-base, recent volcanic materials at high altitudes. Frequently, they are poorly drained, probably Aquods. They are either idle or are managed along with the more fertile adjacent soils.

Histosols, Mollisols, and Oxisols

Histosols do not appear on our map but have local importance in both Australia and New Zealand. In Australia they are found primarily at high altitudes or in cool, moist regions. The Histosols of alpine regions provide some summer grazing. Great emphasis is placed on conservation of these soils, since they are the source of water for most of the streams that feed the hydroelectric plants of the country. They gather, hold, and slowly release the precipitation to feed streams on which the power plants depend. The lower-altitude Histosols generally occur in small patches and are not worth the effort to drain and fertilize. Most are either grazed, forested, or idle. Some peat is harvested for a variety of uses.

Histosols are widespread in New Zealand. Most are found in alpine regions, along coastal plains, in low-lying regions along lake borders, or in alluvial valleys. Although they are not extensive, they are used to produce vegetables for local consumption where they are found in regions large enough to warrant drainage and reclamation.

Mollisols

Mollisols do not appear as a mapping unit on our map. They are found in small patches in both Australia and new Zealand and are of local significance. In Australia, they are limited primarily to subhumid regions along the east coast. Wheat and sorghum are grown extensively. Where they occur on alluvium, they are frequently irrigated, often supporting an intensive dairy industty. Mollisols are less significant in New Zealand.

The Mollisols of Australia tend to be more like those in the southern parts of the United States or Europe. They have less organic matter in the A, slightly more clay movement, and a coarser structure than the Mollisols of the northern prairies of the United States and Canda. Many are Udolls, and their appearance is similar to the Vertisols, which are often found in the same general region. The Udolls develop on parent materials that produce less montmorillonitic clays or in a topographic or climatic zone that results in mild leaching of clays and bases. Many early classification systems did not distinguish between the Mollisols and Vertisols. Areas mapped as Chernozems or Brumizens on older maps of Australia therefore often contain Vertisols, not Mollisols. This leads to confusion when reading older literature.

Oxisols

Oxisols are not a significant soil order but are of local importance in Australia, particularly in the north. They are not found in New Zealand. Soils that do have oxic characteristics or plinthite are usually considered to be remnants of a soil that developed during earlier weathering cycles. In a few areas plinthite-capped landscapes are found but, more often, the only evidence of the earlier weathering cycle is the presence of plinthite gravels in modern soil profiles.

Agricultural Potential

Continued research on Australian soils, particularly in the area of micronutrient fertility and tillage methods, will allow some new lands to be cultivated and some existing farmland to become more productive. But any great expansion in crop production will be limited by the lack of adequate water.

New Zealand is less limited by low rainfall, but the rolliness of the land requires extensive erosion control. Much of the South Island is shallow to bedrock. Fertility and management research should allow some increase in production in the volcanic soil regions. In general, there is little new land to bring under cultivation, and most cultivated lands are producing near their expected potential.

References

Christian, C. S. and G. A. Stewart, *Survey of Katherine-Darwin Region*, Land Research Series 1, CSIRO, Melbourne, 1946.

Gibbs, H. S. *Volcanic Ash Soils in New Zealand*, New Zealand Departmentof Scientific and Industrial Research, Information Series 65, 1968.

Leamy, M. L. and M. Fields, *Soils, Land Classification and Use*. In *New Zealand Atlas*, I. Wards, ed., A. R. Shearer, Government Printer, Wellington, 1976.

Newman J. C. and R. W. Condon, *Land Use and Present Conditions*. In *Arid Lands of Australia*, R. O. Slatyer and R. A. Perry, Ed. Australian National University Press, Canberra, 1969.

Soils Bureau of New Zealand, *Soils of New Zealand*, Bulletin 26, Department of Scientific and Industrial Research, Government Printer, Wellington, 1968.

Stace, H. C. T., G. D. Hubble, R. Brewer, K. H. Northcote, J. R. Sleeman, M. J. Mulcaly, and E. G. Hallsworth, *A Handbook of Australian Soils*, Rellim Technical Publications, Glenside, South Australia, 1968.

Stephens, C. G. *A Manual of Australian Soils*, 3rd Edition, CSIRO, Melbourne, 1962.

Stephens, C. G. *The Soil Landscapes of Australia*, Special Publication 18, CSIRO, Melbourne, 1961.

Stephens C. G. and C. M. Donald, "Australian Soils and Their Responses to Fertilizers" in *Advances in Agronomy*, Vol 10, Academic, New York , 1958.

Europe

The European Environment

Europe lies on the western peninsula of the greater Eurasian continent. Like North America, the northern parts were glaciated during Pleistocene times. Thus the soils of northern Europe are youthful. In addition, the bedrock closest to the surface is frequently calcareous. This slows the rate of acidification and clay movement in these soils, adding to their youthfulness. To the south the land surfaces are older, but much of the area receives less rainfall, which slows the aging of these soils.

A high proportion of Europe lies considerably north of the equator, but it is warmed by the Gulf Stream. Europe is relatively cooler but more humid than the other continents.

Physiography

Europe stretches westward from the low mountain ranges of the Urals to the Atlantic Ocean and includes many islands along the Atlantic coast. It reaches from the Arctic Circle in the north to about 35 degrees north of the equator at the southern tips of Spain, Italy, and Greece, (see Fig. 17-1).

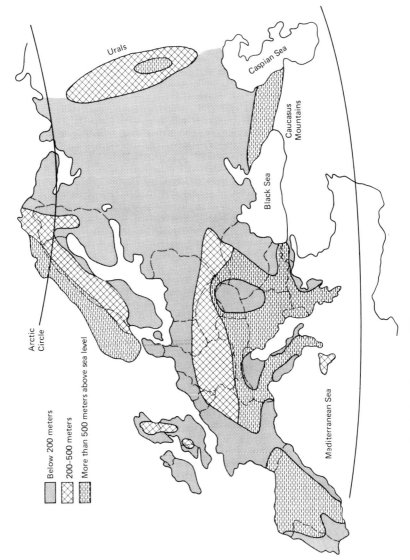

Below 200 meters

200–500 meters

More than 500 meters above sea level

Urals

Caspian Sea

Caucasus
Mountains

Black Sea

Arctic
Circle

Mediterranean Sea

Figure 17-1   Generalized elevation map of Europe.

Two-thirds of Europe is a large plain located between the Urals on the west, the Scandinavian mountains in the northeast, and the complex Alps system to the south. This plain extends through northern Germany and the Benelux countries to western France along the western European coast. In addition, there are several small but important river valleys, including the Danube Basin, primarily in Hungary, and the Po River Valley, in Italy.

The rest of Europe is made up of three mountain ranges and their highland plateaus. The Caledonian mountains extend southwest from Norway and western Sweden to the northern British Isles. Glaciation has worn down much of this area to a collection of broad plateaus. The Hercynian and Alpine folds were formed at different times and intersect each other across southern Europe. Today these folds encompass the Alps and the lower mountain ranges that extend discontinuously across southern Europe. Glacial activity during the Pleistocene altered much of this landscape and produced surfaces typical of areas reshaped by ice, water, and wind. Thus the landscape is young.

Today, a few active glaciers remain in Iceland and in Norway. Volcanic activity was significant in the geologic past in Europe. Now it is significant only in a few small regions, primarily in Italy and Iceland.

Climate

The climate of Europe, particularly western Europe, is very complex and hard to simplify. It lies in the northern part of the temperate zone. It includes part of the Arctic Circle on the north and, at its southern extremes, it is still 12 degrees north of the tropics. The continent contains several large bodies of water, including the Mediterranean, Baltic, and Black seas. These bodies of water as well as the influence of the Gulf Stream along the Atlantic coast have a pronounced tempering influence on climatic extremes. The weather is generally less severe and warmer as a result of this influence. The prevailing western winds bring both warmth and moisture to Europe; except around the Caspian Sea and some isolated mountain regions, the precipitation is generally sufficient for agriculture. The presence of mountains has a strong local influence on precipitation and temperature.

The relatively cool temperatures have resulted in slower soil development compared to the rate that occurs in warmer climates. Although two-thirds of Europe is a plain, most of it lies in Russia, under the influence of a continental climate. Much of western Europe, which does lie in the more favorable climatic zone, is steeper or mountainous. Generally, the precipitation is between 500 to 1500 millimeters, with few areas getting as little as 200 millimeters per year. Figure 17-2 indicates that rainfall less than 500 millimeters generally occurs in three areas of Europe: the high plains and low mountain regions of Spain; northern Scandinavia and adjoining Russia; and along the northern edge of the Black Sea and Caucasus mountain

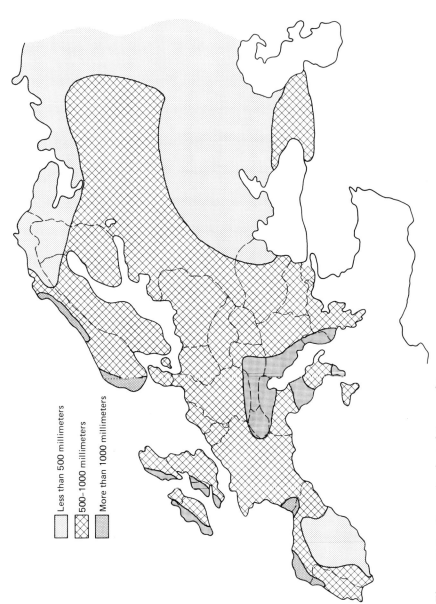

Figure 17-2 Generalized rainfall map of Europe.

Less than 500 millimeters

500–1000 millimeters

More than 1000 millimeters

range. True desert conditions are essentially unknown in Europe, but part of the southern fringe of the continent does approach desert conditions around the Caspian Sea.

Vegetation

Like much of the world that has been inhabited by people for such a long time, much of Europe has lost its native vegetation. Human activities during the development of roads, cities, and agriculture, which frequently were later abandoned, have influenced the present soils of some areas as much as the five soil-forming factors.

The natural vegetation of northern Europe is tundra along the Arctic Circle (Fig. 17-3). Most of west-central Europe was covered by deciduous forests and meadows. To the north, this gives way to extensive coniferous forests. To the southeast, there are grassland prairies. And to the south and southwest a Mediterranean type of vegetation is found. In areas of lower rainfall or regions of steep and rocky soil, the vegetation is a mixture of low trees, shrubs, and grasses that can survive on less moisture. Many of these patterns are interrupted by the influence of altitude, especially in western Europe.

In parts of Europe overgrazing or cultivation has led to severe erosion problems, and efforts are being made to remove this land from agriculture. Programs of reforestation or reestablishment of native grasses have been promoted in areas that are too steep, dry, shallow, rocky, or otherwise unsuited for crop and hay production.

Soil Regions

The generalized soil map of Europe (Fig. 17-4) shows that five soil orders dominate the landscape in Europe except in the regions of high mountains. Alfisols are the most widespread soils in Europe. They are found on a varied range of parent materials and develop under a variety of vegetation types and climatic patterns. The soils are largely Boralfs in the north, Udalfs in the central region, and Xeralfs in the south. The cool summers of the northern Alfisols on young Pleistocene and post-Pleistocene surfaces result in weakly to moderately developed Alfisols. In the United States Alfisols often have a more strongly developed argillic horizon on the same age surface because of a warmer climate.

To the north, Spodosols of northern Russia extend into Scandinavia, northern Germany, and surrounding countries. These have developed on water- and wind-reworked sands of Pleistocene age. South of the Alfisol region Mollisols extend from southern Russia into neighboring countries. These are extensions of the Spodosols and Mollisols of Asiatic Russia, discussed earlier.

Histosols dominate in the western British Isles and are very significant as a secondary soil in the Spodosol and Boralf regions. The presence of these large areas

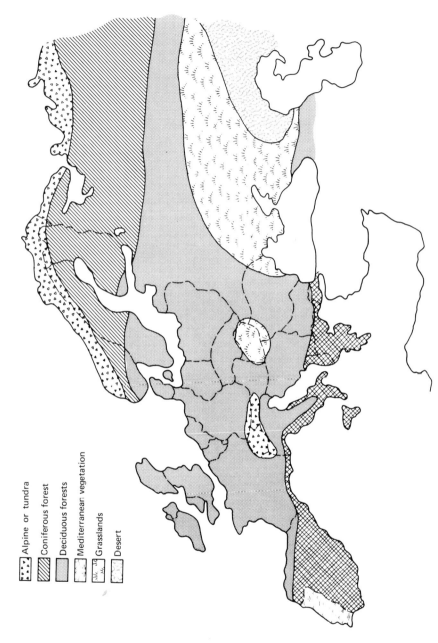

Alpine or tundra
Coniferous forest
Deciduous forests
Mediterranean vegetation
Grasslands
Desert

Figure 17-3 Generalized vegetation map of Europe.

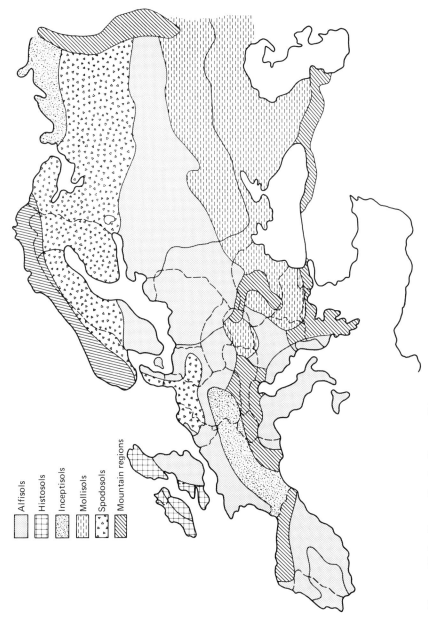

Figure 17-4 Generalized soil map of Europe.

Alfisols
Histosols
Inceptisols
Mollisols
Spodosols
Mountain regions

of Histosols is unique to Europe. Two Inceptisol regions are large enough to appear on the generalized map. The northern Inceptisols are in the tundra zone of the north. The Inceptisol region of France and southwestern Germany is the most extensive grain-growing region in western Europe.

The lack of any significant Aridisol soils is another unique factor of the European map. European Aridisols are found only in a small region around the Caspian Sea. Vertisols are found in small areas scattered throughout Europe. But the percentage of Vertisols on the landscape is the lowest of all the continents. Europe may be the only continent without Oxisols. It also has the smallest proportion of Ultisols.

Approximately 40 percent of the soils of Europe developed on limestone or other calcareous parent materials. This, combined with the cool temperatures of northern Europe and the youthfulness of the glaciated landscapes, results in weakly developed soils. The soils are generally fertile and productive, although many areas are shallow to bedrock or sandy.

Alfisols

Alfisols are the most extensive soil order found in Europe. In addition to being so widespread, they include some of the most productive soils in Europe when fertilized and properly managed. The Alfisols in central and western Europe are found in a discontinuous belt broken by the influence of high plateaus and mountain regions where younger and shallower soils are found. In eastern Europe a large, contiguous area of Alfisol soils is found primarily in Russia. In general, the Alfisols grade from Xeralfs in the Mediterranean climate of southern and western Europe to the Boralfs in the northeast. In between are Udalfs combined with Ochrepts in the west and Aquolls to the east (Fig. 17-5).

Xeralfs

The Xeralfs of Spain and Portugal occur on a rocky plateau and are associated with rock outcrops and lithic soils. This is the driest region of the Mediterranean Alfisols. Precipitation is generally between 500 and 750 millimeters. The soils grade toward Mollisols in this region. The natural vegetation of the area is a mixed tree-grass vegetation that can survive at these lower rainfalls. The soils have good physical properties, are easy to work, and have good drainage. Thus, wherever possible, the trees have been removed and the soils cultivated. A wide variety of crops can be grown. Where the climate and soil depth permit, good yields of wheat, oats, corn, vetch, and some cotton and citrus are expected. The olive production in Europe is generally restricted to the Xeralf soil regions.

The soils of this region are subject to erosion. Local erosion on the steeper slopes has reduced productivity and, in some places, restricted land use to pasture.

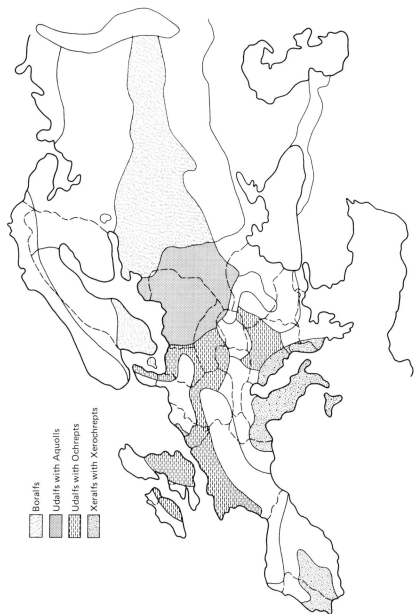

Figure 17-5 Alfisols of Europe.

Boralfs
Udalfs with Aquolls
Udalfs with Ochrepts
Xeralfs with Xerochrepts

Merino sheep are important in the Xeralf region of Spain. One-half of the world's cork production comes from the Xeralf region of Portugal. Wheat is often sown between the cork oak trees. Almonds are also very important in this region of Portugal.

A larger area of Xeralfs encompasses most of Italy south of the Alps, the larger islands in the Tyrrhenian Sea, and the east coast of the Adriatic from Yugoslavia to Greece. The precipitation is higher here than in Spain (750 to 2000 millimeters and above), and the area is somewhat more rugged. The landscape in Italy varies from coastal plains to dissected piedmont and volcanic slopes. In many areas parent materials are basic. Vineyards are extensively in volcanic areas and where parent materials are derived from limestone and other basic rocks.

Over one-half of this region is shallow to bedrock, and the Alfisols are mixed with Inceptisols. The slow rate of weathering of the bedrock plus its high base content retard the formation of an argillic horizon. Xerochrepts are common with Rendolls found on high-lime parent materials. Low rainfall and steeper slopes restrict cultivation, and so forests and grazing are common.

The Xeralfs of the Adriatic coast from Yugoslavia to northern Greece are also shallow. Stones and boulders are at or near the surface. Soil depths vary considerably over short distances, hampering large-scale mechanical field cultivation. The slopes are steep and often calcareous; past erosion of this cleared and cultivated land limits their present use. In some areas where agriculture has been abandoned, forests have taken over. Other areas are stoney and nearly void of any vegetation. Where soil conditions permit, wheat, lentils, vines, figs, dates, and olives are raised.

Udalfs

Farther north the rainfall pattern changes and the Xeralfs give way to Udalfs. Udalfs associated with Ochrepts are found in a discontinuous belt that extends from the British Isles across western France, southern Germany, and into the high slopes above the Danube River.

In Britain the parent materials are of glacial origin. In France and Germany loess is common. In most cases the parent material is moderately deep and unconsolidated, and relief is frequently less than 10 percent. Temperature and rainfall patterns are generally favorable for good crop production. The favorable climate and topography, combined with the high nutrient status and good physical properties of the soils, result in intensive cultivation of these soils (see Fig. 17-6). Wheat and sugar beets are the principle crops; oats, flax, barley, fodder crops, vegetables, and fruit are also grown in abundance. Pastures are much less important in these rich and fertile soils.

Much of the area is underlined by rock at shallow depths. The rock is often closest to the surface along the ridge tops on the landscape. Here the soils are shallow to calcite, dolomite or other calcareous materials. Ochrepts, Rendolls, and

Figure 17-6   A productive Alfisol region in Belgium. The limestone bedrock slows soil development.

other less strongly developed, often calcareous soils are found. The thinner soils of the ridge tops are often forested or in hay. The deeper soils in the lower lying areas of the landscape are in intensive crops. On the steep slopes along river valleys in Germany and France, grapes are important.

The British Isles extension of this region is frequently shallower to bedrock and less well drained than on the mainland. This region has a higher rainfall, giving increased cloud cover, higher humidity, fog, and mists. This reduces both evaporation and solar energy. Thus a higher percentage of the land is in pasture and hay crops. Sheep and cattle grazing are combined with grain production. This emphasis on grass and sheep has led to soils with good fertility status and structure while minimizing erosion.

Above the Hungarian plain in the middle reaches of the Danube River is a large, fertile plateau of Udalfs. It is a hilly area extending from Budapest and widening southward into Yugoslavia. Crop production is hampered in places by the shallowness of the soil and wetness problems. Despite these problems, excellent crop yields are obtained when the soils are properly managed. There are extensive forests on the less suitable slopes, which often contain Ochrepts.

To the north of this area, extending from eastern Germany through Poland into Russia, is another region of Udalfs. The temperatures are cooler and the rainfall higher. Aquolls are associated with the Udalfs. The parent materials tend to be more acid, and the profiles are more strongly developed than in the southern Alfisol regions. The B horizons are heavier in clay, more compact, mottled, and acid.

Many are gleyed from poor internal drainage as well as from unfavorable external drainage patterns. Grasslands may be their best use, although wheat, potatoes, and oats are widely grown. Extensive areas are in forest.

Boralfs

The Boralf region of European Alfisols lies almost entirely within Russia and extends into Asiatic Russia. As in the United States, the Alfisols have been cultivated since the earliest days of settlement. These soils have good to medium fertility levels but often have been subjected to severe local erosion. With lime and fertilizer, excellent yields are possible within the limits of the climate. Moscow is located in the center of this soil region and is about 56 degrees north of the equator (approximately 300 kilometers north of Edmonton, Canada).

The Alfisols of Russia are extensively cultivated for wheat, barley, oats, and rye. Previously, corn was also important, but the acreage of corn has decreased considerably since its peak in the decade around 1960. The total annual solar energy received in the area is less than optimum for the highest corn yields. Sugar beets and flax are important crops in this area.

Spodosols

The soil map of Europe shows that Spodosols, Alfisols, and Mollisols dominate the landscape. This is especially true in Russia, where the soils occur in east-west bands, reflecting the changes in temperature and precipitation from north to south.

About one-fifth of Europe and about one-third of the lowlands of Europe have Spodosols. Spodosols are mapped in two different mapping units (see area S1b and S3a of Fig. 1-11). Both areas were glaciated, and the soils are largely developed on water- and wind-reworked sandy sediments of Pleistocene age. The largest of the two Spodosol areas covers much of northern Russia and extends into Scandinavia. It contains a variety of Spodosols (undifferentiated), mostly in cryic temperature regimes. They are associated with Histosols.

Spodosols Undifferentiated

The Spodosols are found on rolling landscapes of hills and steep mountains with elevation generally below 200 meters but extending to above 500 meters in the Scandinavian areas. The growing season is less than 90 days, and the region is wetter than desired for agriculture more than 50 percent of the time. The short growing season restricts agriculture to root crops such as potatoes and quick-growing, cool-season crops such as flax, barley, and rye. Hay crops help support an animal industry of dairy, beef cattle, and pigs. Buckwheat production has historically been very important in the wetter areas of the Russian Spodosols. In 1971 Russia produced 80 percent of the world's buckwheat, but the total production of buckwheat is declining.

Because of the moderating influence of the ocean on air temperature, agricultural production declines from west to east across this area. The sandy parent materials are acid, low in fertility, and low in water-holding capacity. Additions of lime, fertilizer, and organic matter and the use of grasses to improve soil structure are beneficial. Much of this area remains in forest.

The cultivation and overgrazing of Spodosols in early times caused an invasion of heath. Heath is a low-growing shrub that can survive under conditions of low fertility. It accelerates the podzolization processes, leading to a very acid, nutrient deficient soil. The B horizon is often rich in iron and humus and, in the extreme, becomes cemented. Spodosols can be wet or waterlogged part of the year because of a depressional landscape position or a perched water table. At other times of the year they completely dry out. They can be reclaimed by deep mixing and breaking up of the iron-humus pan. Those at higher elevations are best left in forest, while those with a high water table can grow acid-tolerant crops such as potatoes or rye if tiled to remove excess water.

Humods

A smaller Spodosol region lies in northwestern Europe. The soils here are similar to those of Russia, but the tempering effect of the Gulf Stream lengthens the growing season. The warmer temperatures combined with favorable rainfall patterns have resulted in very productive soils under proper management. Historically, Spodosols in this region were the last of the local soils to be utilized for agriculture. Their acidity, infertility, and low water-holding capacity made them essentially useless for good crop production until the development of chemical fertilizers. When heavily fertilized, they can be extensively used for crop production (see Fig. 17-7). They are especially good in years when fall rains occur during harvest season and cause much greater harvesting losses on the wetter clay soils. The cost of fertilizer is often high, and cleared land is very susceptible to wind erosion. Therefore many of the more marginal Spodosols have been reforested in the last 30 years.

Mollisols

The Mollisols of Europe are found principally in the Danube River Valley in eastern Europe and extend eastward across southern Russia (Fig. 1-11).

Figure 17-7   Sandy Spodosols in the Netherlands can be very productive with heavy fertilization and wind erosion control.

European Mollisols

In the Danube River Valley where Austria, Czechoslovakia, and Hungary meet is a small region of Udolls. It is only in this relatively small valley that soils comparable to the corn belt Mollisols of the United States are found. The rest of the European Mollisols are generally drier and/or cooler. Much of the Udoll region has been under agriculture since prehistoric times. The principle crops are wheat and sugar beets, although corn is also important. Certain areas in Austria are famous for their wine grapes. The vines are usually grown on south-facing slopes where the soils are either very shallow Mollisols or are too weakly developed to be classified as Mollisols. Fruit orchards are also common. Often fruit trees are the only trees on the landscape.

The Danube flows from the Udoll region of Austria through a narrow valley in the hilly region around Budapest. Then the Danube Basin widens out again in the Great Hungarian Plains (see Fig. 17-8). It turns abruptly south to get around the Translvanian Alps and then flows east to the Black Sea. A large region of productive Ustolls is found along the river valley. North and east of the Black Sea lie the Russian Mollisols.

Except for the Caucasus Mountains between the Black and Caspian seas, the area is a large, low-lying plain below 200 meters elevation. Occasionally it extends

Figure 17-8 The Udoll region in the Hungarian Plain is well adapted to large field mechanization on state farms.

up to 500 meters above sea level. The area around the northern perimeter of the Caspian Sea is below sea level. As discussed earlier in the chapter on Asia, the precipitation decreases southward and limits the agricultural productivity of the area.

Russian Mollisols

The Russian Mollisol region is a vast area of very fertile, black, deep soils that developed under prairie vegetation with limited precipitation. Considering both European and Asiatic Russia, the Mollisols account for 13 percent of the land area but 60 percent of the arable land. Their high fertility and high organic matter level makes them suitable for crops that extract high amounts of nutrients from the soil, such as wheat, corn, and sugar beets. The productivity of the area is limited by periodic droughts (especially in the west and south) and the vulnerability of these soils to erosion. The treeless prairies of this region are readily adapted to mechanization. Vast plantings of trees break the otherwise uninterrupted line of sight to the horizon and provide important protection against wind erosion. This is the major wheat-producing area of Russia.

Inceptisols

Inceptisols are found scattered throughout Europe. They are found extensively in only two places—northern Russia, and central France and western Germany (areas I2a and I3a in Fig. 1-11).

Ochrepts in Europe

In western Europe the soils are primarily Dystrochrepts and Fragiochrepts. Thus, at the great group level, they are mapped similarly to the soils of the Appalachian region of the United States. However, in France the topography is more like a level plateau than rolling mountains. This gently undulating area constitutes part of the grain belt of Europe (see Fig. 17-9). The soils developed under hardwood forests and frequently show limited clay movement and acidification. The parent material of the region is varied but includes a variety of calcareous materials, inlcuding limestone. These high-lime soils, when not too steep or stoney, are among the most productive soils of Europe. They have reasonable levels of organic matter and are high in natural fertility. They respond well to nitrogen and phosphorus fertilization.

A wide variety of crops including wheat and corn are produced in the central region on more gentle slopes. At the southeastern and northwestern ends of this area, more acid soils are found on rolling landscapes. Here the areas are more heavily forested. A higher proportion of the cleared land is in pasture than is in crops. In France veal and beef production are important on good pastures receiving lime and fertilizer.

The central part of this soil belt is more gently rolling. It is an important wheat-growing area. It is far enough south that there is enough heat for good yielding corn. North of this region it is too cool for corn for grain, although some corn is grown for silage.

Figure 17-9   The Inceptisol region of France and Germany is the grain belt of Europe.

The steeper and stonier parts of this area are often in vines and include some of the important wine-growing regions of France. In addition, some of the world's more attractive marble is quarried in this region.

Cryaquepts in Russia

To the north, in Russia, lies the other important Inceptisol belt. As the location would suggest, Cryaquepts are common. The climate of this region is somewhat like that of Alaska. The cold temperatures and short growing season of 2 to 3 months severely restrict the productive capacity of the area. The region is also generally poorly drained. Crops are restricted to root crops, rye, and hay, which often support a dairy industry. Cold frames and hothouses can be used for vegetable production. The tundra in the northern part of this region is used as grazing land for reindeer. However, much of the area is uninhabited.

Entisols

As in other parts of the world, Entisols occur extensively in alluvial valleys and on steeper landscapes. There is only one area in Europe where Entisols are extensive. It is a low mountain range separated by high plateaus in central and eastern Spain and northern Portugal (see Fig. 1-11). The elevation of much of this area is between 500 and 1000 meters.

Xerorthents

The soils of this area are primarily Xerorthents. This is one of the driest regions of Europe. It has a Mediterranean climate, and most of the precipitation occurs in the cool season. In much of the area, precipitation is between 200 and 400 millimeters per year. Soils often have good physical properties; if they are level they can be very productive when irrigated. Salinity problems restrict the expansion of irrigation into some areas.

When the annual precipitation increases into the 300 to 500 millemeters range, the soils are somewhat more productive. However, attempts at more intensive agriculture have often led to overgrazing and severe soil erosion. These eroded areas now have a very low carrying capacity. Restoration of the grassland and controlled grazing is often advisable to increase productivity. Some grain farming is possible on the flatter lands with a summer fallow system. Irrigation water is not widely available but, where water is available, high yields of vegetables, wheat, and fruits are obtained. Vines and olives are grown extensively along the eastern coast of Spain.

Figure 17-10   Flower production in Holland on calcareous sands left after the dune was removed. The sands were used to build foundations for buildings and roads.

Entisols of Coastal Areas

In addition to river valleys and steep slopes, Entisols are also found along coastal areas. They frequently include sandy beaches and dunes. Many of these sands have been well sorted by wind and water and are very low in clay, silt, and coarse fragments. In some areas, particularly along the seacoast in the Netherlands, the sands are mined. They are transported many kilometers where roadbeds and foundations are being constructed on sea clay parent materials. After mining, the sandy areas are highly prized for bulb and flower production because of the high water table (see Fig. 17-10). The most desirable sites are those where the remaining sands are calcareous and where the excavation was carried out early in history. The time since excavation has allowed some profile development to occur that seems to be beneficial. Dutch flower growers try to maintain a constant water table at 55 centimeters in these porous sands for optimum flower production. Vegetable production, on the other hand, is better on the non-calcareous sands.

Soils of the Mountainous Areas

Much of Europe was once mountainous. But long periods of weathering have eroded the landscape to high lying plateaus. However, several rugged areas still remain and appear on the soil map as mountainous regions (Fig. 1-11).

The Pyrenees

The Pyrenees of northern Spain include primarily udic great groups of Alfisols, Entisols, Ultisols, and Inceptisols. The soils on the stable landscapes are primarily Alfisols or Ultisols. In the mountain valleys, Entisosl and Inceptisols are common. Entisols are also found on the steep, rocky side slopes.

In the eastern section of the Pyrenees above the timberline, the landscape is covered with shrubs, grass, mosses, and lichens. These thin soils are of little agricultural value. Some grazing occurs, but it is limited by the poor climate, steep topography, and inaccessability created by the high altitude.

Below the timberline, the area is often heavily forested. Where the soils are sufficiently deep, the land may be cleared for fruit or corn. Here in southern Europe the climate is warmer, allowing some cultivation at the higher elevations where soil conditions permit.

In the center of the Pyrenees zone, the parent materials are high in limestone and the ruggedness of the mountains diminishes. The soils here have more organic matter, good structure, and favorable mineral status. Rainfall distribution often restricts the use of these good soils without irrigation. In steeper areas to the east, grazing of sheep is a very important use of the land. The western one-third of this region has more acid parent materials, and both acidity and spotty precipitation limit their use.

The Alps

The Alps stretch from southeastern France across Switzerland, southern Germany, and northern Italy into Austria. The higher altitudes have resulted in the dominance of cryic great groups of Entisols, Inceptisols, and Spodosols.

The western section of the Alps is higher and more rugged. Most of the soils are on steep slopes, so the soils are shallow to stone or rock and the climate is cool. Some of the area is snow covered throughout most of the year. Narrow, steep-sided valleys may have good alluvial soils, but cold air drainage and mountain shadows frequently restrict their productivity. The highest part of the mountains has great scenic beauty. It attracts sportsmen and tourists in all seasons.

During the summer months, there is extensive grazing by cattle and sheep where the land is suitable. The Swiss and Austrians have effectively combined dairy and forestry enterprises. The melting waters from the snow, falling down steep mountain sides, are used to generate hydroelectric power. Industries that can utilize this power, especially where it is not necessary to transport large amounts of raw materials such as coal or iron, have flourished. Switzerland and the surrounding areas are noted for their watchmaking, wood carvings, textiles, and optical instruments.

The southernmost extension of this range in Europe includes the mountainous regions of Greece and Yugoslavia. These areas are largely the xeric great groups of several soil orders. The low summer rainfall in the Mediterranean climate along with the steep, shallow soils restricts their use.

The parent materials are frequently limestone or highly calcareous rocks. The soils are moderately high in organic matter and medium to high in fertility. Use and misuse of these soils since ancient times has resulted in extensive areas of severe erosion. The climate prevents the production of many crops; in some years it is even too dry for good pasture. Much of the area is in forest or in grazing land. Frequently it is overgrazed. Summer dry spells can leave animals without adequate food or water for long periods of time.

Crop production is limited to small strips in alluvial valleys where wheat, barley, and subsistence crops are grown. Vines and olives do well at the lower altitudes. Tree fruits are found in Greece and Turkey.

Northern Mountains

To the north is a mountain region in Scandinavia, principally along the Norwegian coast. The northern latitude, the ruggedness of the fjord landscape, and the isolation of the area result in forestry being the primary land use. Some grazing occurs. In the far north, Laplanders graze reindeer. Because of the ruggedness of the land, only 3 percent of Norway is arable. Most of the farms cultivate less than 4 hectares, and it has been reported that there are only 40 farms in all of Norway that cultivate more than 10 hectares.

The Urals in the Europe-Asia border also have cryic mountain soils. These were discussed in the chapter on Asia.

Histosols

Histosols dominate the landscape along the west coast of Ireland, Scotland, and Great Britain (Fig. 17-11) but, in fact, they are much more widespread than this. Histosols seldom dominate a landscape, but they are found scattered in the lowlands and in the depressions between low hills. They are common throughout northern Europe from the Netherlands, into Germany, Scandinavia, and Russia. Russia contains 60 percent of the world's Histosols, and yet the Histosol mapping unit does not appear in Russia at the scale used on our map. Histosols are found

Figure 17-11   Histosols of Europe.

wherever climate, topography, soil conditions, or human activity trigger a process that supresses the decomposition of organic matter. In much of northern Europe, Histosols are frequently associated with Spodosols on sandy materials, usually of late Pleistocene age.

Basin Peat

Organic soils have been separated according to their mode of formation into blanket peats and basin peats. The formation of basin peats was described under the heading Histosols of North America. In Europe they are generally found in areas enriched with base-rich groundwater under a wide variety of vegetation types. Where incoming waters have carried in mineral sediments, the peat has a high ash content and is less desirable as a fuel. Low-ash peats have been a good source of fuel for hundreds of years, especially in Ireland, the Netherlands, and Scandinavia.

In some areas 10 meters or more of peat have been removed for use as fuel. The mineral material at the base of the peat deposit varies considerably in texture. Where sandy, there is a tendency to leave the area as a lake after the removal of the peat. Where the base material is finer in texture, the area is often drained and reclaimed for use in agriculture. Large areas in Holland have been reclaimed after the removal of peat and are productive agricultural areas today (see Fig. 17-12).

Figure 17-12   Potatoes planted on reclaimed land after the removal of peat. The roadbed behind the brush rests on a strip of unreclaimed peat.

Blanket Peat

Blanket peat generally begins to develop in a wet depression on either a mineral soil or in basin peat. Plants such as sphagnum moss grow and die. As the sphagnum accumulates, it lifts the water table by capillarity, producing a locally wet environment in which the process continues. As the organic deposit rises in the landscape, the water table rises with it. The peat spreads up the side slopes of small hills and, in some cases, completely engulfs them. These soils are generally restricted to very wet climates that lack a dry season and where fog and dew are common. The soil itself never dries out during formation. Since the moisture is from precipitation and upward capillary movement, it is often poor in bases and produces a very acid peat. Since there is no source of mineral sediments, they are low in ash, making them desirable as fuel.

The use of a peat as a fuel is mostly a thing of the past. The best and highest-quality peats have been largely removed, and alternative fuel sources are now readily available. Some peat is still mined, but much of it is used as a soil amendment for horticultural and nursery crops. The low ash content of the peat makes it desirable for the production of activated carbon for purification systems and for medicinal purposes. In many areas, the remaining tracts of peat land have been set aside by local governments to be preserved for science, recreation, and as a living museum.

The bottom few centimeters of blanket peat are often high in ash and therefore are left in place. Because of its susceptibility to wind erosion, the organic residue is often mixed with the underlying mineral soil to produce a highly productive, organic-rich soil. When the water table can be controlled and the pH is good, these reclaimed soils are well suited for many agricultural and horticultural crops.

Where the residual organic layer is thicker or the ash content of the deposit is too high for high-quality fuel, a thick organic layer exists even today. Often the water table is maintained very near the surface. It is too wet for crops but provides subirrigation for optimal grass production and minimizes the problem of oxidation and subsidence of organic material. Such areas often also support an extensive dairy industry.

Vertisols, Aridisols, Ultisols, and Oxisols

Vertisols are practically all under cultivation, except for the steeper slopes, which remain in pasture. They are often difficult to till because they suffer from wide moisture fluctuations within the local climate. Their effective use requires mechanization, erosion control, and water conservancy. The flattest areas are used for wheat, oats, sugar beets, and sunflowers, with some vegetables and hay crops. Some of the more sloping lands are in vines, although reforestation is probably the wisest use of the steeper tracts that have previously been eroded.

Europe is the only continent in which no Aridisols occur on the soil map at the scale used. Rainfall is generally too high to produce true Aridisol soils. A few scattered Aridisols do occur in central Spain and in European Turkey.

Europe is considerably north of the equator, and the presence of true Oxisols would not be expected. Some Ultisols may be found in Europe, but they are not extensive enough to be significant.

Soils Disturbed by Humans

The ordinary activities of drainage, irrigation, fertilization, liming, and soil preparation for domesticated plants have a significant impact on soil properties. Modern classification systems have tried to define criteria so that whenever possible the cultivation of soils by ordinary methods would not alter the classification of that soil compared to its virgin conditions. In some cases, human activities are so extreme that new soils have in fact been created. The soils of the rice paddy were given earlier as examples of this. In Europe human activities have led to two other situations that will be discussed here: the Plaggepts and the land reclaimed from the sea, peat bogs, swamps, lakes, or other bodies of water.

Plaggepts

Plaggepts resulted from human activities in the Middle Ages as people migrated northward into the acid, Spodosol soils. Plaggepts are found primarily on Pleistocene sands in northwestern Europe, principally in the Netherlands and adjacent parts of Germany. The original acid Spodosols were rather unproductive prior to the development of commercial fertilizers. People chose to settle on the hillslope just above the floodplain. Here they built their homes. It was safe from the flooding water and yet low enough in the landscape that a shallow well would provide needed water.

These earlier settlers often settled in villages rather than on isolated farmsteads for protection from predatory animals and thieves. Their animals were in the village. Straw, leaves, heath, sand, soil, seaweed, or any other handy material was used as bedding in the barn to absorb the animal manure. Eventually, the manure-saturated bedding got so deep in the barn that the animals were up against the ceiling. Then the material was hauled out, and the process was repeated. The manure-laden bedding was heavy to handle, so it was usually spread close to the barn where it became mixed with sandy Spodosols. This practice eventually produced a fertile soil that gave high yields compared to the native acid sands. Over time, two areas developed—a bedding source area that provided some grazing and hay as well as the bedding, and a depositional area that produced good yields of adapted crops. The source area was generally 10 to 20 times the size of the depositional area.

Figure 17-13 A Plaggept area in the Netherlands. Many years of adding manure and bedding to the land has raised the site above the surrounding level.

In time, the addition of organic matter to the soil elevated the Plaggepts one to 1½ meters above the landscape (see Fig. 17-13). This practice was eventually abandoned, but the soils remain today as evidence of this former activity. True Plaggepts contain 80 to 200 centimeters of mixed mineral matter enriched with organic matter. They have excellent physical properties, a high water-holding capacity, and may or may not be fertile. Irish Plaggepts tend to be strongly alkaline, with a pH up to 8.4. European Plaggepts tend to be acid reflecting the lack of clacareous materials in the local parent material. They are, however, always high in phosphorus compared to nearby unaltered mineral soils, and this fact is used in identifying Plaggepts. With proper fertilization and management, Plaggepts will usually outproduce the other soils of the region.

The removal of native vegetation for bedding in the source area exposed the sandy soil to wind erosion. Cultivation practices also exposed some areas of the landscape to wind erosion during the year. Thus a second feature common in a Spodosol region is evidence of wind erosion. Areas of loss or deposition of sand are common. In some cases, good agricultural land and even villages were abandoned because of the sand deposition. Today, these areas of blow sand are kept in permanent pasture or have been reforested to prevent further wind erosion problems.

Reclaimed Land

A second feature of this agrarian village life allowed villagers to organize their collective efforts to battle the ravages of floods, storms, and high tides. This led to the reclamation of land from rivers, lakes, or the sea, especially in the Netherlands. Much of the Netherlands is a delta region formed by four major rivers. The alluvial deposits in this delta area are higher in clay than the Pleistocene sand areas. Thus the alluvial soils, usually Fluvents, were much more productive than the upland sandy soils. The villagers built dikes out from the uplands to protect the rich agricultural land at the foot of the village. These dikes were gradually extended further and further into the floodplain. Likewise, dikes were built around coastal marshes and tidal basins. Where there was a supply of fresh water to flush out the salt in these marine sediments, the reclaimed land was often highly productive.

In the fifteenth century, the windmill was developed and utilized to pump water. This allowed shallow lakes and tidal basins to be pumped dry. The areas from which peat had been removed for fuel had flooded to form lakes. These, too, could now be drained and reclaimed as farmland. Lake bottoms were leveled, and ditches and pumps were used to maintain the most desirable water table for the crop to be grown. Today, perhaps 80 percent of the Netherlands would be damaged by high tides and spring floods if it were not for the dikes.

The Zuider Zee Project

In the early 1900s a Dutchman named Lely proposed a project to reclaim a vast area of land from the sea. He suggested that an enclosing dam be built across the mouth of the Zuilder Zee, a large bay of the North Sea (see Fig. 17-14). Sluice gates in the enclosing dike would prevent seawater from entering but allow river water to flow out to the ocean. Upon completion of the enclosing dike, there was no longer any tidal movement in the old Zuider Zee. The saltwater eventually was washed away, and the name Zuider Zee was changed to Issel Meer (Issel Lake) to recognize the change from a saltwater bay to a freshwater lake.

Four polders were proposed behind the enclosing dike (see Fig. 17-15). The first was completed in 1929 and resulted from the consolidation of several small islands and polders reclaimed earlier. The nature of the bottom of the Issel Meer was surveyed by boat, and three large polders were proposed within the Issel Meer. Their location was selected so that they would be developed on the more clayey parts of the lake bottom. Good cropland requires at least 80 centimeters of clay over sand and orchards require at least 100 centimeters of clay. The deeper channels were left as rivers to handle the floodwaters of the rivers emptying into the Issel Meer. The sandy areas of lower productivity were left as lakes for fish and recreation and for storage of fresh water for irrigation. The stored water can also be used during dry spells to fight salt intrusion from the ocean.

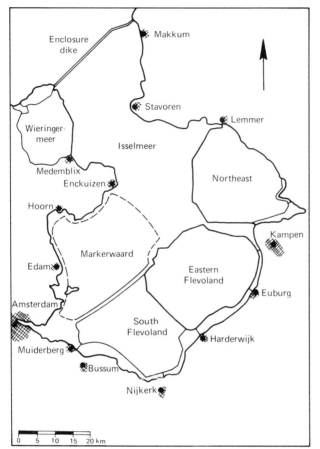

Figure 17-14 The Zuider Zee was dammed in 1924, creating the freshwater Isselmeer. The Wieringermeer polder was diked from 1927 to 1929 and was fully developed by 1940. The northeast polder was diked between 1936 and 1940 and was fully constructed by 1958. The eastern Flevoland was diked between 1950 and 1956 and is nearly fully constructed now. The southern Flevoland was diked between 1959 and 1967 and settlement is now taking place. The dike for the Markewaard polder is now being constructed.

Within each polder an intricate system of canals, ditches, and pumps maintains a water table depth that is constant over the year. However, the depth to the water table varies from one part of the polder to another in recognition of differences in soil texture and type of agriculture. An 80 to 175 centimeter water table is maintained in clayey soils for grain production. A 50 to 75 centimeter water table is

Figure 17-15 The completed portion of the Markewaard dike that also serves as a highway. The water to the left will be drained when the dike is completed.

suited for pastures on sandy soils. The area was systematically laid out so that each farm was on a good road and was of an efficient size. Trees were planted to provide recreational areas, windbreaks, and an artificial horizon on the otherwise flat, monotonous, treeless landscape.

The second polder was completed in 1943. Its construction continued during the World War II German occupancy in recognition of the need for additional land for food production. The northern one-half of the third polder was completed in 1956 and the southern one-half in 1967. During the first few years after reclamation, while roads, ditches, buildings, utilities, and cities are being built, the land is farmed by the government in large blocks. During this period of subsidence and aging of the reclaimed soil, yields are spotty, and the risks involved in crop production are too high to be assumed by individual farmers. Twelve years after the completion of the south one-half of the third polder, the land is still mainly in large blocks. In time, the area will be divided into farmsteads and leased to farmers on long-term contracts. Work is progressing on the dikes that will surround the fourth and final polder of this project.

Farming the Polders

The soil surface usually subsides about 50 centimeters in the first few years after the removal of the water due to the shrinkage of clay and oxidation of organic matter. Initially, the physical properties may not be good because of the high sodium content. Physical properties improve rapidly in the first 3 to 10 years

Figure 17-16   Land reclaimed from the sea in recent times is characterized by large rectangular fields. Many parts of Europe have small irregular fields caused by rolling topography and the division of estates among heirs.

because of natural processes of wetting and drying and the leaching of the soluble sodium. Additions of manure and the growth of crops such as rape also speed structural development.

The soils of the polders are very productive after aging, and most of the area is devoted exclusively to crop production. Wheat, sugar beets, potatoes, onions, and other grains and vegetables are widely grown. The landscape is broken by an occasional road, a row of trees, and drainage ditches. Fields are large and rectangular and are not interrupted by sloping areas, wet spots, or stoney or shallow soils, which are so common in natural areas (see Fig. 17-16). Ditches serve as a barrier to livestock movement, so even fences are rare. Crop yields in the polders are probably higher than in any other part of Europe.

Food Production Potentials

Compared to Asia and Africa, Europe applies more technology to food production and gets a higher return per hectare. Europeans have a high level of technology and generally use fertilizer, pesticides, and machinery very effectively. As a result, their potential of increased yields with current technology is not great.

The Alfisols of western Europe and the more humid Mollisols of central Europe are extensive and productive. The area of additional land that could be brought under cultivation in the future is not great. Most of the potential agricultural land is already being used for pasture or forests, and shifting this land to crop production would simply reduce pasture and forest production.

Russian agriculture has at times been influenced more by politics than by science and technology. New technology and better application of present technology could result in some additional improvement in Russian production. However, weather remains a severe handicap in much of the Soviet Union.

One area where increased production is possible is the southern rim of Europe from Spain to Turkey. Political and economic stability could result in better use of some of the marginal lands. At the same time, many marginal and subsistence lands are now overused and should be removed from crop production to either grazing or forests. The suitable land occurs in small, scattered areas, and considerable work must be done to identify these areas and develop appropriate technology.

In general, European food production is near the maximum that can be expected under current economic and technological conditions. But some shifts in politics could cause important shifts in food production.

Politics, Social Structure, and Agriculture

Agriculture has been historically tightly controlled and regulated by strong, central governments that have recognized the importance of a stable agriculture in maintaining the political stability of the country. The long history of political rivalry in Eruope has led each country to attempt complete self-sufficiency, even where the soil and climate are not well suited for all types of food production. Each country tried to produce all of the grain, sugar, fiber, oil, and other commodities needed to sustain its people. Self-sufficiency was of highest priority. Reliance on trade with a neighbor was considered hazardous; after all, you never know when you might again go to war and be caught with a surplus of sugar and an inadequate supply of flour. Therefore governments have maintained artificially high prices of some commodities to encourage their local production even when yields are low because of less favorable climatic or soil conditions. This leads to mixed farming rather than specialization in the commodity most suited to the local soils and climate.

Crop diversity has the beneficial effect of spreading the work out over the year. A single farm in The Netherlands, for example, might produce potatoes, sugar beets, barley, wheat, onions, flax, peas, and grass seed, plus hay if it has cattle. On a single farm it would be rare for any of these crops to occupy more than 10 hectares, and most occupy considerably less than this. With each crop having slightly different planting, cultivation, and harvesting dates, a farmer is kept constantly busy but is rarely overwhelmed. With increased mechanization, each farmer must have a complete set of tools to prepare a seedbed and to plant, cultivate, harvest, and store this diversity of crops. The machinery investment for all these crops becomes prohibitive. Therefore informal cooperatives of 2 to 10 farmers have been formed. Under this system, one farmer might own a grain combine and another a potato digger, a third a sugar beet harvester, and so forth. They would then share equipment and labor to harvest the crops.

With the development of the European Common Market, the political necessity of complete independence and crop diversity may diminish, but the concept of subsidies and the labor advantage of a mixed farming system may make change very slow.

An interesting aspect of Swiss subsidies has a more significant bearing on tourism than agricultural needs. Historically, many of the steep slopes of Switzerland were cleared and used to produce hay. Availability of jobs in the cities, the advantages of mechanized production on the flat lands, and the amount of hard work involved in harvesting hay on steep slopes have all led to a decline in hay production on these slopes. If this trend were to continue, many of the slopes in the highlands of Switzerland would be abandoned from agriculture and eventually revert to forest. Solid forests are not nearly as picturesque as lands that are broken by small fields and pastures. Therefore the Swiss government subsidizes small landowners to produce hay either by hand or with small machinery on these steep slopes (see Fig. 17-17). The goal is not to promote hay or livestock or to support the related cheese and milk industries but to maintain the landscape that will continue to attract tourists.

Figure 17-17 Harvesting hay on steep mountainside slopes in Switzerland.

Thus advances in food production in Europe, as on the other continents, depends on many factors. Social structures, educational levels, availability of capital, chemicals, machinery, seeds, markets, and transportation facilities as well as what is perceived to be in the best national interest of the local government may be as important in determining future world food supplies as soil and climatic conditions.

References

Central Intelligence Agency, *USSR Agriculture Atlas.* U.S. Government Printing Office, Washington, D.C., 1974.

Conry, M. J., *Irish Plaggen Soils—Their Distribution, Origin and Properties Journ. Soil Sci., 22,* 1971.

Curtis, L. F., F. M. Courtney and S. T. Trudgill, *Soils in the British Isles,* Longman, London, 1976.

Dudal, R., R. Tavernier, and D. Osmond, *Soil Map of Europe 1:2,500,000,* FAO/U.N., Rome, 1966.

Eshuis, J. A., *Reclamation of Heathland and its Cultivation, Soil Sci., 74,* 1952.

Farnham, R. S., and H. R. Finney, *Classification and Properties of Organic Soils, in Advances in Agronomy,* Vol. 17, Academic, New York, 1965.

Hellinga, F., *Water Control, Soil Sci., 74,* 1952.

Ministry of Agriculture, Fisheries and Food, *Agriculture of the Netherlands,* The Hague, 1962.

Moore, P. D., *Origin of Blanket Mires, Nature, 256,* July 24, 1975.

Pape, J. C., Plaggen Soils in the Netherlands, *Geoderma, 4,* 1970.

Van Der Meer, K., *Reclamation of Dune Sand Soils, Soil Sci., 74,* 1952.

18

South America

The South American Environment

South America contains approximately 12 percent of the earth's land surface. It is the fourth largest continent, but it is more compact and has the shortest coastline per area of landmass. It lies almost entirely east of the United States. The west is dominated by the Andean range, which is relatively narrow and transverses the whole continent from north to south. Much of the continent lies on a vast central valley whose eastern margins are at a higher elevation than the center. Only a few rivers penetrate this rim, and they bear the burden of draining the vast interior to the sea. South America is second only to Europe in the amount of water it deposits daily into the ocean. Only Europe contains fewer desert regions. No other continent has more forests than South America. Except for the deserts and a few grass prairies, South America was originally one continuous forest.

Physiography

South America extends further from north to south than any of the continents. It reaches from 12 degrees north of the equator to 56 degrees south of the equator. And yet, because of its shape, 80 percent of its land lies in the tropics. In contrast to Africa, whose tropical zone contains a vast desert, much of the tropics of South America are at higher altitudes.

South America compares more to Europe in elevation features. A good proportion of the land surface is below 200 meters in elevation. The Andes range is about the same age as the Rockies, and its elevation is almost continuously above 1000 meters from tip to tip. The mountains rise suddenly out of the Pacific with almost no shoreline, and only two suitable natural harbors occur along the whole west coast. The mountains soar to great heights and drop back to less than 1000 meters in elevation within 600 kilometers from the west coast. In places, rain falling only 150 kilometers away from the Pacific must flow 4000 kilometers east before draining into the Atlantic (see Fig. 18-1).

The Andes rank second to the Himalayas in height, but they are more populated. The mountains contain high, flat plateaus between 3500 and 4500 meters in elevation. Except for Tibet, there is no other area like it in the world. It is cold but humid enough to support some agriculture.

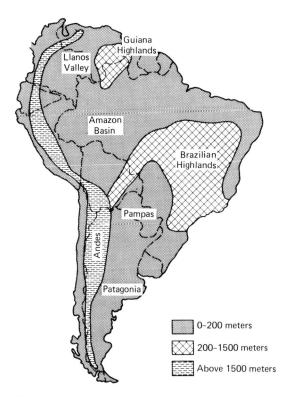

Figure 18-1   The physiographic regions and general elevation of South America.

Two separate smaller ranges also reach to 1000 meters: The Guiana Highlands, located primarily in southern Venezuela, and the Brazilian Highlands. The Brazilian Highlands slope westward and northwest, forming a vast plateau between 600 and 1000 meters in elevation. This separates the lowlands of South America into two regions. To the north is the vast Amazon Basin, much of it less than 200 meters above sea level. The area is so flat and low that the Amazon drops only 44 meters in the last 1400 kilometers and only 73 meters in the preceding 1600 kilometers.

The Amazon is second only to the Nile in length and handles the largest volume of water of all of the world's rivers. One-fifth of all flowing water in the world flows down the Amazon. It handles 10 times the volume of the Mississippi. It has seven tributaries that are over 1600 kilometers long. In places, the river level rises 15 meters from the driest to the wettest parts of the year. Floodwaters sometimes extend almost 100 kilometers from the banks of the river. It is so flat that tides can be detected nearly 1000 kilometers from the coast. It is navigable farther from the ocean than any other river in the world. However, climate and other factors have rebuffed efforts to use it to open up the interior.

North of the Guiana Highlands is the Llanos Valley. The headwaters butt against the Amazon Basin, but the Guiana Highlands separate their drainage systems.

To the south of the Brazilian Highlands are the Argentine Lowlands or the *pampas*. The area is drained by several rivers flowing into the Rio de la Plata between Montevideo and Buenos Aires. Although it is overshadowed by the Amazon in size, it handles one-twelfth of the world's flowing water. Much of this area is below 250 meters. Ancient rivers have deposited sands and clays over an old rock plateau. Over time, the elevation has dropped as the rocks sank.

South of the *pampa* is Patagonia. It sits on a plateau that rises above the *pampas* and merges with the Andes as they diminish from the west. This is a cool desert region.

Climate

Winds from several different systems bring moisture to South America. In the summer a large low-pressure system sits over the interior Amazon Valley. Winds from the northeast bring moisture-laden winds across much of the valley and the western Brazilian Highlands. More than 1000 millimeters of moisture falls over much of this area from November through April. The *pampas* receives 250 to 750 millimeters of moisture during this same period from winds blowing off the ocean. West of the Andes, winds blow eastward from the South Pacific, bringing over 500 millimeters of moisture to the southern coast of Chile. These winds are deflected northward by the Andes and flow parallel to the coast. As in Australia, these parallel winds bring little precipitation to the rest of the coast. In the Patagonian region in the south, only dry winds make it across the lower-lying southern Andes, and much of the interior is desert.

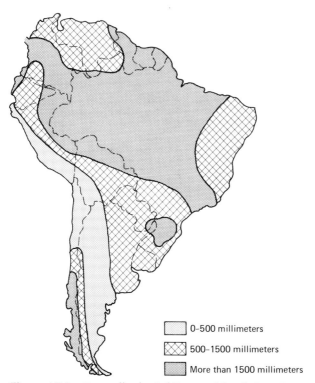

0–500 millimeters

500–1500 millimeters

More than 1500 millimeters

Figure 18-2   Generalized rainfall map of South America.

In the winter the size of the low-pressure system shrinks and moves northwest-ward. Much of the Amazon Basin receives another 750 millimeters or more of moisture in the winter season, from May to October.

The *pampas* of Uraguay and southern Brazil also receive additional winter mois-ture, but the amount and extent is diminished. On the other hand, the Pacific rain belt in southern Chile extends further northward in the winter.

The average rainfall in much of the Amazon Basin is around 2500 millimeters each year. Extremes of 1800 to 9000 millimeters have been recorded. In parts of Colombia, the rainfall averages over 2.5 centimeters per day for 300 days each year. A generalized rainfall map is given in Fig. 18-2.

While wind systems do bring the rains, the wind velocities in the Amazon Basin are generally low. Typhoon-type storms are virtually unknown in the interior. The relative humidity is high, with the mean usually above 75 to 80 percent.

Three-fourths of the continent lies in the tropics, parts of it at very high altitudes. However, over one-half of the continent still has a tropical or subtropical climate. The tropics are not known so much for their hot temperatures as for their con-stantly warm temperatures. In addition, the fact that the sun is more directly overhead allows 56 to 59 percent of the solar energy to reach the ground, while 46 percent reaches the earth's surface at 40° latitude and only 33 percent at 60° latitude. Thus the crop yield potential of the tropics, based on total solar radiation reaching the surface, is nearly twice that of the temperate region. There are, of course, other limiting factors to production, including moisture and soil conditions, but the utilization of this high solar energy level challenges researchers in their quest to increase yields in the tropics.

Vegetation

South America is dominated by forests (see Fig. 18-3). The Amazon Basin is a large tropical rain forest. It is said that one-fourth of the trees growing on the earth today are in this huge valley. Forty percent of them are inaccessible. Temperate forests are located in the Brazilian range and in the more humid Chilean Pacific coast.

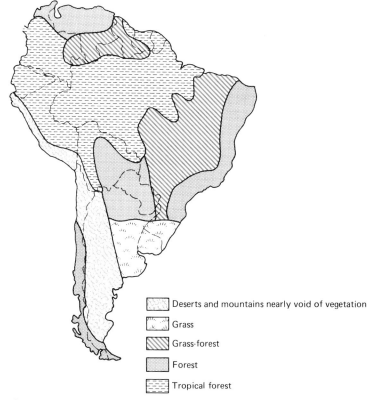

Deserts and mountains nearly void of vegetation

Grass

Grass-forest

Forest

Tropical forest

Figure 18-3 Generalized vegetation map of South America.

The grasslands are largely in the *pampas* area of Uruguay and Argentina. The Guiana Highlands and Brazilian Highlands contain areas of mixed trees and grass. The deserts are located in the temperate region rain shadow of the Andes. Much of the Andes in the tropical region are also arid.

Soil Regions

Approximately one-half of South America's soils are nutrient-poor soils on unconsolidated sediments. They range from Oxisols to sandy or clayey Entisols in river valleys. An additional 20 percent of them are too dry for crop production. About 10 percent are too steep, rocky, or cold, and another 10 percent are too wet. But some of these wet soils can be productive if drained and properly managed. Only 10 percent are fertile soils in a suitable climate for highest productivity. These include Mollisols, Inceptisols, Alfisols, and Ultisols.

Figure 18-4 shows a generalized soil map of South America. The Oxisols of the

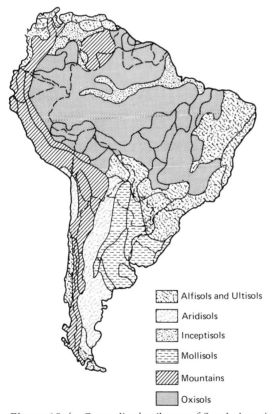

Figure 18-4   Generalized soil map of South America.

Amazon Basin and surrounding highlands dominate this map. The humid climate and the relatively flat and stable landscape have resulted in very old and highly developed soils.

Alfisols and Ultisols are shown together on this map. They are often found adjacent to each other or intermingled on the landscape. Subtle variations in parent material, climate, topography, and other factors on the landscape are responsible for this. Parts of these areas are more rolling, and erosion has prevented the formation of the more highly weathered Oxisols.

The Mollisols are confined largely to the *pampas* south of the Brazilian Highlands. The soils are fertile, but over much of the region low or undependable rainfall has limited the use of these soils to grazing.

The Andes and the Guiana Highlands in the north are steep. Altitude is the single most significant factor in determining soil properties. The Aridisols of South America are confined largely to the rain shadows of the southern Andes and to the coastal area of northern Chile.

Inceptisols appear on the generalized map in the Amazon Basin, the Guiana Highlands, and the cold region of the southern tip of South America. This is the southernmost land area in the world, excluding Antarctica.

Entisols appear on the more detailed world map (Fig. 1-11) but not on the generalized map. Histosols, Spodosols, and Vertisols are of minor importance in South America.

Oxisols

Figure 18-5 shows the specific Oxisol regions of South America contained in the detailed soil map of Fig. 1-11. Most of South America's Oxisols are in the Amazon Valley and the surrounding uplands. The valley was once part of an ancient sea that was later uplifted. It is now a low-lying, flat area that has been weathered for a long time. It is the largest continuous area of tropical, humid climate in the world, and much of it receives rain almost daily. All of this combined to give soils of great age and degree of weathering.

The climate causes the area to be somewhat inhospitable to humans. It is not only uncomfortably hot and humid, but many tropical diseases flourish here; however, it is estimated that one-half of the land potential for expanded cultivation lies in the tropics. Therefore much of this potential would be expected to lie in the Amazon Basin.

One of the problems with developing this potential is identifying and locating soils with the greatest potential. Partly because of their age, soils in Oxisol regions are quite variable across the landscape. Small differences in parent material composition, after long periods of weathering, cause significant variation in soil properties. Without fertilizer these small variations can result in very significant differ-

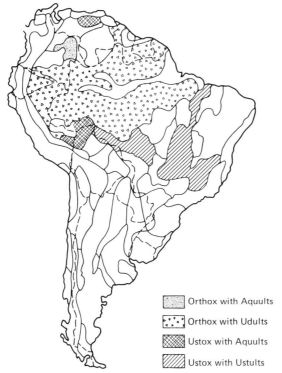

Figure 18-5   Oxisols of South America.

ences in yields. This is especially true where a specific micronutrient is absent or present in inadequate amounts.

The activities of people, termites, and other animals over long periods have also caused considerable local variability in soil properties. These small differences have often been ignored. In experimental plots they cause enough variability to overshadow the influence of the treatment under study, and they hinder the transfer of knowledge from one site to another.

Least Understood Soil Order

Our understanding of Oxisols and tropical soils in general is somewhat tentative. Until after World War II, studies of soils in tropical areas were spotty. Today, we are just beginning to identify the significant features of soils in tropical regions and to understand their genesis and classification. Oxisols have so gradual a change in properties with depth, as shown in Fig. 18-6, that horizon boundaries are generally arbitrary.

Early classification systems stressed the red color, absence of weatherable minerals, supposed absence of organic matter as evidenced by lack of dark-colored horizons, and the presence of laterite or plinthite when classifying tropical soils. These criteria proved to be weak. More recently, the $SiO_2:R_2O_3$ (silicon dioxide:sesquioxide) ratio has been used. Ratios less than two are common in the tropics as silicate minerals weather and sesquioxides (iron and aluminum oxides) accumulate. This has lead to the concept of the oxic horizon and the order Oxisols. The soil taxonomy classification system is admittedly weak in the classification of Oxisols. The handling of tropical soils in other classification systems has not been much better. Because of our lack of understanding, criteria for classifying soils in various systems are quite different. Until we understand enough to develop a good system of classification, our progress will be limited. In the meantime, the variance between systems often makes the transfer of knowledge from one system to another next to impossible.

Figure 18-6 Oxisol profile in the Brazilian Highlands. It lacks the striking horizon differentiation common in most of the other soil orders.

While we conclude that our level of knowledge is not what we would like it to be, we also note that much good research is underway. Research is revealing new ideas and questioning the validity of earlier concepts. Some of the discussion contained here about Oxisols will very likely be quickly outdated as our knowledge expands.

The two most important aspects of soil productivity is the ability of the soil to store water and to provide nutrients to plants. It seems that the transfer to the tropics of our temperate region knowledge in both of these areas is being challenged.

Water Storage

Tropical soils generally have a lower water-holding capacity than temperate soils. Recent work has shown that the temperate region rule of thumb that available water is stored between 0.3 and 15 bars tension in the soil does not apply in the tropics. It seems that the nature of clay minerals and organic matter in the tropics causes much of the available water to be held at lower tensions. The strong structure of the clay results in sand-sized aggregates with large macropores that drain easily and hold little water. Thus the water retention characteristics of fine-textured soils in the tropics are more like those of the coarse-textured soils in the temperate zone. Approximately 50 to 75 percent of the water held between 0.1 and 15 bars tension is actually held between 0.1 and 1 bar tension. Much of the water that remains in the soil at 1 bar tension is held above 15 bars and is unavailable to plants.

Oxisols generally can store only one-third to two-thirds as much water between 0.3 and 15 bars as Alfisols or Ultisols of similar texture. Thus Oxisols are more dependent on a steady water supply. In Brazil, corn can wilt on Oxisols after six rainless days. A 10-day period without rain in the rainy season is expected one year in two in Brazil. A 14-day dry spell occurs one year in five. These brief rainless periods occur when evaporation rates are very high. Thus short-term periods of water stress are much more detrimental to yields than has been previously recognized. Irrigation of crops in the tropical rain forest regions may be an economical investment in some cases.

Furthermore, the warmer soil temperatures reduce the viscosity of water. This results in substantially more unsaturated flow of water within the soil. This can aid in moving water to the roots, and it is undoubtedly significant in ion transport in the soil. It can also increase leaching.

Aluminum Toxicity

The problem of low water storage is further complicated by the fact that many tropical Oxisols (and Ultisols) have toxic levels of aluminum in the B horizon. This restricts rooting depth of plants and further accentuates the moisture problem.

The acidic nature of the lower part of the profile causes aluminum toxicity to be the biggest single factor reducing crop growth in the tropics. Manganese toxicity is also common. Soluble aluminum begins to appear in the soil solution when the pH drops below 5.4. By the time the pH reaches 4.5, the exchange surface is 50 percent aluminum saturated. At 4.2, this has increased to 60 percent.

At a given pH level, Oxisols and Ultisols can be expected to have similar percent aluminum saturation on the clay surface. But the higher cation-exchange capacities of Ultisols can result in up to 10 times the total aluminum concentration in solution. Thus, although aluminum toxicity is more widespread in Oxisols, it is more difficult to correct when it occurs in Ultisols.

In general, liming to pH values between 5.3 and 5.6 will solve yield problems as aluminum and manganese toxicities disappear. Yields of corn, sorghum, and soybeans have been increased 7 to 140 percent by liming. But the best response occurs when lime is incorporated to a depth of 30 centimeters instead of the more usual 15 centimeters. Even deeper incorporation is probably beneficial, but this is not easy to accomplish.

Early attempts to lime soils in the humid tropics into the 6.5 to 7 pH range met with disaster. At these high values, micronutrient deficiencies appeared and phosphorus availability declined. It is now recommended that tropical soils be limed only sufficiently to eliminate aluminum toxicity problems (pH 5 to 5.5) and provide adequate levels of calcium and magnesium. Some legumes will show responses when limed up to 6 or slightly above. This is due mainly to increased molybdenum availability (see Fig. 18-7).

Figure 18-7 Influence of lime on soybeans on the second plot back on an Ustox. Note the small stature of the trees in the native vegetation at this site near Brazilia.

Soil Color

We have mentioned throughout this text that most soil classification systems use color as an important classification criterion. In the tropics the distinction between red and yellow soils has frequently been considered significant. American soil classification has tended to consider soil color more incidental, preferring to concentrate on more basic soil properties such as type of clay, soil pH, and other properties. Now work by Uehara and his associates and the application of this work by Sanchez has suggested that soil color in the tropics may be a quick, visible way to approximate the clay system in a soil. Specifically, the clay system in red Alfisols and Ultisols is often more comparable to Oxisols. The clays in yellow Alfisol and Ultisol soils are more like those of temperate region Mollisols and Inceptisols. Recent work has deepened our understanding of the difference between the two basic clay systems found in soils—the silicate clays and the oxide clays.

Clay Systems

The silicate clays, more specifically the 2:1 silicate clays, derive most of their exchange capacity from isomorphous substitution. Because of this, the cation-exchange capacity of the 2:1 clays is relatively constant over most of the normal soil pH range.

The 1:1 silicate clays, oxide clays, and organic matter have a highly pH-dependent cation-exchange capacity. This is because, to a large extent, they derive their exchange capacity from surface ionization. As the pH increases, the cation-exchange capacity increases (see Fig. 18-8). As the soil becomes acid, the cation capacity decreases. When pH values fall in the 4 to 5 range, the cation-exchange capacity reaches a zero point. Below this point the soil has a net anion-exchange capacity.

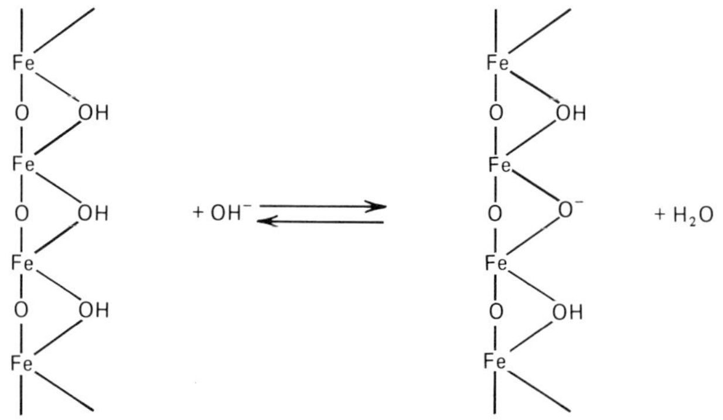

Figure 18-8   Increasing soil pH causes an increase in the cation-exchange capacity of oxide clays. As the pH increases, hydroxide ions force this reaction to the right, increasing cation-exchange sites.

Because the oxide clays are amorphous, they can coat the crystalline clays. This physically blocks their pH-independent exchange sites. The red Alfisols and Ultisols of the tropics often react more like the Oxisol soils because the properties of the 2:1 clays are masked by the properties of the oxide coatings. The yellow Alfisols and Ultisols of the tropics lack this amorphous coating and react more like the Mollisols, Inceptisols, and other soils in which the 2:1 silicate clays dominate. Therefore, in the yellow soils of the tropics, theories established for the silicate clay systems of the temperate regions seem to apply. But an understanding of the red soils of the tropics may require some new thinking.

This leads to problems when developing classification systems for tropical soils. Our methods of determining cation-exchange capacity need careful examination. Should cation-exchange capacity be determined at an arbitrary pH value or at the pH of the natural soil? The former makes comparisons between soils more meaningful, while the latter is more meaningful in determining needed management practices.

Several practical problems follow from this. First, lime requirement determinations may give erroneous recommendations. A portion of the lime is used to increase the cation-exchange capacity of the soil. This fraction of the added lime does not appreciably alter soil pH. Second, only small amounts of cationic fertilizer elements can be held on the exchange when the cation-exchange capacity is low. Thus heavy applications of fertilizer may result in large leaching losses. Finally, phosphorus is very reactive with soluble aluminum. The resulting compound is unavailable, so much of the phosphorus fertilizer applied on acid soils never gets to plants.

But the picture is not all bad. Both the oxide clays and the oxide-coated silicate clays have well-developed structures. This aids in water infiltration and gaseous exchange and reduces erosion problems. Because of its strong structure, some Oxisols that are 80 percent clay can be plowed the day after a heavy rain. The best structure seems to be in ustic rather than in udic regions, suggesting that a good drying out of the soil is important in structural development. Sandy Oxisols and Ultisols often lack the iron that flocculated clays needed for good structure. Therefore sandy soils may be less permeable and may erode and compact more easily than soils that are higher in clay.

However, one must again be careful about making broad generalizations. First, although they are important, Oxisols are not the only soil in the tropics. Second, although Oxisols are less erodible than soils of other orders, it cannot be implied that erosion is not a problem in tropical regions. As in temperate regions, management systems strongly influence soil erosion. Shifting agricultural systems tends to be soil conserving. Converting to more intensive plantation type of management systems is usually accompanied by a marked increase in soil erosion, primarily because large tracts of land are left exposed to the devastating effects of high-energy monsoon rainstorms. Furthermore, since soil fertility is concentrated in the organic surface materials and the subsoils may contain toxic levels of aluminum, yields can be reduced as much as 40 percent with the loss of as little as 2.5 centimeters of topsoil.

Organic Matter Levels

Sanchez also discusses another significant, heretofore unappreciated fact about Oxisols. Extensive data are not available and conclusions must be considered somewhat tentative at this time, but Oxisols seem to be much higher in organic matter than previously realized. Sanchez suggests that the organic matter levels in the tropics are often similar to temperate region soils. Over one-half of the Oxisols in an East African study contained over 4 percent organic matter. Another comparison showed Oxisols to contain more organic carbon in the top 15 centimeters than nearby Ultisols and Alfisols in both Zaire and Brazil.

The reasons for these high levels is not clear. Evidently, the relationship between color and organic matter level of the temperate region does not hold in the tropics. Many red Oxisols and Ultisols have been shown to contain more organic matter than adjacent black Vertisols. Some reseachers have tentatively suggested that organic matter in the tropics may be "colorless."

Two factors may contribute to these higher than expected organic matter levels. The organic matter may form very stable complexes with the oxide clays and allophane in the clay fraction. This could either provide a physical barrier to the microbe, or the complex could prevent the microbes from decomposing the organic part of the complex. A second possibility is that the extreme phosphorus deficiencies that commonly occur inhibit the decomposing organisms. Whatever the reason, we need to reconsider our previous picture of tropical soils being very low in organic matter.

Oxisols of the Amazon Basin

By far the largest area of Oxisols in South America and in the world is in the Amazon Basin. These are primarily Orthox soils, with Udults on the younger land surfaces. Because of the extent of Oxisols in the Amazon Valley, we are beginning to accumulate some knowledge about these soils. They have a deep, porous solum with little horizon differentiation. They tend to be low in silt, and clay skins are usually absent in the B horizon. They are porous, friable when wet, and hard when dry, and gully erosion is not extensive. There may be some increase in clay from the A to the B that sometimes makes it difficult to distinguish them from Ultisols and Alfisols. They are acid, with a percent base saturation commonly less than 15 percent. Most of the cation exchange is from the soil humus.

A few of these soils are in shifting agriculture, producing low yields of cassava, rice, beans, pineapple, and cotton. The low fertility of the area requires a long fallow period.

Figure 18-9 Shifting agriculture experiment station in the upper Amazon Valley. Research will allow increased yields using traditional methods.

Pepper and rubber are produced on plantations in the area. Good response of pepper to fertilizer has been noted. A continued shift to more permanent agriculture can be expected with time as the area is settled by the efforts of the Brazilian government and as better management systems are developed (see Fig. 18-9). But, at the moment, nearly one-half of the fertilizer in this region is used on the relatively few hectares devoted to peppers.

The need for fertilizer is generally quite high due to the low levels of fertility. But the low exchange capacities of the soil require careful management to avoid overfertilization, which leads to unnecessary leaching or nutrient imbalance. Oxisols are generally low in all nutrients. Oxide clays have an especially high phosphorus fixation capacity. Many of the soils of the Amazon region are low in clay relative to other Oxisol regions, and they have lower fixation rates. They are generally low in sulfur, potassium, calcium, and magnesium. Most of the sulfur is in the organic matter, and up to 75 percent of the sulfur in the vegetative cover can be lost when plant residues are burned.

Zinc, molybdenum, and boron deficiencies are frequent problems. As yet there is no good zinc test for tropical soils. Molybdenum problems often improve when the soils are limed. Liming also reduces the chances of manganese toxicity. When copper fungicides are used on bananas, it can induce an iron deficiency on a succeeding crop of rice. Copper toxicities have shown up in Chile, where irrigation water was previously used in copper mining operations.

The main problem to development in this area remains its general inacessibility and the inhospitable nature of the environment for humans. The vast majority of the area is covered by a wide variety of tropical trees, many of which have no commercial value for lumber but may support a paper industry.

Other Oxisol Regions

The Oxisols of the Brazilian Highlands are largely Ustox, with Ustults on the younger surfaces. As the soil names imply, this is an area with a more distinct dry season. These soils tend to be redder, which is characteristic of soils that dry out some time during the year. These soils generally lack plinthite and have a low cation-exchange capacity (less than 1.5 milliequivalents per 100 grams of clay from NH_4Cl).

Their immediate potential is no better than the Orthox soils in the valley below. They generally have an even lower fertility level, and some have essentially no calcium at all in the subsoil. The percent base saturation is usually less than 20 percent, and the aluminum saturation is over 50 percent. The topography of the area is generally level, and their strong structure minimizes the risk of erosion. Because Brazilia, the new captial of Brazil, lies in this soil region, much effort is now being directed toward these soils. Brazilian soil scientists have developed management systems that have resulted in higher production from some of these soils (see Fig. 18-10).

Two small areas of Ustox with Plinthaquults occur north and south of the Amazon Basin. This mapping unit would imply that the plinthite horizon is more common in the Ultisols than in the Oxisols of the area. The southern Ustox region is located above the Amazon Basin. Deficiencies of phosphorus, potassium, copper, cobalt, and sodium have been reported. This not only limits crop yields but seriously limits the rate of gain of cattle grazing on these natural grasslands. Plinthite limits depth of rooting. It can cause waterlogging in the wet season, thus limiting root growth. This further accentuates the problems in the dry periods.

The northern area lies on the Guiana Shield and is mostly uninhabited. In addition to the fertility problem normal to these soils, the hilly landscape limits the potential for mechanization. Where plinthite reduces infiltration rates in steep soils, erosion is a very serious problem.

The final region of Oxisols is a small area of Orthox with Plinthaquults. This area lies on the northeast edge of the Amazon Basin in a wetter area. This area, too, is largely uninhabited and is in native vegetation. In other more densely populated regions small areas of similar soils have been cleared for continuous production of coffee, citrus, cotton, bananas, pineapple, and cassava. But most have later reverted to pasture as they lose their natural fertility. Although the soil fertility level is low, the nutrient status of the vegetation is higher and shifting agricultural systems can be supported.

Figure 18-10 Intensive agriculture on Oxisol soils in the southern highlands of Brazil. When the local economy allows the implementation of agronomic research high yields of a wide variety of crops are possible.

Ultisols

The Ultisols are the second most extensive soil in South America and are mapped in 10 separate units on our map (Fig. 18-11). Many of these soils are intermingled with Oxisols on the landscape, and comments about the red Ultisols in the previous section also apply to them. Other Ultisols are mixed with Alfisols on the landscape. These tend to be in temperate regions and reflect parent material differences. Alfisols are found on the more basic parent materials and Ultisols on the more acid ones. Where Ultisols are mapped with Inceptisols, one can generally expect the Inceptisols on the steeper, more eroded sites and the Ultisols on the more stable slopes.

The two Ultisol regions along the northwest coast are Ultisols mixed with Inceptisols. These are in a rolling area in the foothills of the Andes. The northern region in Colombia contains Tropudults with Tropepts. This region lies just north of the equator in an area receiving 8 to 12 months of rainfall. The southern region in Equador lies on and south of the equator. The Ustults and Ustochrepts reflect a longer dry season.

Both areas are in a marine climate and receive a high rainfall. Much of the area receives 2000 millimeters per year and is known to exceed 8000 millimeters per

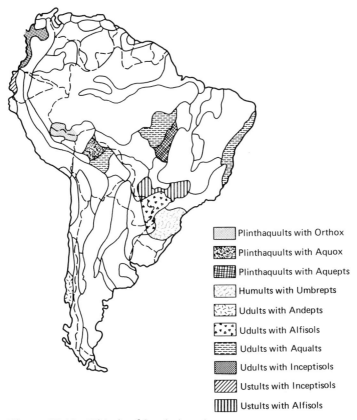

Plinthaquults with Orthox

Plinthaquults with Aquox

Plinthaquults with Aquepts

Humults with Umbrepts

Udults with Andepts

Udults with Alfisols

Udults with Aqualts

Udults with Inceptisols

Ustults with Inceptisols

Ustults with Alfisols

Figure 18-11 Ultisols of South America.

year in some parts. This leads to a dense, lush jungle that contains as many tree species as the jungles of the Amazon Basin. Some volcanic activity has deposited ash in the area. This is more extensive in the southern Ustult region. The population density of these areas is not high.

Along South America's southwest coast is a zone of Udults with Andepts. This is the only humid region south of the equator on the west coast. These soils are found in a narrow region of foothills. Volcanic soils (Andepts) are high in allophane which, like oxide clays, have a pH-dependent cation-exchange capacity. In some places the volcanic ash is 4 meters deep.

These soils are often low in fertility, with a base saturation well below 50 percent. Slope, shallowness, and stoniness further hinder their use. The high rainfall of this area results in a forest cover. The commercial value of these trees in the temperate region is higher than those of the tropical forests. Along parts of the seacoast, the landscape is often rugged fjords.

Ultisols of the Amazon Basin

The rest of the Ultisol soils lie east of the Andes, primarily along the edge of the Brazilian Shield. Four separate Ultisol units are mapped on the north edge of the shield and three on the south.

On the north, furthest west, are Plithaquults with Orthox soils. This area is surrounded by Oxisols, and the soils here are not too dissimilar from them. The surrounding forested Oxisol area is better drained. This region is flatter, low-lying, poorly drained, and contains savanna.

To the southeast, the savanna again reappears and two Ultisol areas are mapped: Plinthaquults and Tropudults with Aquults. This, too, is a flat, marshy area that is the headwaters for several tributaries of the Amazon. The soils vary primarily in their drainage and the presence or absence of plinthite. These areas are largely uninhabited. Low fertility and occasional water saturation of the soils also restrict their potential. Deficiencies of phosphorus, potassium, copper, and cobalt reduce their potential for further crop production and as grazing land for cattle. Lowering of the water table has sometimes led to hardening of the plinthite, which restricts the root zone.

Further east along the Brazilian rim of the Amazon Basin is another region of Tropudults and a region of Plinthaquults. These regions are similar to those just described. Another area of Tropudults is mapped along the coast. This is a rolling, dissected area known as the Sea of Hills. Much of the original forests have been removed by humans and are now in grassland.

Population densities are somewhat higher along the coast and there is more agriculture. Perhaps 20 to 30 percent is in crops and 40 to 50 percent is in pasture. Coffee, sugarcane, citrus, and corn are extensive. Pineapple, tea, and rice are also grown. These soils have a low level of fertility. This, combined with the rolling topography, limits the potential for mechanization. Erosion is a serious hazard, and there is a great need for conservation.

Ultisols of the Brazilian Highlands

To the south of the Brazilian Highlands, three Ultisol regions grade from the highland to the coast. The soils grade from Ustults to Udults to Humults. The Ustult region has a 5 to 7 month dry season and is largely in savanna with some forests. The soils contain considerable amounts of unweathered minerals, generally from more basic rocks.

The Udults form on similar rocks but are in a more humid area. The population density is somewhat higher and, therefore, Udults are more extensively used, but more for pasture rather than crops. Wheat, oats, potatoes, corn, beans, and cassava are raised. The slopes are more gentle and erosion, while still common, is less severe than in the other Ultisol regions. While some soils are fertile, others may contain toxic levels of aluminum and require lime. The landscape lends itself to mechanization on those soils of moderate natural fertility. If the aluminum toxicity problem can be solved, these soils have a good potential for extensive wheat production for export markets.

The coastal area is mapped as Humults. These soils have the highest organic matter level of the freely drained Ultisols. The productivity potential of Udults is even higher here.

Inceptisols

Inceptisols are found scattered throughout South America, mostly in river basins, on steeper slopes, or in cold regions (see Fig. 1-11). The largest Inceptisol region is the floodplain of the Amazon River. These soils are primarily Tropaquepts, with Plinthaquults on the more stable highlands. The properties of the soils of the region vary enormously, but they have in common the limitation of flooding and poor drainage. Seldom is it economical to establish the drainage system necessary to utilize them. Except for some local rice production, they are rarely used.

An area in the Llanos Valley north of the Guiana Highlands is mapped similarly. It occurs near sea level, and extensive drainage systems have been developed. Sugarcane, rice, bananas, cocoa, coffee, cassava, corn, and horticultural crops are produced. But most of this region is grazed, with a carrying capacity of two cows per square kilometer. Many of these soils have developed on young marine sediment, which is often high in natural fertility. These soils are usually basic, while the Amazon Basin soils are more acid.

A third region of similar soils is located on the edge of the Brazilian Highlands. While classified here as Inceptisols, they are closely related to soils in the albic or palic great groups of Alfisols and Ultisols. Many soils in the region have a heavy clay B horizon, usually at a shallow depth that restricts water movement and root growth. It is a flat, swampy area and produces excellent rice crops in Brazil and neighboring Argentina. In the dry season sugarcane is produced or the land is grazed.

A small, swampy coastal region of these same soils occurs at the mouth of the Courantyne River.

A small region of Dystrochepts is found higher up the Llanos Valley. This is an area of steeper slopes and more erosion. Use is limited by shallowness to rock and the steeper landscape.

A larger Inceptisol region is found along the eastern coast in southern Brazil. This is part of the Sea of Hills area. But here the landscape is steeper, and Tropepts have developed as this eroded landscape begins to stabilize. This coastal area is more heavily populated and includes the city of Rio de Janiero.

Further south along the coast, a Humaquept region lies in southern Brazil and Uruguay. This is a region of heavier marine clays and some salt problems.

Umbrepts with Aqualfs are found in the Tierra del Fuego region on the southern tip of South America. It is a very cold and wet area in which organic matter decomposition is very slow. Many of the soils are Histosols; others are very shallow to bedrock. The cold weather severely limits land use. This region is as far south of the equator as the southern tip of Alaska is north of the equator.

Mountainous Areas

The Andes are the highest mountain range in the Western Hemisphere. They rose from the sea during the Tertiary period. They were glaciated during Pleistocene times, and some glaciers still exist today above 3800 meters. In many areas there has been extensive volcanic activity leading to locally significant regions of Andepts.

The areas mapped as mountain regions (Fig. 1-11) lie in the continuous belt of the Andes along the west coast, plus the Guiana Highlands. The soils of the Andes are cryic in the south and aridic in the central region. To the north, in the tropics, the soils are ustic and cryic on the coastal side of the Andes. On the inland side, where the altitude drops and the area receives the benefits of the moisture from the Atlantic, the area is udic.

The soils of the Guiana Highlands are also mapped as udic. Like most mountain areas, it is a region of contrasting rock formations, tilted and eroded to various degrees. Within this range are flat, high plateaus and alluvial valleys. Their development is hindered by their isolation, but local productivity is good where topography, soils, and climate are found in productive combinations. Much of this area is in tropical forest, and many weathered, infertile soils occur on the flatter landscapes.

There is a small ustic mountain range along the north coast. The mountains rise abruptly from the sea. This is a more densely populated area and includes Caracas, the capital of Venezuela. Many of these soils are of volcanic origin and are low in bases.

The northern region of the Andes is the most humid region of these mountains. The area receives rainfall primarily from the storm systems that move across the Amazon Basin from the Atlantic. It is a steep area that rises sharply from the valley. In Ecuador the valleys are filled with volcanic ash that is not found very far north or south of Ecuador.

The highest parts of the mountain ranges in the tropical region are ustic and cryic. The slopes are steep, and climate and vegetation change rapidly over short distances. Volcanic materials are interspersed with sedimentary and metamorphic rocks. In the south the udic area shifts eastward.

The aridic and cryic zones straddle the Tropic of Capricorn. In the north these peaks are west of the mountains, and a high valley is located between them. During the formation of the Andes, high basins without external drainage were formed.

These were later, partially filled by local sediment. This area is known as the Altiplano. The subhumid part of the Altiplano receives 3 to 4 months of summer rainfall. Wet and salty soils are found here. The semiarid parts are primarily large salt flats. The flats are 3750 meters above sea level. Volcanic peaks rise in places to over 6000 meters. These volcanoes are the source of Andepts.

The most productive area is probably the Altiplano of Bolivia. It includes Lake Titicaca, which is the highest large lake in the world. The elevation of the Altiplano ranges between 3500 and 4500 meters, and it is about 200 kilometers across. Water from the surrounding mountains flows into these closed basins.

The most arid regions are in the south. Many are void of vegetation. This dry region grades into the cryic region of the southern Andes of Chile. The elevation of the peaks gradually drops from 4000 meters in the north to 2000 meters in the south. The seacoast is largely a fjord landscape except where the Alfisol and Ultisol regions are mapped. Again, volcanoes are found throughout this area and were a source of ash. They are most numerous in the central region, diminishing to the north and to the south.

Alfisols

Four Alfisol areas are mapped in South America. The largest is in the Brazilian Highlands and contains Ustalfs with Tropustults (see Fig. 1-11). In the western part of the Brazilian Highlands are Ustalfs with Ustolls. Two small areas are located west of the Andes. Ustalfs with Ochrepts are found in Columbia, and Xeralfs are in Chile.

A large Ustalf area is on the eastern edge of the Brazilian Highlands. The climate is semiarid, and much of the area is grazed. Cotton, peanuts, and sisal are raised as cash crops along with cassava, corn, and beans for local consumption. The fertility level of these soils is generally high, but the dry period in the summer limits their productivity.

The landscape is gently undulating, with isolated mountains, hills, and steeper areas. In these areas the soils are very erodible and shallow. Along the coast and at higher elevations where rainfall is higher, forests are found. The rest of the area is covered by scrub typical of areas of lower rainfall.

The western Ustalf area is smaller and mixed with Ustoll soils. The rainfall is slightly higher here. The population density is fairly low. The native vegetation is grass or savanna, and much of the area is grazed.

Crop production in the Ustalf area of Columbia is limited by a yearly dry season, but it is still an important crop production area for Columbia. It is in a gentle to moderately sloping area along the coast and includes some good farmland.

The Xeralf area of Chile also has a dry period. In this case, a Mediterranean climate exists. This region stretches north from Santiago. Because of the rainfall restriction, it is less intensively used than the Ultisol region to the south. But the cropping systems are not unlike those of the Mediterranean area: wheat, oats, grapes, peaches, and olives. Irrigation is common in many areas where water is available.

Mollisols

The Mollisols of South America are found primarily in two areas. A large area of Mollisols dominates the *pampas* between the Brazilian Highlands and the desert, and a small region of Mollisols is found in Patagonia in southern Argentina (see Fig. 18-12). Only in the *pampas* of southern Brazil and in Australia can one find farms in the tropics that compare with the U.S. family farm.

Three separate mapping units are found in the *pampas:* Udolls along the coast, Albolls in two smaller regions in the north and south, and Ustolls in the drier west. The northern Udoll region is in Uruguay and Brazil. About one-half of the land is flat; the rest is more rolling and consists of outcrops of older rocks. Loess deposits are the parent material of most of the Mollisols. Some are high enough in carbonates to allow Rendolls to develop. Others are depressional soils and are wet and/or salty.

South of the Rio de la Plata are the Argentine *pampas.* This is the largest area of deep, fertile soils in South America. The soils develop in loess, whose texture becomes finer from west to east. The unique feature of this loess is its volcanic origin. The eastern area is more humid and this, combined with the finer textures, results in aquic soils.

The region is a natural prairie that has been converted to rangeland and mixed farming in the last 100 years. It is one of the world's most important beef-produc-

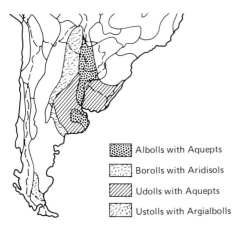

Albolls with Aquepts

Borolls with Aridisols

Udolls with Aquepts

Ustolls with Argialbolls

Figure 18-12 Mollisols of South America.

ing areas. Wheat is also grown extensively on Ustolls and exported all over the world. Corn is important in the Udoll area. A wide variety of crops can be grown here; however, moisture is not always predictable, and periodic drought can be devastating.

The Alboll region is primarily in the north-central *pampas,* with a small coastal area south of Buenos Aires. This northern area is the main drainage basin of the *pampas.* The basin is flat, filled with sediments, and poorly drained. The northern-most parts are drier, and saline problems are frequent. The fluctuating water table throughout the area seems to be responsible for the formation of the albic horizon. The high water table combined with a less permeable argillic B and, in some cases, a natric horizon results in soils very susceptible to drought.

The western edge of the *pampas* contains Ustolls with Argialbolls. The region is a large, flat, sedimentary plain showing some volcanic influence. Winds off the desert to the west have redeposited salts and some soil particles. Soils with salt problems are rarely used. Where irrigation is possible, alfalfa, vegetables, fruit, and grapes are produced. Without water the land is limited to grazing, and the area is extensively grazed. However, limited water in the dry season combined with local salinity problems results in areas that have seasonably inadequate water supplies for livestock. Frequently, a heavy-textured B horizon causes serious wetness prob-lems in the rainy season.

The southern Mollisol region is mapped as Borolls, indicating the cold climate that limits the use of the soil in Patagonia.

Aridisols

A region of Aridisols with Orthents of South America lies along the western boundary of the Andes south of the equator. The largest Aridisol area lies to the east of the Andes in the rain shadow. Two smaller areas lie along the coast north of the humid part of the Chilean coast. A small area of Aridisols with Ustalfs lies along the Gulf of Venezuela (see Fig. 1-11).

The Venezuela region lies mostly below 75 meters above sea level. The rainfall of the area varies over short distances from humid to semiarid. The Guajira Pen-insula has a large area of shifting sand dunes. The rainfall is below 500 millimeters, and sisal and pineapples are raised without irrigation. Extensive goat grazing in the area has removed much of the ground cover.

An arid region is found on the westernmost protrusion of South America on the Peru-Ecuador border. This is an area of calcareous coastal dunes. The low rainfall (50 to 300 millimeters) severely limits their use. A larger Aridisol area is found along the Chile-Peru border. This region is frequently sodic or saline, which restricts their use even where water is available.

Not all the coastal desert region is shown in Fig. 1-11. The desert extends nearly 2000 kilometers along the coast but rarely extends more than 150 kilometers from the coast. About 52 rivers cross this desert; not all of them flow all year round. This gives a complex soil pattern that reflects largely parent material patterns. Most of the agriculture is on the intermediate terraces with deep, medium- to fine-textured sediments. These irrigated valleys produce vegetables, cereals, alfalfa, flax, cotton, and tropical fruits. It produces the food for 50 percent of the population of Peru.

The largest desert region of South America is in Argentina in the rain shadow of the southern Andes. The rainfall in the northern region is 100 to 200 millimeters per year. Strong winds desiccate the plants and suppress plant growth. Desert pavements are common, and many soils are calcareous, saline, or sodic.

The rainfall in the central region is higher, approaching 500 millimeters in some areas. The area contains extensive colluvial materials from the Andes. The southern region is colder. Rainfall is lower, but lower evaporation rates partially compensate for this. Soils are shallow and stoney. Histosols occur in the depressions, where organic matter decomposition is slow.

Entisols

The Entisol regions of the north are primarily in river valleys (see Fig. 18-13). Aquents are found at the mouth of the Orinoco River. The soil properties vary with respect to texture, depth, organic matter content, and many other features, but they all are poorly drained. They are frequently basic. Near sea level, in the tropics, good production of sugarcane, rice, bananas, and cocoa can be expected when the soils are drained.

The second Entisol area along the northern coast is crescent shaped. Several rivers flow through here to the sea. Much of the area is sandy, and parts are shallow to ancient rocks. The soils are mainly Psamments. Orthox soils are found on the stable upland positions. The Psamments in the Brazilian Highlands are also sandy soils combined with Ustox soils on the shield. The two regions astride the desert in the south are Psamments with Aridisols.

Agricultural Potential of South America

Most of the agriculture of the tropics is on Alfisols, Mollisols, and Vertisols with some additional production from higher fertility Entisols and Inceptisols. These occupy about 18 percent of the tropics.

It seems that there are at least 500 million hectares of arable land in South

Figure 18-13 Irrigation of Torriorthents along the coast of Peru. Narrow coastal areas receive little rainfall but are very productive when irrigation water is available from nearby rivers.

America which, today, are in their original, natural state. Our objective must be to identify the remaining land that is suitable for production and to apply proper management techniques to it.

The infertility of tropical Ultisols and Oxisols and the associated low-fertility Entisols and Inceptisols, which occupy one-half of the tropics, will cause this expansion of production to be slow. But there is no other area in the world that is as large and as thinly populated and that also receives so much rainfall and solar energy. The only comparable soils are in the tropics of western Africa. The advantage in South America is that this region lies almost entirely in one country, Brazil. This greatly simplifies the social and political barriers to development. If the Brazilian government decides to place a high priority on the development of this region, it is relatively easier to organize and mobilize the expertise to develop new management techniques; to organize so that the necessary inputs are available at an acceptable cost; to teach the local farmers about the new management systems and to properly utilize the inputs; and to handle the marketing and distribution of the produce. The problems in undertaking this development in Africa are much more complex because the area is shared by many smaller countries. Local rivalries and differences in economic resources, educational levels, national priorities, ethnic backgrounds, and social structure slow the potential for developing regions of similar resources.

The Brazilian Highlands is another area of potential. Rainfall and the solar energy levels are lower than in the Amazon Basin, but they are still relatively high. Soil problems here are sufficiently unique that management systems different from those of the Amazon Valley must be developed. But, again, the potential is there, and enough of the problems are solvable that one can expect significant increases in agricultural productivity in the future.

Outside the tropics most of South America is too rugged, dry, or cold for increased production. The one exception to this is the *pampas.* Shrader estimates that only one-third of the potentially cultivated land is now being cultivated.

Expansion in this area will be hampered by claypans and poor drainage, but the techniques for solving these problems are known. In some areas social, political, and economic problems exist. This is one of the few areas of the world where very significant increases in grain exports can occur based on current technology. However, to do so will be at the expense of the beef production in the *pampas.*

Agricultural Potential of the World

Vast areas on this planet are suited to expansion in food production. But this increase will not be easy. Most of the best soils of the world, especially those in the temperate regions, are now under cultivation. In most of the temperate regions of the world, with the possible exception of the *pampas,* production is about as high as current technology and economic structures will allow. Increased yields are to be expected, but only after considerable research and time to test the economics of new technology.

It is true that in some areas there remains some additional land to drain, some areas to irrigate, and some problem soils to reclaim. Plant breeders are making some progress in the development of varieties that will expand production into areas of restrictive environments. The possibility of developing cold-tolerant varieties holds considerable promise. Nevertheless, the expansion of food production into wet, dry, and cold areas will come at a high price. The river reversal projects of Russia, the land reclamation of the Netherlands, and the efforts of the People's Republic of China to rebuild lands are examples of what can be done where peoples are willing and able to direct their efforts and financial reserves into such projects. But it must always be remembered that altitude, cold, and aridity will dictate that vast areas of the world will always be unuseable. Deserts are deserts because they lack water. There are very few regions of the desert that have replenishable sources of good-quality water for irrigation that are not already in use.

The story in the tropics is somewhat different. It is true that altitude and aridity also severely limit expansion over vast areas. Some potentially productive areas are small and isolated. But the two biggest limitations to expanded production in the tropics are the soils and people themselves. The true nature of the limitations inherent in tropical soils is only now being unlocked. And we can expect that our knowledge of tropical soils will expand as rapidly in the next century as our understanding of temperate region soils in the past century. This knowledge will not come easily and will require much hard work.

Even as we accumulate this knowledge, progress may be very slow. The limiting factors will probably be human in origin: people's social, political, educational, and economic structures. In order to apply knowledge as it is gained, it must be carried to the farmers. They must be educated enough to understand more complex management systems. Farmers cannot properly follow the instructions on an insecticide label if they cannot read. The label must be in their language, and the units of measurement must be adapted to their system for measuring weights, volumes, and areas.

The materials must be available when and where they are needed, in adequate amounts, and at a reasonable cost. They require an infrastructure to transport and market produce at a reasonable price. And the producers must be able to take pride, within the local social structure, in their contribution to society. Their financial rewards must be able to provide them with comfortable homes that are comparable to the nonfarming segment of their society. They must be able to see that the future for their children, whether on or off the farm, will be better. And they must have medical care to insure that most days they feel good enough to tackle their tasks with enthusiasm and comfort.

While these goals may seem reasonable to many, they are not realized and, at the moment, show no evidence of being realized in the short run in many areas of the world. Many social, political, educational, and economical barriers must fall before such goals can be realized. Given the present reality, these barriers will fall very slowly and not without much effort and possible bloodshed.

Finally, we must always remember that there exists on our planet an equation that can be stated as follows.

$$\text{needed production} = \text{number of people} \times \text{standard of living}$$

The remarks in this book are confined largely to factors that influence the left side of the equation. The right side of the equation poses even more complex problems that cannot be ignored.

The task before us is enormous; there is no alternative but to face it and succeed. We hope that in some small way we have made a positive contribution to its solution by helping you to understand better the soil input into the solution.

References

Bornemisza, E., and A. Alvarado, Ed. *Soil Management in Tropical America,* Soil Science Department, North Carolina State University, Raleigh, 1975.

Buol, S. W., "Soil Genesis, Morphology and Classification," *N.C. Agri. Exp. Sta. Tech. Bull. 219,* 1973.

Buringh, P., *Introduction to the Study of Soils in Tropical and Sub-tropical Regions,* Centre for Agricultural Publishing and Documentation, Wageningen, 1968.

Cox, F. R., "Micronutrients," *N.C. Agri. Exp. Sta. Tech. Bull. 219,* 1973.

Drosdoff, M., Ed., *Soils of the Humid Tropics,* National Academy of Sciences, Washington, D.C. 1972.

FAO/UNESCO, *Soil Map of the World Volume IV South America,* 1971. Available through Unipub, New York.

Kamprath, E. T., "Soil Acidity and Liming," *N.C. Agri. Exp. Sta. Bull. 219,* 1973.

Sanchez, P. A., *Properties and Management of Soil in the Tropics,* Wiley, New York, 1976.

Sanchez, P. A., and S. W. Buol, *Soils of the Tropics and the World Food Crisis, Science 188:*598–603, May 9, 1975.

Shrader, W. D., *Soil Resources Characteristics, Potential and Limitations.* In *Dimensions of World Food Problems.* Iowa State University Press, Ames, 1977.

Sombroek, W. G., *Amazon Soils: A Reconnaissance of the Soils of the Brazilian Amazon Region,* Centre for Agricultural Publications and Documentation, Wageningen, 1966.

Uehara, G., and J. Keng, "Management Implications of Soil Mineralogy in Latin America," in *Soil Management in Tropical America,* E. Bornemisza and A. Alvarado, Eds., Soil Science Department, North Carolina State University, Raleigh, 1975.

Appendix

Table 1 Lowercase Suffixes and Their Meaning in the American and Canadian Systems

Description or Meaning	Suffix	
	United States	Canada
Accumulation of carbonates of alkaline earths	ca	ca
Accumulation of calcium sulfate	cs	—
Accumulation of concretions	cn	cc
Accumulation of salts more soluble than calcium sulfate	sa	—
Accumulation of salts more soluble than calcium and magnesium carbonate	—	sa
Buried soil horizon	b	b
Cementation by silica	si	—
Cementation, strong	m	c
Cryoturbation	—	y
Eluviated clay, iron, organic matter	—	e (as in Ae)
Fragipan	x	x
Frozen soil, permafrost	f	z
Gleying, strong	g	g
Illuvial clay	t	t
Illuvial humus	h	—
Illuvial iron	ir	f
Humus enriched	—	h
Plowing or other disturbance	p	p
Presence of carbonates	—	k
Slight alteration in B horizon	—	m

Table 2 Approximate Area and Extent of Orders and Suborders in the United States

Soil Order and Suborder	Area, square miles	Percent Extent[a]
Alfisols	483,450	13.4
Aqualfs	36,400	1.0
Boralfs	107,500	3.0
Udalfs	211,550	5.9
Ustalfs	94,650	2.6
Xeralfs	33,350	.9
Aridisols	414,400	11.5
Argids	310,900	8.6
Orthids	103,500	2.9
Entisols	284,250	7.9
Aquents	8,200	.2
Fluvents	10,750	.3
Orthents	188,450	5.2
Histosols	18,750	.5
Fibrists	8,300	.2
Hemists	7,050	.2
Saprists	3,400	.1
Inceptisols	655,700	18.2
Andepts	66,950	1.9
Aquepts	411,100	11.4
Ochrepts	153,450	4.3
Umbrepts	24,100	.7
Mollisols	896,000	24.6
Aquolls	46,100	1.3
Borolls	176,800	4.9
Udolls	170,450	4.7
Ustolls	318,400	8.8
Xerolls	184,250	4.8
Oxisols	500	<.02
Spodosols	183,050	5.1
Aquods	25,900	.7
Orthods	157,150	4.4
Ultisols	463,050	12.9
Aquults	41,250	1.1
Humults	27,550	.8
Udults	357,650	10.0
Xerults	36,600	1.0
Vertisols	35,300	1.0
Uderts	13,500	.4
Usterts	21,500	.6

From Soil Taxonomy, 1975.

[a]*Areas not included are barren rock, lava, or salt and make up 4.5 percent. These areas are mainly the Alaska and Brooks ranges in Alaska, the area adjacent to the Great Salt Lake in Utah, and a few small areas of Hawaii. Inland bodies of water of 50 square miles or more are also excluded. Soils that occur in very small areas that are not included are Albolls, Folists, Humods, Humox, Torrerts, Torrox, Tropepts, Ustults, and Xererts.*

Index